Application of Multi-Omics Analysis in Cancer Diagnosis, Treatment and Prognosis

Application of Multi-Omics Analysis in Cancer Diagnosis, Treatment and Prognosis

Editors

Wei Wu

Trever G. Bivona

MDPI

Basel • Beijing • Wuhan • Barcelona • Belgrade • Novi Sad • Cluj • Manchester

Editors
Wei Wu
The University of California,
San Francisco (UCSF)
San Francisco
USA

Trever G. Bivona
The University of California,
San Francisco (UCSF)
San Francisco
USA

Editorial Office
MDPI AG
Grosspeteranlage 5
4052 Basel, Switzerland

This is a reprint of articles from the Special Issue published online in the open access journal *Cancers* (ISSN 2072-6694) (available at: https://www.mdpi.com/journal/cancers/special_issues/Application_Multi_Omics_Analysis_Cancer_Diagnosis_Treatment_Prognosis).

For citation purposes, cite each article independently as indicated on the article page online and as indicated below:

Lastname, A.A.; Lastname, B.B. Article Title. *Journal Name* **Year**, *Volume Number*, Page Range.

ISBN 978-3-7258-1999-7 (Hbk)
ISBN 978-3-7258-2000-9 (PDF)
doi.org/10.3390/books978-3-7258-2000-9

Contents

About the Editors vii

Preface ix

Aubhishek Zaman and Trever G. Bivona
Quantitative Framework for Bench-to-Bedside Cancer Research
Reprinted from: *Cancers* **2022**, *14*, 5254, doi:10.3390/cancers14215254 1

Jovanka Gencel-Augusto, Wei Wu and Trever G. Bivona
Long Non-Coding RNAs as Emerging Targets in Lung Cancer
Reprinted from: *Cancers* **2023**, *15*, 3135, doi:10.3390/cancers15123135 19

Ihsan Ullah, Le Yang, Feng-Ting Yin, Ye Sun, Xing-Hua Li, Jing Li and Xi-Jun Wang
Multi-Omics Approaches in Colorectal Cancer Screening and Diagnosis, Recent Updates and Future Perspectives
Reprinted from: *Cancers* **2022**, *14*, 5545, doi:10.3390/cancers14225545 37

Lok-Man Ho, Sai-Kit Lam, Jiang Zhang, Chi-Leung Chiang, Albert Chi-Yan Chan and Jing Cai
Association of Multi-Phasic MR-Based Radiomic and Dosimetric Features with Treatment Response in Unresectable Hepatocellular Carcinoma Patients following Novel Sequential TACE-SBRT-Immunotherapy
Reprinted from: *Cancers* **2023**, *15*, 1105, doi:10.3390/cancers15041105 59

Garrett Winkelmaier, Brandon Koch, Skylar Bogardus, Alexander Borowsky and Bahram Parvin
Biomarkers of Tumor Heterogeneity in Glioblastoma Multiforme Cohort of TCGA
Reprinted from: *Cancers* **2023**, *15*, 2387, doi:10.3390/cancers15082387 77

Zainab A. H. Alebady, Mahsa Azizyan, Sirintra Nakjang, Emma Lishman-Walker, Dhuha Al-Kharaif, Scott Walker, et al.
CDC20 Is Regulated by the Histone Methyltransferase, KMT5A, in Castration-Resistant Prostate Cancer
Reprinted from: *Cancers* **2023**, *15*, 3597, doi:10.3390/cancers15143597 92

Yanjing Dong, Jiang Zhang, Saikt Lam, Xinyu Zhang, Anran Liu, Xinzhi Teng, et al.
Multimodal Data Integration to Predict Severe Acute Oral Mucositis of Nasopharyngeal Carcinoma Patients Following Radiation Therapy
Reprinted from: *Cancers* **2023**, *15*, 2032, doi:10.3390/cancers15072032 111

Jonghyun Lee, Seunghyun Cha, Jiwon Kim, Jung Joo Kim, Namkug Kim, Seong Gyu Jae Gal, et al.
Ensemble Deep Learning Model to Predict Lymphovascular Invasion in Gastric Cancer
Reprinted from: *Cancers* **2024**, *16*, 430, doi:10.3390/cancers16020430 127

Jiang Zhang, Xinzhi Teng, Saikit Lam, Jiachen Sun, Andy Lai-Yin Cheung, Sherry Chor-Yi NG, et al.
Quantitative Spatial Characterization of Lymph Node Tumor for N Stage Improvement of Nasopharyngeal Carcinoma Patients
Reprinted from: *Cancers* **2023**, *15*, 230, doi:10.3390/cancers15010230 141

Heng Jia, Ruzhi Li, Yawei Liu, Tian Zhan, Yuan Li and Jianping Zhang
Preoperative Prediction of Perineural Invasion and Prognosis in Gastric Cancer Based on Machine Learning through a Radiomics–Clinicopathological Nomogram
Reprinted from: *Cancers* **2024**, *16*, 614, doi:10.3390/cancers16030614 . . . . **155**

About the Editors

Wei Wu

Wei Wu obtained his medical and doctoral degrees in China. He received postdoctoral training in molecular biology and cancer genomics at the University of Chicago, USA. Dr. Wu worked on systems and computational cancer biology at the University of Calgary, Canada, before joining Dr. Trever Bivona's cancer translational research team at UCSF in 2015. His current research is focused on understanding the underlying complexity of genomic- and epigenomic-mediated targeted therapy-induced drug-tolerant persister cell states, using big data and artificial intelligence, and deep learning approaches in oncogene-driven non-small cell lung cancer, as well as in the development of effective treatments for aggressive cancers.

Trever G. Bivona

Trever Bivona is a board-certified medical oncologist with a PhD in cell and molecular biology. Dr. Bivona maintains an active academic clinical practice and clinical research program while also leading a basic and translational research laboratory focused on cancer signal transduction and genetics and precision medicine. One of his major research interests is enhancing the understanding of the molecular basis of targeted cancer therapy response and resistance in lung cancer and other solid cancers. He leads a multi-disciplinary team of investigators in laboratory-based, patient-focused investigations and is a principal investigator in clinical trials, enabling a bench-to-bedside research program. The overall goal of these efforts is to improve survival in the molecular subclasses of cancer patients through novel, effective precision medicine approaches.

Preface

Cancer is prevailingly viewed as a disease of evolving genomic instability and abnormal epigenomic modifications. Pioneering work has discovered and chromosomally mapped the genomic locations of oncogenes and tumor suppressor genes responsible for cancer initiation, progression, and metastasis. Through the efforts of cancer genome projects in the United States (e.g., The Cancer Genome Atlas research network, TCGA) and worldwide (International Cancer Genome Consortium, ICGC) and many other research groups, abundant genomic, transcriptomic, and proteomic data have been generated using state-of-the-art high-throughput sequencing technologies. As such, we are witnessing mutation signatures, gene copy number alterations, and aberrant gene expression profiles in specific types of cancer. Advances in sequencing technology and dramatic decreases in its cost offer the potential to accurately inspect the cancer genome at the level of single cells and with spatial resolution to understand cancer heterogeneity, the tumor microenvironment, spatial relationships, and the mechanisms of evolving drug resistance. Cutting-edge computational approaches and bioinformatics algorithms, such as artificial intelligence, machine learning, and deep learning methods, provide powerful toolkits to systematically identify clinically relevant biomarkers for early cancer diagnosis, prognosis, and stratification for precision cancer therapy.

This Special Issue reprint includes reviews and original research articles written by leading research scientists who are experts in multidisciplinary fields pertaining to a variety of cancer types, including lung cancer, gastric cancer, breast cancer, and brain tumors. This reprint covers a range of basic cancer research to translational and clinical research and is suited for a broad readership in the cancer research community, including graduate students, postdoctoral fellows, and physicians–scientists.

We would like to thank all of the authors who made excellent contributions to this Special Issue. Thanks to Ms. Zella Zhu from MDPI for her editorial assistance

Wei Wu and Trever G. Bivona
Editors

cancers

Review

Quantitative Framework for Bench-to-Bedside Cancer Research

Aubhishek Zaman [1,2,*] and Trever G. Bivona [1,2,3,*]

1 Department of Medicine, University of California, San Francisco, CA 94158, USA
2 UCSF Helen Diller Comprehensive Cancer Center, University of California, San Francisco, CA 94158, USA
3 Chan-Zuckerberg Biohub, San Francisco, CA 94158, USA
* Correspondence: aubhishek.zaman@ucsf.edu (A.Z.); trever.bivona@ucsf.edu (T.G.B.)

Simple Summary: Technological advancements and emerging high throughput molecular data have transformed biology into a more quantitative and multidisciplinary discipline. This has accelerated the translation of laboratory based findings into applied and clinically relevant applications and therapeutics. A shared practice for quantifying and statistical rank-ordering the effects of such translational applications and for understanding their underlying mode-of-action is now critical. In this manuscript, we discuss some of the major types of quantitative translational research and the best practices. We propose that adherence to these guidelines will improve assay design and reduce missteps in translational biomarker and therapeutics clinical application and adoption

Abstract: Bioscience is an interdisciplinary venture. Driven by a quantum shift in the volume of high throughput data and in ready availability of data-intensive technologies, mathematical and quantitative approaches have become increasingly common in bioscience. For instance, a recent shift towards a quantitative description of cells and phenotypes, which is supplanting conventional qualitative descriptions, has generated immense promise and opportunities in the field of bench-to-bedside cancer OMICS, chemical biology and pharmacology. Nevertheless, like any burgeoning field, there remains a lack of shared and standardized framework for quantitative cancer research. Here, in the context of cancer, we present a basic framework and guidelines for bench-to-bedside quantitative research and therapy. We outline some of the basic concepts and their parallel use cases for chemical–protein interactions. Along with several recommendations for assay setup and conditions, we also catalog applications of these quantitative techniques in some of the most widespread discovery pipeline and analytical methods in the field. We believe adherence to these guidelines will improve experimental design, reduce variabilities and standardize quantitative datasets.

Keywords: quantitative biology; chemical biology; bench-to-bedside; OMICS; IC50; high throughput screen (HTS)

Citation: Zaman, A.; Bivona, T.G. Quantitative Framework for Bench-to-Bedside Cancer Research. *Cancers* **2022**, *14*, 5254. https://doi.org/10.3390/cancers14215254

Academic Editors: Serge Roche and John J. Turchi

Received: 1 September 2022
Accepted: 24 October 2022
Published: 26 October 2022

1. Introduction

"What is life?", once asked quantum physicist and Nobel laureate Erwin Schrödinger, when he prophesized that behind the mystical nature of life there must remain a quantifiable mathematical elegance [1]. Not long after, James Watson and Francis Crick presented the first mathematical model for the 'molecule of life'-DNA [2]. Ever since then, from a systems biology point of view, the quest to generate mathematical models to quantify biological processes as reactions, to quantify stimuli and response as input and output has emerged. Over the years, due to a remarkable expansion in high throughput data acquisition and in our ability to analyze, biological research has become vastly more quantitative. In vitro research coupled with powerful statistical analysis have successfully recapitulated patient and in vivo biology indicating the power of quantitative biology in the field of biochemistry, molecular and cellular biology and cancer research and treatment. This has also caused a paradigm shift towards massively automated and computation-heavy annotation and analysis in many disease contexts such as cancer biology [3]. For example, quantitative chemical biology research has transformed our understanding of tumor genomics, prediction of novel candidate therapies against cancer and improved personalized targeted therapy for cancer patients. In last decade alone, worldwide chemical biology efforts have resulted in an avalanche of clinical trials and FDA-approved therapies against many different types of tumors, subsequently increasing patient longevity and decreasing therapy toxicity [4].

The broader goal of quantitative chemical biology research is to model the pharmacokinetics and pharmacodynamics of chemicals in diseased patients, but in a much more conducive and tractable in vitro system initially. Modern chemical biology research takes advantage of our ability to screen unbiasedly through a wide array of compounds in systems such as in patient-derived cell lines, patient derived xenografts (PDXs) or purified protein-ligand binding assays [5]. Furthermore, modeling drug response by high throughput screening (HTS) is followed by identifying candidate biomarkers, signaling pathways and molecular targets. Chemical biology research also focuses on improving the response of the drugs through modifications of chemical structure followed by quantitative structure activity relationship (QSAR) studies or through drug synergy assays. Quantitative chemical biology is making a reality what once was science fiction [6]. Yet, there are gaps that must be addressed to better translate quantitative cancer research into clinical implementation. Here, in this review, we outline individual aspects of chemical biology research and their quantitative frameworks. We also summarize progress in this area of cancer research, highlight key gaps, and propose concrete steps forward.

2. Modeling Drug Dose Response

The goal of quantitative biology is to quantify biological processes and chemical biology to discern the effect of chemicals on biological systems. Proteins are the functional molecules of life, important for carrying out most biological reactions and in return for steering biological processes. Hence, a major focus of chemical systems biology is to model the response of different doses and kinetics of chemical perturbagens on enzymes.

For example, enzyme inhibitors are often pharmacological agents that competitively and reversibly inhibits substrate binding and enzyme activity [7]. The kinetic behavior for many enzymes can be explained with a Michaelis-Menten (MM) model for enzyme-ligand or enzyme-substrate binding and catalysis:

$$\mathrm{E} + \mathrm{S} \underset{\mathrm{k}_{-1}}{\overset{\mathrm{k}_1}{\rightleftharpoons}} \mathrm{ES} \overset{\mathrm{k}_2}{\rightarrow} \mathrm{E} + \mathrm{P}$$

where E stands for enzyme, S stands for substrate and P stands for product. ES is an enzyme-substrate complex that is formed prior to the catalysis. Formation of ES requires only binding and hence is reversible, indicated by equivalent rate constant k_1 and k_{-1} for forward and reverse rate for the event. On the other hand, the overall rate-limiting and irreversible step in the reaction is the breakdown of the ES complex to yield product, which can proceed with rate constant k_2.

Reaction velocity from this reaction can be described as a function of substrate concentration using the following formula, which is typically referred to as the Michaelis-Menten (MM) equation:

$$v = ([S]Vmax)/([S] + Km)$$

where, v = rate of reaction during initial velocity condition; Vmax = maximal reaction rate; S = substrate concentration; Km = Michaelis-Menten constant. Interestingly, since the value of Km at 0.5 Vmax condition equals to [S], Km is often termed as the substrate concentration at half maximal velocity (Figure 1a).

A MM equation is different from an enzyme substrate reaction progress curve that often describes a kinetic equilibrium. It is important to note that the MM model plot in Figure 1a does not describe activity of the enzyme under a continuous time variable;

instead, velocity, v is calculated separately for corresponding individual substrate values (Figure 1b) [8]. Hence, many different measurements of v are calculated each under initial velocity conditions with varying substrate concentrations at or below the Km value (Figure 1a). Measurement at the initial velocity conditions ensures that the equation is insensitive to the effect of velocity variation during reaction progression. However, the equation predicts saturation of reaction rate at Vmax and an initial logarithmic increase in velocity as a function of substrate concentration (Figure 1a) [7].

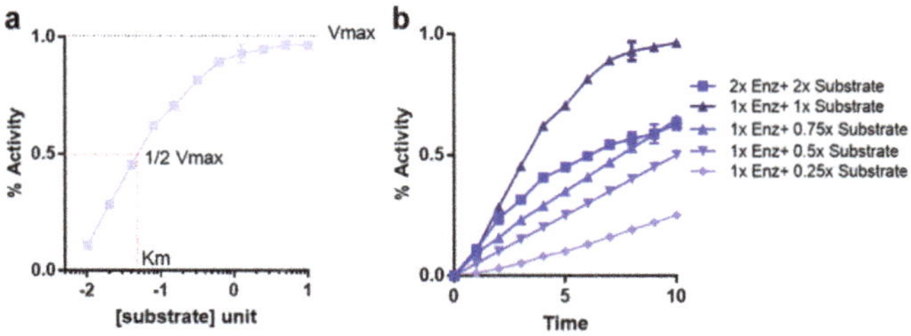

Figure 1. Reaction kinetics and velocity plots. (**a**) Michaelis-Menten reaction rate is plotted as a function of substrate concentration. Reaction rate is saturated at Vmax. (**b**) Reaction progress curve for initial velocity measurement under a varying concentration of substrates and enzyme as indicated.

Formation of product, post-enzyme-substrate mixture is the quantitative estimate of reaction rate and when tracked and plotted over a period indicates the reaction progress rate (Figure 1b). However, initial velocity of the enzymatic reaction is separate from the velocity of reaction during reaction progression. Initial velocity represents the reaction rate when less than 10% of the substrate has been depleted or less than 10% of the product has formed. Under these conditions, it can be safely assumed that the substrate concentration does not significantly change and does not reach saturation limit for enzymatic activity. Furthermore, it can also be assumed that, in such conditions, the contribution of reverse reaction is minimal [8].

The MM equation is a first order reaction which results in a linear increase in v until the reaction saturates due to maximal occupancy of enzyme with substrate. This is a classic example of 1:1 substrate enzyme interaction consistent with the 'lock and key' model. However, for enzymes with multiple substrates, cooperativity amongst the substrates for enzyme binding and requirement of maximal enzyme site occupancy for enzyme activity results in an initial lag time for reaction initial velocity followed by an exponential increase in enzyme activity and subsequent saturation. In this case, the reaction follows a multi-order non-linear and sigmoidal reaction kinetics, often consistent with a so-called 'induced fit' model, determined by what is called a hill coefficient. The higher the hill coefficient the sharper the inflexion of the hill curve (Figure 2) [9].

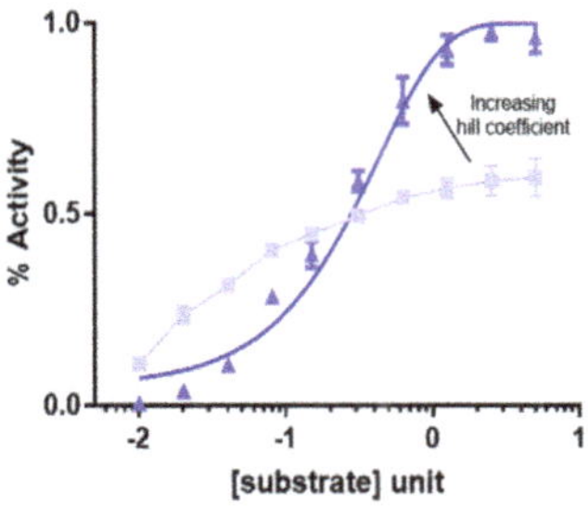

Figure 2. Hill co-efficient and sigmoidal curve. The light blue curve represents a classical Michaelis-Menten reaction kinetics, whereas the darker blue curve represents transformation of the Michaelis-Menten reaction kinetics in presence of cooperativity and hill coefficient. The steeper slope that the inflexion of the sigmoidal curve has the higher the hill coefficient for co-operativity.

3. Determination of IC50 for Inhibitors

The measurement of enzyme activity in the linear initial velocity condition is often time not possible due to multiple reasons, lack of a measurable assay read out being one of them. In these cases, the enzyme or a linked biomarker response is measured at a fixed time point as a function of varying substrate concentration. This concentration-response plot is similar to the Michaelis-Menten plot except that response is not calculated at initial velocity condition (Figure 3a). Furthermore, for enzyme inhibitors the curve simply follows an upside-down version of the concentration response plot for an enzyme substrate (Figure 3b) [10]. Effects of an inhibitor on enzymatic reaction can often be determined by plotting a dose/concentration response plot where *x*-axis represents varying doses of the inhibitor and *y*-axis represents corresponding reaction rate estimates. Notably, in this setting, the enzyme and substrate concentration are kept constant. The dose response plots are widely applied in pharmacology and chemical biology. For example, it is often the first assay to be performed to identify a candidate list of active/lead compounds by screening a library of known or unknown ones. It is also applied for determination of ideal dose range and therapeutic window as well as for structure activity relationship (SAR) assay for chemical/functional group variation of a desired compound's backbone for activity enhancements. C The 4-parameter logistic nonlinear regression model (4PL) for data fitting, that describes the sigmoid-shaped response pattern, is an example of this type of model (Figure 3a,b) [10]. For example, in Figure 3a light blue curve represents a concentration response plot for an enzyme inhibitor.

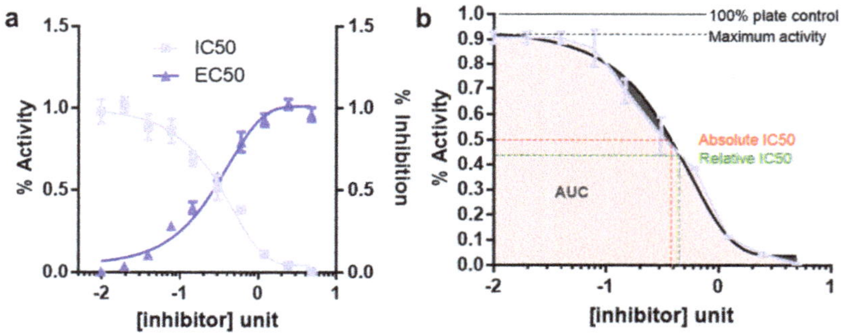

Figure 3. IC50, EC50 and AUC measurements from dose response plots. (**a**) Dose response plots for AC50 (dark blue) and IC50 (light blue). (**b**) AUC calculation of a dose response plot. The figure also represents the theoretical (red) and relative/tested (green) IC50 measurements from the dose response curves. The theoretical (solid black line) activity represented as plate control and maximum activity tested (broken black line) can also be different.

Analogous to MM constant (Km), substrate concentration required to result in 50% activity is called EC50 (effective concentration to reach 50% activity). On the contrary, the concentration of compound that results in 50% inhibition of maximal activity is termed the IC50 (inhibitor concentration yielding 50% inhibition) [11]. In this review, we discuss the IC50 calculation often deployed in cancer biology, where instead of inhibitor binding to the target of interest (target-based), biologists measure the inhibitor response on cellular viability (phenotype/cell based).

Some criteria for successful concentration-response curves are listed in the discussion below.

1. Well defined top and bottom plateau values need to be established. To do so, it is important to use sufficient range of inhibitor concentrations. These parameters are critical for the mathematical models used to fit the data
2. A minimum of 8–10 inhibitor concentration data points for an accurate IC50 determination should be used

3. Concentration ranges for the inhibitors should be spaced equally
4. The concentration data point counts and the range should be chosen so that half the data points on the IC50 curve are above the IC50 value and half are below the IC50 value. This is difficult for IC50 measurements for compounds for which there exist no prior knowledge. In this case, the inhibitors should be tested for response using a broader range of doses followed by final IC50 estimation using narrower range of doses
5. Enzyme concentration should always be kept constant and the lower limit for determining an IC50 is half of the enzyme concentration
6. Well readable and quantifiable screening strategies for measuring the response should be employed. The quantification should be benchmarked under different experimental conditions. For example, cellular viability can be measured by viable cell adenosine triphosphate (ATP) level using the reagent cell titer glo (CTG)
7. At least three replicates for each data point should be collected. For cellular viabilities these replicates need to be biological replicates
8. Criteria for reporting IC50′s are the maximum % inhibition should be greater than 50%; top and bottom values should be within 15% of theory; the 95% confidence limits for the IC50 should be within a 2–5-fold range. Relative and absolute IC50 and EC50 is described in Figure 3b.

Depending on necessity, a wide array of reagents can be used as a replacement of CTG for cell viability measurements. For example, dye coating and dye exclusion -based experimental setup requires use of crystal violet, trypan blue, eosin, congo red and erythrosine B staining. For non-ATP and nicotinamide adenine dinucleotide phosphate (NADPH) -based colorimetric assay MTT, MTS, XTT, WST staining reagents are used (PMID: 9869118, 28470513) [12,13]. For cellular protein and enzyme level as a proxy for viability LDH and SRB assays are widely accepted. On the other hand, alamar blue and CFDA-AM are two commonly used fluorometric cell viability assay (PMID: 28470513) [13]. Additionally, immunofluorescence and flow cytometry-based assays (e.g., Brdu, annexin 5) are also commonly used for the purpose of determining viable cells (PMID: 28573164) [14].

The concentration-response curve response does not plateau at the baseline (e.g., 0%) or does not saturate at the highest point (e.g., 100%). This may happen due to inter sample heterogeneity (e.g., bimodal response samples) or due to technical issues. In this case, the theoretical IC50 values is different from the test IC50 values calculated, giving rise to inaccurate IC50 estimates. In those cases, area under curve (AUC) calculation can offer a more accurate estimate of the response [15].

4. HTS Using Pharmaco-Chemical Library

A holy grail in oncology is so-called 'magic bullet' therapies that perturb only diseased cells/proteins but leave normal healthy cells/proteins untouched. Over the course of time, our understanding of proteins, as the functional molecules of cells, has immensely improved and has resulted in interest to target them in pathophysiological conditions [16]. For example, we have undertaken technologically sophisticated high throughput screening for pharmacological compounds that perturb or ameliorate the activity of a protein molecule and thereby correct a disease phenotype.

In such a drug screening experiment, the efficacy of several pharmacological agents are evaluated either against a disease phenotype or against an enzyme activity. The former is called a phenotypic screen whereas the latter is known as the target-based screen. The goal of these screening approaches is to identify, from a wide an array of initial compound list, a smaller and tractable number of candidate compounds (often called 'leads') [17].

To find novel therapeutics against a disease, phenotypic screening, where a myriad compounds are tested for reduction of a disease phenotype, is most commonly employed. Phenotypic screening is unbiased and agnostic about the mechanism of action (MOA) or the molecular target for the tested agents. Therefore, subsequent analysis for target deconvolution is required for comprehensive understanding of the effect of the compound.

In this regard, a more focused version of screening called target-based screening can be applied. wherein this case, pharmacological agents are tested against a single or handful of molecular targets; targets that have already been identified as a causal mechanism of disease pathology. Target-based screenings are usually less time consuming, but difficult to design. Moreover, target-based screening approaches are a non-starter for diseases for which a knowledge deficiency exist (Table 1) [18]. Both approaches have their pros and cons. Below is a comparison of phenotypic screenings and target-based screening (Table 1):

Table 1. Comparison between phenotypic and target-based screening.

	Phenotypic Screening	**Target Based Screening**
Molecular targets	Not known	Known
MOA	Not known, but can be targeted based on signaling pathways	Known
Assay type	Cell viability (e.g., luminescence read out live cells)	Direct binding assays (e.g., fluorescence read out in FRET)
Assay scale	Relatively difficult to scale up	Easily scalable into high throughput
Biological relevance	Highly relevant to biology	May not be relevant to functional biology
Quantification methods	Not available	Structure activity relationship (SAR)
Novel target scope	High	Low

For HTS, the concept of combinatorial chemistry was developed in the mid 1980′s, with Geysen's multi-in technology where hundreds of thousands of peptides were synthesized on solid support in parallel [19,20]. Subsequently, one-bead one- compound (OBOC) combinatorial peptide libraries and solution-phase mixtures of combinatorial peptide libraries and phage display libraries were introduced. However, it was not until mid-90 s, when the first example of a small-molecule combinatorial library was reported [21].

Combinatorial chemistry has been used for both drug lead discovery and optimization. The highly focused parallel synthesis of small-molecule libraries (hundreds to thousands of compounds), when developed in conjunction with computational chemistry, are particularly useful for optimization of drug leads [10].

Recommendations and challenges for IC50 calculations in HTS:

Doubling time: IC50 is best calculated in an isogenic setting, where response of the cells to a particular perturbation is best compared with that of a response without the perturbation. However, in absence of an isogenic system, classifying IC50 spectrum of many different cell lines into high and low, leads to a possibility that the difference in IC50 is due to doubling time differences (Figure 3a). For example, cells with higher metabolic activity and doubling time, are prone to up taking the compound faster and hence will be killed faster, resulting in an IC50 smaller than cells that grow slower. Hence, regression analysis of IC50 and doubling time is required to rule out this phenomenon where, ideally, no significant correlation between the doubling time and IC50 is preferred.

Number of cells or seeding confluency: To balance for the doubling time often cells are seeded in a manner so that by the time they are ready for measuring the effect of the drug they are of around 90% confluency. Quite intuitively, it has been observed that, higher confluency of seeding requires a higher dose of compound to kill 100% of the cells. The effect is often described as a drug sync/sponge effect. Hence, the initial seeding densities of the cells required, needs to be accurately estimated by empirical trials [10].

Edge effect: Screening platforms often use small multi-well formats (e.g., 384 and 96-wells). It has been reported that the wells in the plate that are situated at the edge of the plates are exposed to external stimulus such as temperature un-uniformly than that of the wells in the middle. Hence, edge wells are usually exempted from using during the IC50 calculation in these plates.

Vehicle effect: Each perturbation must be compared with a vehicle treatment cohort. Often the Vehicle treatments can result in some response alone (Figure 3b). At long as the response is below 10%, the response is considered acceptable. The vehicle can also result in some confounding cellular effect above a certain dose; hence it is very important to keep the vehicle dose within the limits of acceptable range. The lead compounds being tested should have the same amount of vehicle in volume for comparative analysis. If compounds involved in a screening assay were dissolved in various vehicles, the screen must consist of many different 'vehicles alone' controls for comparative analysis as well.

5. Biomarker Prediction

One of the major limitations of phenotypic screens is the lack of understanding of any molecular targets for the drug itself. Hence, target discovery from phenotypic screen has been a major challenge in the field of chemical biology. Since the advent of high throughput genomics technologies, many computational approaches have been undertaken to correlate drug phenotype response to cellular genomic, epigenetic, transcriptomic, proteomic and metabolic features [22–24].

National Cancer Institute (NCI) initiated a Drug sensitivity Dialogue for Reverse Engineering Assessment and Methods (DREAM7; Available online: http://dreamchallenges.org/ (accessed on 12 August 2022) project to gather momentum and bolster enthusiasm for this very important challenge of predicting biomarkers from drug sensitivity and vice versa. The Challenges is one of the first of kind- a community-based collaborative competition oriented towards crowdsourcing solution and open-data sharing [22,23]. The DREAM7 Challenge also benchmarks many drug sensitivity prediction methods. For example, kernel-based prediction methods, which depend on machine learning algorithm for pattern matching (e.g., support vector machine or SVM), differ from feature-based methods, which depends on a feature map generated by training dataset, in terms of utilization of the user-defined feature map [22,23,25,26].

In NCI-DREAM7 Challenge, for training datasets a multi-OMICS (e.g., copy number variation, DNA methylation, point mutations, transcriptional and protein level estimates) approach was pursued. Interestingly, the predictive models that used multi-omics profiles outperformed a single-OMICS prediction model, which suggests genomic, epigenomic, and proteomic profiles provide complementary signal for drug response prediction [24,27,28]. Importantly, prediction algorithms validated previous biological knowledge for breast cancer and provided insight into non-linear feature relationships during modeling [22,23].

One useful approach is a regularized regression model known as elastic net [29]. One of the major problems of these biological datasets is the asymmetry of the matrices. The columns of the matrices containing various treatments (<200) were much too small in number than the rows of the matrices that contain genomic features (>5000), which poses a computational challenge for regression model often known as '$p >> n$ ratio problem' [24,25,30]. Ordinary regression models due to this asymmetric in matrices generates overfitting solutions resulting in false positive/type1 errors. To solve this, the elastic net generates sparser biomarkers based on a regularized regression model where the equation balances between lasso and ridge regressions. Furthermore, the resulting solutions can be represented as a heatmap (Figure 4a) [15,30–32].

The best-performing algorithm was based on the Bayesian efficient multiple kernel learning (BEMKL) model. BKMEL uses a kernelized regression model that makes use of both multi-task and multi-view learning algorithms [23,26]. In Multiple kernel learning (MKL) algorithm, pairwise similarities of cell line OMICS profile constitute an initial kernel and are subsequently combined into a compound kernel. In multi-task learning (MTL), on the other hand, the model is trained simultaneously for all the drugs and thus differ from the stepwise kernel generation strategy employed in MKL. BKMEL introduces hyper-parameters and an error term/bias to account for poor intersection of multi-OMICS datasets [22,23].

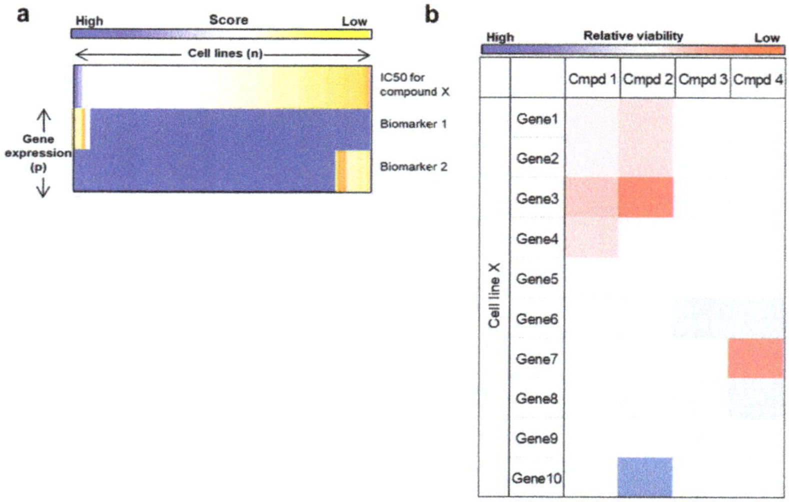

Figure 4. From heatmap to actionable OMICS feature selection. (**a**) Heatmap representation of a typical regularized regression (elastic net) driven dose response versus biomarker plot. IC50 of many different drugs in a panel of cell lines is plotted on the first row and the other row 2 represents anti-correlative biomarkers and row 3 represents correlative biomarkers. Biomarker solutions are rank ordered based on their score. (**b**) Heatmap representation of chemigenomic interaction in an isogenic setting. Many genetic alterations in 'Cell line X' is presented on rows whereas columns represent viability of the genetically modified cells under different compounds (columns).

6. IC50 Measurements in Isogenic Settings

Biomarker prediction depend on a correlation between features and drug response. However, these methods do not essentially establish causality. To address this, incorporating tumor associated alterations in an isogenic system is increasingly being pursued for comprehensive chemogenomic analysis [33]. In this setting, a patient derived cell line or organoid of interest is genetically subjected to very specific genetic modifications and subsequently drug responses are measured across the board to determine the effect of the genetic alterations. The resulting viabilities can be represented as a heatmap (Figure 4b). On many occasions, these have contributed crucial understanding of oncogenic addiction, specificity of crosstalk between pathways and genetic interactions in cancer. Recently, the ease of activating or perturbing genetic alterations using CRISPR based technologies have paved the way for new opportunities for high throughput chemogenomic interaction analyses [34–40].

7. Signaling Pathway Analysis and Target Discovery

In biological response versus feature correlation analyses, instead of enrichment of a single biomarkers, enrichment of a list of functionally related group of genes is more informative. As a result, increasingly the classical gene-based approaches that ignore the modular nature of most human traits is being replaced with a more functionally holistic pathway enrichment approach. In this regard, statistically computing overlapping between experimental OMICS datasets (such as exome, methylome, RNAseq, quantitative proteomic, metabolomic, etc.) and curated pathway databases (e.g., GO, ENCODE, KEGG, REACTOME, etc.) have become routine [41–45]. The computational analyses depend on either hypergeometric tests (ENRICHR) or a signal to noise based (S2N) analysis (BROAD institute GSEA) [45–47].

In pathway analysis, a set of candidate/query genes are compared against a library of curated 'gene sets' each of which includes genes that are bundled due to their participation in a signaling pathway or biological function). The candidate/query gene sets are usually composed of genes that are differentially upregulated or downregulated in an OMICS dataset such as gene expression, proteomics, etc. (Figure 5). The prototypical enrichment or overrepresentation (ORA) analysis is usually performed via comparing the test gene set with that of the curated gene set using hypergeometric test, where null hypothesis represents a baseline or random-chance representation probability [48–51]. Overlapping is considered significant if the hypergeometric test produces a significant p value (Figure 5). Hypergeometric test can be done in presence or absence of weighted or ranked gene sets as well. However, this method does not consider the topology of a signaling network. In biological pathways, genes/proteins tend to perform in a network where each gene/protein can be thought to be as a 'node' (drawn as circles) and their regulation between one another is signified by 'edges' (drawn as lines) (Figure 5). From this network point of view, minor variation in gene/protein neighborhood and directionality of reaction contributes to a vastly different biological function. Hence, while considering overrepresentation, curated databases that incorporate directionality and neighborhood information (such as Reactome, KEGG, WikiPathways, etc.) produces better signaling pathway analysis (Figure 5) [42,43,52]. More recently, even more granular context specific sub-circuitry and subnetwork based over representation analysis have become increasingly useful and is called mechanistic pathway activity (MPA) based pathway analysis. For this analysis, the curated database not only has the topology information it also has positivity and negativity information for the nodes in play (Figure 5) [53–56].

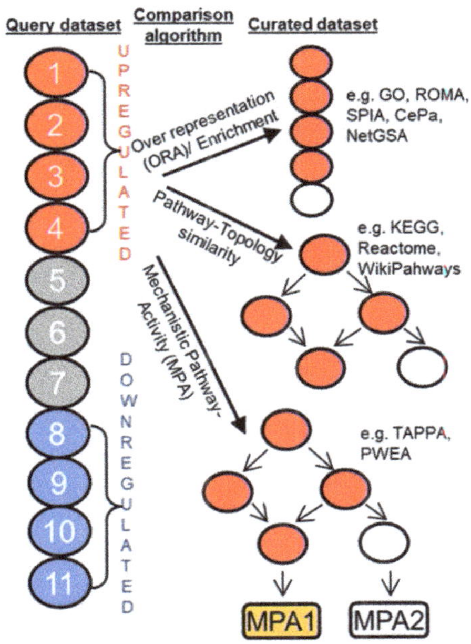

Figure 5. Pathway analysis methods. Comparison of test dataset with curated databases can be performed in many ways- based on over representation, pathway topology-based, mechanistic-pathway activity.

8. Form Pathway to Target Discovery

Pathway analysis, based on correlation, generates a single or a few candidates signaling pathways as target of the drug. However, it neither establish causality nor pinpoint a single gene/protein as a target. Hence, for target discovery the goal is to home in on a single protein–protein interaction (PPI) from protein interaction signaling networks; a single enzyme target for cascade of enzymatic reaction; a single gene target for a gene regulatory network [34,57,58]. Causal relationships of a perturbagen often needs to be established by genetic manipulation of the candidate genes/proteins one at a time [59].

Recent technological advances, including genomics, proteomics, small interfering RNA CRISPR, and mouse knockout models have allowed us to measure the effect of shortlisted candidate pathway regulators on the cellular phenotype, which allows us to identify a targetable protein [18,59].

Computational biology and structural biology have been instrumental in deciphering drug-protein and protein–protein interactions in absence and presence of the perturbagens. If a single protein is thought as a hub, then the protein that interacts with it and forms a PPI network constitute the interactome for the protein of interest [60,61].

Structural biology is crucial for PPI research. X-ray crystallography, protein-based nuclear magnetic resonance (NMR) spectroscopy has made it possible to generate and study 3D structure and active site pockets of protein molecules. Computer simulated docking of perturbagens, often known as in silico docking, often faithfully recapitulates the biological ligand/inhibitor binding pocket on the protein [58,61]. However, 3D structures generated through these methods often are static and fails to recapitulate the dynamic nature of the protein-ligand interaction. Furthermore, crystallization of the protein itself requires many modifications of the protein molecule such as truncation and/or mutations [58]. Although far from perfect, in the last few years, significant improvement in dynamic structural simulations such as monte carlo simulations have raised remarkable promise for in silico simulations for protein-ligand/inhibitor interaction [62].

9. Quantitative Structure Activity Relationship (QSAR) and Physicochemical Properties of Drugs

Quantitative Structure-Activity Relationship modeling is one of the major computational tools employed in medicinal chemistry [63,64]. In QSAR analysis a structural element (called a molecular descriptor) of the lead chemical compound is modified and the response in activity is measured [65]. The goal, this way, is to generate an array of activity response and curate and finetune the best response. Modification of molecular descriptor of the compound can be based on its chemical 2D structure as well as 3D topography [63,66].

One of the major challenges with QSAR equations is that of faithfully predicting the effect of multiple modifications at once. For example, two colinear molecular descriptors independently may result in improvement in QSAR response; however, when introduced together may result in antagonistic response. This indicates the importance of experimental validation of QSAR response to avoid such confounder effect.

Moreover, applications of the concept of drug-likeness, which compares physical properties of candidate pharmacological chemicals (such as lipophilicity) with that of other validated compounds and approved drugs to predict pharmacodynamics and pharmacokinetics of the candidate drug [66]. These in silico predictions help both the final in vivo preclinical and clinical validation experiments, by helping decide the range of doses and time-period to be tested.

10. Drug Synergy

For any complex disorders, rational design of multi-targeted drug combinations is a promising strategy not only to improve individual drug potency and efficacy, but also to tackle resistance to individual drugs. A drug combination is usually classified as synergistic or antagonistic, depending on the deviation of the observed combination response from the expected effect calculated based on a reference model of non-interaction [67]. There are many metrics for drug combination measurements. Combination effect measurements can vary due to the experimental design. For example, before the advent of high throughput platforms single dose combination therapy was widespread [15,40,68]. However, the field has moved towards a much more sophisticated and comprehensive methods of combining the doses of the drugs (Figure 6) [68–71].

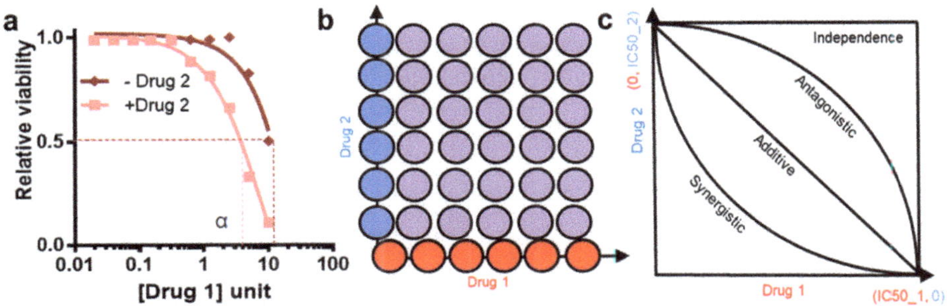

Figure 6. Drug synergy measurement methods. (**a**) Synergy measured by visualizing left-shifting of dose response curve of Drug 1 in presence of the Drug 2. (**b**) Cartoon illustrating comprehensive checkerboard- style experimental setup where combinatorial doses are marked as purple and only Drug 1 as red and only Drug 2 as blue. (**c**) Isobologram analysis indicates multiple possible outcomes for drug interaction analysis- for additivity diagonal, independence rectangular, for synergy concave and for antagonistic convex isobolograms.

Often, synergy is calculated by generating comprehensive dose response under a varying level of Drug 1 and Drug 2 (Figure 6a) [72]. These generate a symmetric matrix and often called checkerboard dose combinations (Figure 6b). However, due to effort intensive nature of this setup, synergy is often calculated by measuring dose response curve of cells under Drug 2, in the presence of a few doses of a test compound, Drug 1 [68,72,73]. The graph looks like few dose–response curves in the same plot and they superimpose in top of each other in absence of synergy or antagonism and shift left or right in presence of synergy and antagonism, respectively, (often called a multiple-ray plot). Synergy can also be measured using just a single dose of Drug 1 and Drug 2. However, quantification of synergy under those condition is difficult [72,73]. For any two compounds Drug 1 and Drug 2, following is a summary of ways for quantifying the synergy.

Statistical independence/Bliss: This quantification is applicable for synergy calculation even using a single dose. However, Synergy calculation using a few doses, or a single dose is also possible. This can be calculated by calculating the probability/percentage of killing under each individual drug treatment-

$$P_{a+b} = 1 - (P_a \times P_b)$$

where P_a = probability of killing cells by drug 1; P_b = probability of killing cells by drug 1 and P_{a+b} = probability of killing cells together.

The other quantification methods for synergy are quantified by measuring and adding maximum response by each drug alone and then measuring the effect of the combined dose. If combined effect of maximum dose is more than that of the additive effect, the interaction is called synergistic.

Gaddum pharmacological interaction: In presence of the Drug B; the dose response curve of Drug 1 shifts on the left and the new IC50 value is α, lower than that of the IC50 in absence of Drug 2. Gaddum pharmacological interaction measures difference in this IC50 as a measure of synergy (Figure 6a) [74].

Isobologram analysis: In checkerboard comprehensive synergy analysis, a response-surface plot generated as such allows generation of isobologram graphs (such as contour plots in geography) by connecting identical level of toxicities in the different drug combinations [75,76]. For example, if Drug 1 alone causes 20% toxicity and Drug 2 causes 20%; Drug 1 and 2 together causes 40% toxicity then the isobologram would go through a rectangle line as all the 20% toxicity values would fall in that isobologram line. However, this visualization does not aptly explain the additive effect of two drugs (Figure 6b,c).

Lowe additive model analysis: In comprehensive analysis, Drug 1 will interact with Drug 1 as in an additive manner [77,78]. From this expectation, If Drug 1 and Drug 2 behaves as though their effect is additive the isobologram would go through a diagonal line. Similarly, if the combination results in better response than each compound alone then the isobologram would go through a concave line and similarly for antagonistic interaction, a convex line.

The Chou Talay combination index from this can be calculated using the formula for this is (Figure 6c):

$$\text{Combination index (CI)} = \frac{a}{IC50\_A} + \frac{b}{IC50\_B}$$

CI > 1 means antagonism; CI < 1 means synergistic interaction and CI = 0 means additive interaction (Figure 6c) [79]

11. Case Study

In this section, we now describe an experimental example that illustrates the concepts discussed above. In this example, an investigator found a cellular receptor tyrosine kinase called Fibroblast Growth Factor Receptor 1 (FGFR1) and its downstream pathway to be upregulated in a subset of non-small cell lung cancer (NSCLC) samples (Figure 5). Additionally, the investigator found that genetic suppression of FGFR1 reduces the viability of NSCLC cells by performing crystal violet and cell titer glo (CTG)-based viability assays (described in Section 3). Hence, the investigator concluded that this particular subset of NSCLCs required FGFR1 for cell proliferation.

Establishment of this causal relationship between the FGFR1 pathway and cell viability motivated the investigator to next test existing chemical perturbagens of the FGFR1 pathway, such as Infigratinib, using the CTG assay (Figure 3). Unfortunately, the investigator found that although the effect is specific, the response to the FGFR inhibitor has a narrow therapeutic window indicated by some toxicity even in normal cells and also requires a high dose for therapeutic effects. Hence, the investigator decided that one next step would be to design an improved inhibitor for this pathway of interest.

The next goal was to test a library of chemical compounds which contain both structurally diverse as well as structurally similar compounds with slight functional group variation compared to Infigratinib. The library was tested in a panel of ~100 NSCLC patient-derived cell lines with 3 replicates for each data point and 12 doses of 3 times serial dilution from the highest dose. The highest dose used for any compound is empirically determined (e.g., 33 μM for Infigratinib). The compound response patterns were also clustered in an unbiased manner. In an ideal case, this cluster should reflect biologically relevant information. For example, small molecules with shared MOA should cluster together. As expected, AZD4547, another FGFR inhibitor was in the same functional clade as Infigratinib. A candidate list of 10 compounds was rank ordered from this experiment and further validated in vitro. These 10 compounds include candidate leads from both the same (such as other FGFR inhibitors) and different functional clades (such as AKT inhibitor MK-2206) as Infigratinib.

Leads that parsed into a different functional clade (and hence employ different MOAs), and yet worked potently against FGFR1 upregulated NSCLC provided novel insights. For example, these small molecules might exhibit on-target blockade of the FGFR pathway up or downstream of the FGFR receptor tyrosine kinase (e.g., AKT inhibitor MK-2206). Additionally, some small molecules could also function in a synthetic lethal-like manner, where they exhibit potent responses against FGFR1 upregulated NSCLC by acting indirectly against additional pathways required for NSCLC tumor survival in this molecular context (e.g., PARP inhibitors).

Although the compound screen was informative in identifying novel leads, no single agent perturbation was sufficiently potent to elicit a complete response (near 100% cell death). Hence, the investigator next applied these 10 lead compounds in dual combinations (10 × 10 = 100 unique combinations) and then measured the viability response again in a high throughput manner. The compound combination effect was then mathematically classified as synergistic, antagonistic, or additive as explained above (Figure 6). The investigator identified potent synergy between FGFR and Mitogen-activated protein kinase (MAPK) pathway inhibitors (e.g., Trametinib). Combined application of these small molecules exhibited a more complete effect against FGFR1 upregulated NSCLC cells.

Interestingly, this compound screen could also be carried out in an unbiased manner in the absence of the pathway information mentioned above. Given the availability of molecular feature dataset for each individual cell lines (such as genomic, transcriptomic, proteomic and/or other), mathematical models can then be applied to uncover associations between cell viability response and genomic feature variables (Figure 4). In this way, it is also possible to identify therapeutic and MOA hypotheses in an unbiased manner even in the absence of any candidate pathway a priori.

We use this example to demonstrate scenarios where these tools can be directly applied. There are many possible combinations of ways these techniques can further be used in the field.

12. Bench-to-Bedside Translation

Preclinical research, despite certain limitations, is a powerful approach to predict in vivo clinical responses [5,80]. However, translating observations identified in preclincial systems into actionable therapies in vivo involves a range of additional validations. For example, pharmacogenomic analysis carried out in patient-derived cell lines are often validated in patient derived xenografts (PDXs) and preclinical animal models (a process often called 'T1-T4 stage' validations) [81,82]. Once a concordant observation is achieved in many different such models, the therapy is advanced into clinical trials. However, despite this hierarchy of validation process, clinical trials often show discouraging outcomes [83]. Hence, a goal of translational research is to sequentially narrow down as many candidate therapies as possible. In this regard, as our ability to test thousands of compounds and natural products in hundreds of cell lines and animal models has proven useful [15,84].

Preclinical research has also paved the way for predicting pharmacokinetics and pharmacodynamics of drugs. For example, the effective dose of a drug required in the patient can be measured accurately by comparing QSAR and IC50 properties of drugs with other optimized drugs. Moreover, doses of a wide variety of drugs measured in isogenic cell lines settings can successfully recapitulate the dose spectrum required in syngeneic mouse models and even in human patients [15,71,85].

Although, the link between IC50 at the cellular level and in vivo at the plasma concentrations may sometimes be complicated, a general equation linking in vivo doses and effective concentrations is:

$$(D/\tau) = (CL/F) \times C_{target}$$

where: D = Dose, τ = dosing interval, CL = body clearance of the drug, F = bioavailabilty (fractions absorbed) of the drug by the selected route of administration and C_{target} = total plasma concentration required for desired effect [86]. The C_{target} is usually greater than IC50, since desired response often is more than just 50% of inhibition of the system.

Furthermore, precise in vitro genetic modification of patient derived T-cells has resulted in breakthrough efficacy in the form of immunotherapy [87]. In this regard, the efficacy of immunotherapeutic agents- 'checkpoint inhibitors', have also been widely facilitated by mechanistic insight gained by in vitro and preclinical research [88].

13. Challenges and Scopes

Quantitative chemical biology research is not without limitations. For example, HTS performed in cell lines has not often faithfully recapitulated the complexity of the microenvironment, cell-tissue heterogeneity and host-microbiome interactions in mouse models or human patients [89,90]. To address this, the field is increasingly experimenting with 3D and co-culture driven models [91,92]. As mentioned before, a caveat of biomarker and candidate target prediction approaches is that it often is correlative and lacks causality. With the advent of artificial intelligence, the use of machine learning is proving useful to predict causal biomarkers more accurately for therapy [23]. However, machine learning is limited in its ability to nullify batch artifacts as well as to combine different OMICS data—challenges to be addressed in the future.

For causality analysis, the recent advent in CRISPR technology has made it possible to perform high throughput loss of function genetic screens and couple that information with chemical perturbagen screens in order to improve target discovery [34,59]. Variations of CRISPR technologies, such CRISPRa, has made it possible to conduct gain of function screens in this context as well [37,38,59].

One of the major limitations of current cancer therapies is the emergence of resistance. Accumulating evidence indicates not only mutation of the drug target but also the presence of co-occurring mutations and heterogenous genetic and epigenetic background of tumors as causative factors promoting drug resistance [93,94]. Acquisition and analysis of -OMICS data serially to characterize tumor evolution during treatment has the potential to provide target-driven therapeutic approaches. One goal would be to predict the trajectory of tumor evolution during treatment based on preclinical and clinical data and deploy and adapt the therapy regimen accordingly [83,95].

14. Conclusions

In this review, we have attempted to summarize recent developments in quantitative chemical biology and outlined parameters for such quantitative assays. Successful implementation of cancer therapeutics requires a comprehensive understanding and analysis of both -OMICS data and pharmacodynamic-pharmacokinetic responses [83]. We hope that an improved quantitative understanding of chemical biology will transform aggressive cancers into chronic or curable conditions through more accurate clinical use of current and future systemic therapies.

Author Contributions: Conceptualization, A.Z. and T.G.B.; writing—original draft preparation, A.Z.; writing—review and editing, T.G.B.; visualization, A.Z.; supervision, T.G.B.; funding acquisition, T.G.B. All authors have read and agreed to the published version of the manuscript.

Funding: The authors acknowledge the Chan-Zuckerberg Biohub for research support.

Conflicts of Interest: The authors declare no conflict of interest related to this article.

References

1. Schrödinger, E. *What Is Life? The Physical Aspect of the Living Cell*; The University Press: Cambridge, UK; The Macmillan Company: New York, NY, USA, 1945; Volume viii, p. 91.
2. Watson, J.D.; Crick, F.H. The structure of DNA. *Cold Spring Harb. Symp. Quant. Biol.* **1953**, *18*, 123–131. [CrossRef] [PubMed]
3. Werner, H.M.; Mills, G.B.; Ram, P.T. Cancer Systems Biology: A peek into the future of patient care? *Nat. Rev. Clin. Oncol.* **2014**, *11*, 167–716. [CrossRef] [PubMed]
4. Viktorsson, K.; Lewensohn, R.; Zhivotovsky, B. Systems biology approaches to develop innovative strategies for lung cancer therapy. *Cell Death Dis.* **2014**, *5*, e1260. [CrossRef] [PubMed]
5. Barretina, J.; Caponigro, G.; Stransky, N.; Venkatesan, K.; Margolin, A.A.; Kim, S.; Wilson, C.J.; Lehár, J.; Kryukov, G.V.; Sonkin, D.; et al. The Cancer Cell Line Encyclopedia enables predictive modelling of anticancer drug sensitivity. *Nature* **2012**, *483*, 603–607. [CrossRef]
6. Kirschner, M.W. The Meaning of Systems Biology. *Cell* **2005**, *121*, 503–504. [CrossRef]
7. Reuveni, S.; Urbakh, M.; Klafter, J. Role of substrate unbinding in Michaelis–Menten enzymatic reactions. *Proc. Natl. Acad. Sci. USA* **2014**, *111*, 4391–4396. [CrossRef]
8. Chen, W.W.; Niepel, M.; Sorger, P.K. Classic and contemporary approaches to modeling biochemical reactions. *Genes Dev.* **2010**, *24*, 1861–1875. [CrossRef]
9. Neubig, R.R.; Michael, S.; Terry, K.; Arthur, C.; International Union of Pharmacology Committee on Receptor Nomenclature and Drug Classification. XXXVIII. Update on terms and symbols in quantitative pharmacology. *Pharmacol. Rev.* **2003**, *55*, 597–606. [CrossRef]
10. Markossian, S.; Grossman, A.; Brimacombe, K. (Eds.) *Assay Guidance Manual*; Bethesda: Rockville, MD, USA, 2004.

11. Stewart, M.; Watson, I. Standard units for expressing drug concentrations in biological fluids. *Br. J. Clin. Pharmacol.* **1983**, *16*, 3–7. [CrossRef]
12. Abate, G.; Mshana, R.N.; Miörner, H. Evaluation of a colorimetric assay based on 3-(4, 5-dimethylthiazol-2-yl)-2, 5-diphenyl tetrazolium bromide (MTT) for rapid detection of rifampicin resistance in *Mycobacterium tuberculosis*. *Int. J. Tuberc. Lung Dis.* **1998**, *2*, 1011–1016.
13. Präbst, K.; Engelhardt, H.; Ringgeler, S.; Hübner, H. Basic Colorimetric Proliferation Assays: MTT, WST, and Resazurin. In *Cell Viability Assay*; Methods in Molecular Biology; Humana Press: New York, NY, USA, 2017; Volume 1601, pp. 1–17.
14. Orellana, E.A.; Kasinski, A.L. Sulforhodamine B (SRB) Assay in Cell Culture to Investigate Cell Proliferation. *Bio-Protocol* **2016**, *6*, e1984. [CrossRef] [PubMed]
15. McMillan, E.A.; Ryu, M.; Diep, C.H.; Mendiratta, S.; Clemenceau, J.R.; Vaden, R.M.; Kim, J.; Motoyaji, T.; Covington, K.R.; Peyton, M.; et al. Chemistry-First Approach for Nomination of Personalized Treatment in Lung Cancer. *Cell* **2018**, *173*, 864–878. [CrossRef] [PubMed]
16. D'Errico, G.; Machado, H.L.; Sainz, B. A current perspective on cancer immune therapy: Step-by-step approach to constructing the magic bullet. *Clin. Transl. Med.* **2017**, *6*, 3. [CrossRef]
17. Swinney, D.C. Phenotypic vs. Target-Based Drug Discovery for First-in-Class Medicines. *Clin. Pharmacol. Ther.* **2013**, *93*, 299–301. [CrossRef]
18. Zheng, W.; Thorne, N.; McKew, J.C. Phenotypic screens as a renewed approach for drug discovery. *Drug Discov. Today* **2013**, *18*, 1067–1073. [CrossRef] [PubMed]
19. Geysen, H.M.; Meloen, R.H.; Barteling, S.J. Use of peptide synthesis to probe viral antigens for epitopes to a resolution of a single amino acid. *Proc. Natl. Acad. Sci. USA* **1984**, *81*, 3998–4002. [CrossRef]
20. Houghten, R.A. General method for the rapid solid-phase synthesis of large numbers of peptides: Specificity of antigen-antibody interaction at the level of individual amino acids. *Proc. Natl. Acad. Sci. USA* **1985**, *82*, 5131–5135. [CrossRef]
21. Liu, R.; Li, X.; Lam, K.S. Combinatorial chemistry in drug discovery. *Curr. Opin. Chem. Biol.* **2017**, *38*, 117–126. [CrossRef]
22. Ali, M.; Aittokallio, T. Machine learning and feature selection for drug response prediction in precision oncology applications. *Biophys. Rev.* **2019**, *11*, 31–39. [CrossRef]
23. Costello, J.C.; Heiser, L.M.; Georgii, E.; Gönen, M.; Menden, M.P.; Wang, N.J.; Bansal, M.; Ammad-ud-din, M.; Hintsanen, P.; Khan, S.A.; et al. A community effort to assess and improve drug sensitivity prediction algorithms. *Nat. Biotechnol.* **2014**, *32*, 1202–1212. [CrossRef]
24. Jang, I.S.; Neto, E.C.; Guinney, J.; Friend, S.H.; Margolin, A.A. Systematic assessment of analytical methods for drug sensitivity prediction from cancer cell line data. *Biocomputing* **2014**, 63–74. [CrossRef]
25. Dong, Z.; Zhang, N.; Li, C.; Wang, H.; Fang, Y.; Wang, J.; Zheng, X. Anticancer drug sensitivity prediction in cell lines from baseline gene expression through recursive feature selection. *BMC Cancer* **2015**, *15*, 489. [CrossRef] [PubMed]
26. Hejase, H.; Chan, C. Improving Drug Sensitivity Prediction Using Different Types of Data. *CPT Pharmacomet. Syst. Pharmacol.* **2015**, *4*, 98–105. [CrossRef] [PubMed]
27. LaCroix, B.; Gamazon, E.R.; Lenkala, D.; Im, H.K.; Geeleher, P.; Ziliak, D.; Cox, N.J.; Huang, R.S. Integrative analyses of genetic variation, epigenetic regulation, and the transcriptome to elucidate the biology of platinum sensitivity. *BMC Genom.* **2014**, *15*, 292. [CrossRef] [PubMed]
28. Weng, L.; Ziliak, D.; Lacroix, B.; Geeleher, P.; Huang, R.S. Integrative "omic" analysis for tamoxifen sensitivity through cell based models. *PLoS ONE* **2014**, *9*, e93420.
29. Eskiocak, B.; McMillan, E.A.; Mendiratta, S.; Kollipara, R.K.; Zhang, H.; Humphries, C.G.; Wang, C.; Garcia-Rodriguez, J.; Ding, M.; Zaman, A.; et al. Biomarker Accessible and Chemically Addressable Mechanistic Subtypes of BRAF Melanoma. *Cancer Discov.* **2017**, *7*, 832–851. [CrossRef]
30. Ding, Z.; Zu, S.; Gu, J. Evaluating the molecule-based prediction of clinical drug responses in cancer. *Bioinformatics* **2016**, *32*, 2891–2895. [CrossRef]
31. Geeleher, P.; Cox, N.J.; Huang, R.S. Clinical drug response can be predicted using baseline gene expression levels and in vitro drug sensitivity in cell lines. *Genome Biol.* **2014**, *15*, R47. [CrossRef]
32. Geeleher, P.; Zhang, Z.; Wang, F.; Gruener, R.F.; Nath, A.; Morrison, G.; Bhutra, S.; Grossman, R.L.; Huang, R.S. Discovering novel pharmacogenomic biomarkers by imputing drug response in cancer patients from large genomics studies. *Genome Res.* **2017**, *27*, 1743–1751. [CrossRef]
33. Gainor, J.F.; Dardaei, L.; Yoda, S.; Friboulet, L.; Leshchiner, I.; Katayama, R.; Dagogo-Jack, I.; Gadgeel, S.; Schultz, K.; Singh, M.; et al. Molecular Mechanisms of Resistance to First- and Second-Generation ALK Inhibitors in ALK-Rearranged Lung Cancer. *Cancer Discov.* **2016**, *6*, 1118–1133. [CrossRef]
34. Dixit, A.; Parnas, O.; Li, B.; Chen, J.; Fulco, C.P.; Jerby-Arnon, L.; Marjanovic, N.D.; Dionne, D.; Burks, T.; Raychowdhury, R.; et al. Perturb-Seq: Dissecting molecular circuits with scalable single-cell RNA profiling of pooled genetic screens. *Cell* **2016**, *167*, 1853–1866. [CrossRef] [PubMed]
35. Adamson, B.; Norman, T.M.; Jost, M.; Cho, M.Y.; Nuñez, J.K.; Chen, Y.; Villalta, J.E.; Gilbert, L.A.; Horlbeck, M.A.; Hein, M.Y.; et al. A Multiplexed Single-Cell CRISPR Screening Platform Enables Systematic Dissection of the Unfolded Protein Response. *Cell* **2016**, *167*, 1867–1882. [CrossRef] [PubMed]

36. Liu, X.S.; Wu, H.; Ji, X.; Stelzer, Y.; Wu, X.; Czauderna, S.; Shu, J.; Dadon, D.; Young, R.A.; Jaenisch, R.; et al. Editing DNA Methylation in the Mammalian Genome. *Cell* **2016**, *167*, 233–247. [CrossRef] [PubMed]
37. Chen, S.; Sanjana, N.E.; Zheng, K.; Shalem, O.; Lee, K.; Shi, X.; Scott, D.A.; Song, J.; Pan, J.Q.; Weisslederet, R.; et al. Genome-wide CRISPR screen in a mouse model of tumor growth and metastasis. *Cell* **2015**, *160*, 1246–1260. [CrossRef]
38. Shalem, O.; Sanjana, N.E.; Hartenian, E.; Shi, X.; Scott, D.A.; Mikkelson, T.; Heckl, D.; Ebert, B.L.; Root, D.E.; Doench, J.G.; et al. Genome-scale CRISPR-Cas9 knockout screening in human cells. *Science* **2014**, *343*, 84–87. [CrossRef]
39. Kampmann, M.; Horlbeck, M.A.; Chen, Y.; Tsai, J.C.; Bassik, M.C.; Gilbert, L.A.; Villalta, J.E.; Kwon, S.C.; Chang, H.; Kim, V.N.; et al. Next-generation libraries for robust RNA interference-based genome-wide screens. *Proc. Natl. Acad. Sci. USA* **2015**, *112*, E3384–E3391. [CrossRef]
40. Kampmann, M.; Bassik, M.C.; Weissman, J.S. Functional genomics platform for pooled screening and generation of mammalian genetic interaction maps. *Nat. Protoc.* **2014**, *9*, 1825–1847. [CrossRef]
41. Mi, H.; Muruganujan, A.; Ebert, D.; Huang, X.; Thomas, P.D. PANTHER version 14: More genomes, a new PANTHER GO-slim and improvements in enrichment analysis tools. *Nucleic Acids Res.* **2019**, *47*, D419–D426. [CrossRef]
42. Kanehisa, M.; Sato, Y.; Kawashima, M.; Furumichi, M.; Tanabe, M. KEGG as a reference resource for gene and protein annotation. *Nucleic Acids Res.* **2016**, *44*, D457–D462. [CrossRef]
43. Jassal, B.; Matthews, L.; Viteri, G.; Gong, C.; Lorente, P.; Fabregat, A.; Sidiropoulos, K.; Cook, J.; Gillespie, M.; Haw, R.; et al. The reactome pathway knowledgebase. *Nucleic Acids Res.* **2020**, *48*, D498–D503. [CrossRef]
44. Fabregat, A.; Jupe, S.; Matthews, L.; Sidiropoulos, K.; Gillespie, M.; Garapati, P.; Haw, R.; Jassal, B.; Korninger, F.; May, B.; et al. The Reactome Pathway Knowledgebase. *Nucleic Acids Res.* **2018**, *46*, D649–D655. [CrossRef] [PubMed]
45. Chen, E.Y.; Tan, C.M.; Kou, Y.; Duan, Q.; Wang, Z.; Meirelles, G.V.; Clark, N.R.; Ma'ayan, A. Enrichr: Interactive and collaborative HTML5 gene list enrichment analysis tool. *BMC Bioinform.* **2013**, *14*, 128. [CrossRef] [PubMed]
46. Liberzon, A.; Birger, C.; Thorvaldsdóttir, H.; Ghandi, M.; Mesirov, J.P.; Tamayo, P. The Molecular Signatures Database (MSigDB) hallmark gene set collection. *Cell Syst.* **2015**, *1*, 417–425. [CrossRef] [PubMed]
47. Liberzon, A.; Subramanian, A.; Pinchback, R.; Thorvaldsdóttir, H.; Tamayo, P.; Mesirov, J.P. Molecular signatures database (MSigDB) 3.0. *Bioinformatics* **2011**, *27*, 1739–1740. [CrossRef] [PubMed]
48. Martignetti, L.; Calzone, L.; Bonnet, E.; Barillot, E.; Zinovyev, A. ROMA: Representation and Quantification of Module Activity from Target Expression Data. *Front. Genet.* **2016**, *7*, 18. [CrossRef]
49. Tarca, A.L.; Draghici, S.; Khatri, P.; Hassan, S.S.; Mittal, P.; Kim, J.; Kim, C.J.; Kusanovic, J.P.; Romero, R. A novel signaling pathway impact analysis. *Bioinformatics* **2009**, *25*, 75–82. [CrossRef]
50. Gu, Z.; Wang, J. CePa: An R package for finding significant pathways weighted by multiple network centralities. *Bioinformatics* **2013**, *29*, 658–660. [CrossRef]
51. Ma, J.; Shojaie, A.; Michailidis, G. Network-based pathway enrichment analysis with incomplete network information. *Bioinformatics* **2016**, *32*, 3165–3174. [CrossRef]
52. Kutmon, M.; Riutta, A.; Nunes, N.; Hanspers, K.; Willighagen, E.L.; Bohler, A.; Mélius, J.; Waagmeester, A.; Sinha, S.R.; Miller, R.; et al. WikiPathways: Capturing the full diversity of pathway knowledge. *Nucleic Acids Res.* **2016**, *44*, D488–D494. [CrossRef]
53. Gao, S.; Wang, X. TAPPA: Topological analysis of pathway phenotype association. *Bioinformatics* **2007**, *23*, 3100–3102. [CrossRef]
54. Hung, J.-H.; Whitfield, T.W.; Yang, T.-H.; Hu, Z.; Weng, Z.; DeLisi, C. Identification of functional modules that correlate with phenotypic difference: The influence of network topology. *Genome Biol.* **2010**, *11*, R23. [CrossRef] [PubMed]
55. Nguyen, T.M.; Shafi, A.; Nguyen, T.; Draghici, S. Identifying significantly impacted pathways: A comprehensive review and assessment. *Genome Biol.* **2019**, *20*, 203. [CrossRef] [PubMed]
56. Amadoz, A.; Hidalgo, M.R.; Çubuk, C.; Carbonell-Caballero, J.; Dopazo, J. A comparison of mechanistic signaling pathway activity analysis methods. *Brief. Bioinform.* **2019**, *20*, 1655–1668. [CrossRef]
57. Szklarczyk, D.; Gable, A.L.; Lyon, D.; Junge, A.; Wyder, S.; Huerta-Cepas, J.; Simonovic, M.; Doncheva, N.T.; Morris, J.H.; Bork, P.; et al. STRING v11: Protein-protein association networks with increased coverage, supporting functional discovery in genome-wide experimental datasets. *Nucleic Acids Res.* **2019**, *47*, D607–D613. [CrossRef]
58. Meng, X.Y.; Zhang, H.; Mezei, M.; Cui, M. Molecular docking: A powerful approach for structure-based drug discovery. *Curr. Comput. Aided Drug Des.* **2011**, *7*, 146–157. [CrossRef] [PubMed]
59. Gilbert, L.A.; Horlbeck, M.A.; Adamson, B.; Villalta, J.E.; Chen, Y.; Whitehead, E.H.; Guimaraes, C.; Panning, B.; Ploegh, H.L.; Bassik, M.C.; et al. Genome-Scale CRISPR-Mediated Control of Gene Repression and Activation. *Cell* **2014**, *159*, 647–661. [CrossRef] [PubMed]
60. Zaman, A. Docking studies and network analyses reveal capacity of compounds from Kandelia rheedii to strengthen cellular immunity by interacting with host proteins during tuberculosis infection. *Bioinformation* **2012**, *8*, 1012–1020. [CrossRef]
61. Lindsay, M.A. Target discovery. *Nat. Rev. Drug Discov.* **2003**, *2*, 831–838. [CrossRef]
62. Pagadala, N.S.; Syed, K.; Tuszynski, J. Software for molecular docking: A review. *Biophys. Rev.* **2017**, *9*, 91–102. [CrossRef]
63. Cherkasov, A.; Muratov, N.A.; Fourches, D.; Varnek, A.; Baskin, I.I.; Cronin, M.; Dearden, J.; Gramatica, P.; Martin, Y.C.; Todeschini, R.; et al. QSAR modeling: Where have you been? Where are you going to? *J. Med. Chem.* **2014**, *57*, 4977–5010. [CrossRef]
64. Wolff, M.E.; Hansch, C. Correlation of physicochemical parameters and biological activity in steroids 9α-substituted cortisol derivatives. *Experientia* **1973**, *29*, 1111–1113. [CrossRef] [PubMed]

65. Katayama, M.; Gautam, R.K. Synthesis and Biological Activities of Substituted 4,4,4-Trifluoro-3-(indole-3-) butyric Acids, Novel Fluorinated Plant Growth Regulators. *Biosci. Biotechnol. Biochem.* **1996**, *60*, 755–759. [CrossRef] [PubMed]
66. Islam, M.R.; Zaman, A.; Jahan, I.; Chakravorty, R.; Chakraborty, S. In silico QSAR analysis of quercetin reveals its potential as therapeutic drug for Alzheimer's disease. *J. Young Pharm.* **2013**, *5*, 173–179. [CrossRef]
67. Foucquier, J.; Guedj, M. Analysis of drug combinations: Current methodological landscape. *Pharmacol. Res. Perspect.* **2015**, *3*, e00149. [CrossRef]
68. Palmer, A.C.; Sorger, P.K. Combination Cancer Therapy Can Confer Benefit via Patient-to-Patient Variability without Drug Additivity or Synergy. *Cell* **2017**, *171*, 1678–1691. [CrossRef] [PubMed]
69. Shah, K.N.; Bhatt, R.; Rotow, J.; Rohrberg, J.; Olivas, V.; Wang, V.E.; Hemmati, G.; Martins, M.M.; Maynard, A.; Kuhn, J.; et al. Aurora kinase A drives the evolution of resistance to third-generation EGFR inhibitors in lung cancer. *Nat. Med.* **2019**, *25*, 111–118. [CrossRef]
70. Wang, V.E.; Xue, J.Y.; Frederick, D.T.; Cao, Y.; Lin, E.; Wilson, C.; Urisman, A.; Carbone, D.P.; Flaherty, K.T.; Bernards, R.; et al. Adaptive Resistance to Dual BRAF/MEK Inhibition in BRAF-Driven Tumors through Autocrine FGFR Pathway Activation. *Clin. Cancer. Res.* **2019**, *25*, 7202–7217. [CrossRef]
71. Lin, L.; Sabnis, A.J.; Chan, E.; Olivas, V.; Cade, L.; Pazarentzos, E.; Asthana, S.; Neel, D.; Yan, J.J.; Lu, X.; et al. The Hippo effector YAP promotes resistance to RAF- and MEK-targeted cancer therapies. *Nat. Genet.* **2015**, *47*, 250–256. [CrossRef]
72. Yadav, B.; Wennerberg, K.; Aittokallio, T.; Tang, J. Searching for Drug Synergy in Complex Dose-Response Landscapes Using an Interaction Potency Model. *Comput. Struct. Biotechnol. J.* **2015**, *13*, 504–513. [CrossRef]
73. Palmer, A.C.; Chidley, C.; Sorger, P.K. A curative combination cancer therapy achieves high fractional cell killing through low cross-resistance and drug additivity. *eLife* **2019**, *8*, e50036. [CrossRef]
74. Gaddum, J.H. Discoveries in therapeutics. *J. Pharm. Pharmacol.* **1954**, *6*, 497–512. [CrossRef]
75. Grabovsky, Y.; Tallarida, R. Isobolographic Analysis for Combinations of a Full and Partial Agonist: Curved Isoboles. *J. Pharmacol. Exp. Ther.* **2004**, *310*, 981–986. [CrossRef] [PubMed]
76. Chou, T.-C. Theoretical Basis, Experimental Design, and Computerized Simulation of Synergism and Antagonism in Drug Combination Studies. *Pharmacol. Rev.* **2006**, *58*, 621–681. [CrossRef]
77. Greco, W.R.; Bravo, G.; Parsons, J.C. The search for synergy: A critical review from a response surface perspective. *Pharmacol. Rev.* **1995**, *47*, 331–385. [PubMed]
78. Loewe, S. Effect of combinations: Mathematical basis of problem. *Arch. Exp. Pathol. Pharmakol.* **1926**, *114*, 313–326. [CrossRef]
79. Chou, T.C.; Talalay, P. Quantitative analysis of dose-effect relationships: The combined effects of multiple drugs or enzyme inhibitors. *Adv. Enzyme Regul.* **1984**, *22*, 27–55. [CrossRef]
80. Garnett, M.J.; Edelman, E.J.; Heidorn, S.J.; Greenman, C.D.; Dastur, A.; Lau, K.W.; Greninger, P.; Thompson, I.R.; Luo, X.; Soares, J.; et al. Systematic identification of genomic markers of drug sensitivity in cancer cells. *Nature* **2012**, *483*, 570–575. [CrossRef]
81. DuPage, M.; Dooley, A.L.; Jacks, T. Conditional mouse lung cancer models using adenoviral or lentiviral delivery of Cre recombinase. *Nat. Protoc.* **2009**, *4*, 1064–1072. [CrossRef]
82. Byrne, A.T.; Alférez, D.G.; Amant, F.; Annibali, D.; Arribas, J.; Biankin, A.V.; Bruna, A.; Budinská, E.; Caldas, C.; Chang, D.K.; et al. Interrogating open issues in cancer precision medicine with patient-derived xenografts. *Nat. Rev. Cancer* **2017**, *17*, 254–268. [CrossRef]
83. Zaman, A.; Bivona, T.G. Emerging application of genomics-guided therapeutics in personalized lung cancer treatment. *Ann. Transl. Med.* **2018**, *6*, 160. [CrossRef]
84. Kim, H.S.; Mendiratta, S.; Kim, J.; Pecot, C.V.; Larsen, J.E.; Zubovych, I.; Seo, B.Y.; Kim, J.; Eskiocak, B. Systematic identification of molecular subtype-selective vulnerabilities in non-small-cell lung cancer. *Cell* **2013**, *155*, 552–566. [CrossRef] [PubMed]
85. Nichols, R.J.; Haderk, F.; Stahlhut, C.; Schulze, C.J.; Hemmati, G.; Wildes, D.; Tzitzilonis, C.; Mordec, K.; Marquez, A.; Romero, J.; et al. RAS nucleotide cycling underlies the SHP2 phosphatase dependence of mutant BRAF-, NF1- and RAS-driven cancers. *Nat. Cell Biol.* **2018**, *20*, 1064–1073. [CrossRef] [PubMed]
86. Lux, L.J.; Posey, R.E.; Daniels, L.S.; Henke, D.C.; Durham, C.; Jonas, D.E.; Lohr, K.N. *Pharmacokinetic/Pharmacodynamic Measures for Guiding Antibiotic Treatment for Hospital-Acquired Pneumonia*; Agency for Healthcare Research and Quality: Rockville, MD, USA, 2014.
87. Leach, D.R.; Krummel, M.F.; Allison, J.P. Enhancement of Antitumor Immunity by CTLA-4 Blockade. *Science* **1996**, *271*, 1734–1736. [CrossRef] [PubMed]
88. McIntyre, B.W.; Allison, J. The mouse T cell receptor: Structural heterogeneity of molecules of normal T cells defined by Xenoantiserum. *Cell* **1983**, *34*, 739–746. [CrossRef]
89. Haibe-Kains, B.; El-Hachem, N.; Birkbak, N.J.; Jin, A.C.; Beck, A.H.; Aerts, H.; Quackenbush, J. Inconsistency in large pharmacogenomic studies. *Nature* **2013**, *504*, 389–393. [CrossRef]
90. Better support translational research. *Nat. Microbiol.* **2017**, *2*, 1333. [CrossRef]
91. Dijkstra, K.K.; Cattaneo, C.M.; Weeber, F.; Chalabi, M.; Haar, J.V.; Fanchi, L.F.; Slagter, M.; Velden, D.L.; Kaing, S.; Kelderman, S.; et al. Generation of Tumor-Reactive T Cells by Co-culture of Peripheral Blood Lymphocytes and Tumor Organoids. *Cell* **2018**, *174*, 1586–1598. [CrossRef]
92. Bivona, T.G.; Doebele, R.C. A framework for understanding and targeting residual disease in oncogene-driven solid cancers. *Nat. Med.* **2016**, *22*, 472–478. [CrossRef]

93. Blakely, C.M.; Watkins, T.B.K.; Wu, W.; Gini, B.; Chabon, J.J.; McCoach, C.E.; McGranahan, N.; Wilson, G.A.; Birkbak, N.J.; Olivas, V.; et al. Evolution and clinical impact of co-occurring genetic alterations in advanced-stage EGFR-mutant lung cancers. *Nat. Genet.* **2017**, *49*, 1693–1704. [CrossRef]
94. Gerlinger, M.; Rowan, A.J.; Horswell, S.; Math, M.; Larkin, J.; Endesfelder, D.; Gronroos, E.; Martinez, P.; Matthews, N.; Stewart, A.; et al. Intratumor heterogeneity and branched evolution revealed by multiregion sequencing. *N. Engl. J. Med.* **2012**, *366*, 883–892. [CrossRef]
95. Zaman, A.; Wu, W.; Bivona, T.G. Targeting Oncogenic BRAF: Past, Present, and Future. *Cancers* **2019**, *11*, 1197. [CrossRef] [PubMed]

cancers

MDPI

Review

Long Non-Coding RNAs as Emerging Targets in Lung Cancer

Jovanka Gencel-Augusto [1,2], Wei Wu [1,2,*] and Trever G. Bivona [1,2,3,*]

1 Department of Medicine, University of California San Francisco (UCSF), San Francisco, CA 94158, USA; jovanka.gencelaugusto@ucsf.edu
2 UCSF Hellen Diller Comprehensive Cancer Center, San Francisco, CA 94158, USA
3 Chan-Zuckerberg Biohub, San Francisco, CA 94158, USA
* Correspondence: wei.wu@ucsf.edu (W.W.); trever.bivona@ucsf.edu (T.G.B.)

Simple Summary: Long non-coding RNAs (LncRNAs) are non-protein coding molecules longer than 200 nucleotides. They play essential roles in normal cell function and development, and can contribute to diseases such as cancer when dysregulated. Although lncRNAs have oncogenic or tumor-suppressive properties in lung cancer and can serve as stable biomarkers, this is still an understudied field. Here, we discuss recent evidence for lncRNAs role in lung cancer development, therapy resistance, biomarker potential, and therapeutic strategies. We conclude that understanding the interplay between non-coding and coding molecules in lung cancer should be explored further and may open up new avenues for treatment.

Abstract: Long non-coding RNAs (LncRNAs) are mRNA-like molecules that do not encode for proteins and that are longer than 200 nucleotides. LncRNAs play important biological roles in normal cell physiology and organism development. Therefore, deregulation of their activities is involved in disease processes such as cancer. Lung cancer is the leading cause of cancer-related deaths due to late stage at diagnosis, distant metastasis, and high rates of therapeutic failure. LncRNAs are emerging as important molecules in lung cancer for their oncogenic or tumor-suppressive functions. LncRNAs are highly stable in circulation, presenting an opportunity for use as non-invasive and early-stage cancer diagnostic tools. Here, we summarize the latest works providing in vivo evidence available for lncRNAs role in cancer development, therapy-induced resistance, and their potential as biomarkers for diagnosis and prognosis, with a focus on lung cancer. Additionally, we discuss current therapeutic approaches to target lncRNAs. The evidence discussed here strongly suggests that investigation of lncRNAs in lung cancer in addition to protein-coding genes will provide a holistic view of molecular mechanisms of cancer initiation, development, and progression, and could open up a new avenue for cancer treatment.

Keywords: LncRNAs; lung cancer; metastasis; therapy resistance; biomarkers; alternative therapies

Citation: Gencel-Augusto, J.; Wu, W.; Bivona, T.G. Long Non-Coding RNAs as Emerging Targets in Lung Cancer. *Cancers* **2023**, *15*, 3135. https://doi.org/10.3390/cancers15123135

Academic Editor: Aamir Ahmad

Received: 9 May 2023
Revised: 4 June 2023
Accepted: 8 June 2023
Published: 10 June 2023

1. Introduction

Long non-coding RNAs (LncRNAs) are broadly defined as RNAs that usually do not encode for proteins and that are longer than 200 nucleotides. These are messenger RNA (mRNA)-like molecules that are transcribed by polymerase II, 5′ capped, and have a 3′ poly-A tail. Because they can form complex secondary structures, they often have functions. Many lncRNAs are preferentially found in the nucleus, where they participate in the regulation of chromatin organization and transcription, often by forming lncRNA-DNA triplex [1–3], as well as through the formation of nuclear speckles and regulation of splicing. In the cytoplasm, lncRNAs can regulate mRNA stability, bind to other non-coding RNAs, and modulate protein post-translational modifications and protein function [4–8].

LncRNAs have been studied in mammals since the early 1990s due to their involvement in developmental processes. For example, the *Xist* (X-inactive specific transcript)

lncRNA contributes to reshaping the architecture of chromatin to achieve X chromosome-silencing in early embryonic development [9]. The *H19* lncRNA is involved in genomic imprinting and regulation of the insulin growth factor 2 (*IGF2*) and other genes involved in embryonic growth [10,11]. The study of *HOX* genes, master regulators of embryonic development, led to the discovery of the lncRNA *HOTAIR* (Homeobox transcript antisense intergenic RNA). *HOTAIR* is transcribed from the antisense strand of the *HOXC* gene.

HOTAIR has been reported to repress the transcription of the *HOXD* loci via interaction with PRC2 (polycomb repressive complex 2) [12], although in vivo models developed later report conflicting results regarding *HOXC* or *HOXD* genes regulation by *HOTAIR* [13,14]. Because of the importance of physiological context to understand lncRNA molecular function, more recently, other groups have reported exclusively in vivo approaches using animal models of lncRNA genetic ablation. For example, a comprehensive study developed 18 knock-out (KO) mouse models for less well-known lncRNAs with human orthologs. Certain lncRNAs' expression was highly tissue-specific (such as *Fendrr*, *Manr* and *linc–Cox2* expressed mainly in lung), supporting a unique physiological role, while other lncRNAs were more ubiquitously expressed. Three of the analyzed lncRNAs were required for embryonic development (*Fendrr, Mdgt, Peril*), speaking to their fundamental functions [15]. Thus, the involvement of lncRNAs in normal cell physiology and organism development suggests that these may also control disease-related processes such as cancer.

Lung cancer is the leading cause of cancer-related mortality in the U.S., and non-small cell lung cancer (NSCLC) is the most common subtype. Lung cancer is commonly diagnosed in late stages, where patients present distant metastasis with 9% having a 5-year survival rate [16]. Although the development of targeted therapies (e.g., tyrosine kinase inhibitors, TKIs) has improved patient outcomes, their clinical efficacy is often limited by both innate and acquired resistance, permitting tumor progression and recurrence leading to poor survival rates [17]. It is imperative to better understand molecular drivers of tumorigenesis, metastasis, and therapy resistance in lung cancer to develop improved therapeutic strategies. LncRNAs are emerging as important molecules in cancer due to their oncogenic or tumor-suppressive functions. Here, we summarize the latest works providing in vivo evidence available for the role of lncRNAs in lung cancer, therapy resistance, and their potential as biomarkers.

2. Role of lncRNAs in Lung Cancer

Because of the role of lncRNAs in regulating a diverse array of cellular functions, deregulation of their activities is involved in cancer. Some mechanisms reported for lncRNAs functions in cancer are as follows: acting as miRNA sponges to modulate activity on their targets; interacting with histone-modifier enzymes to modulate known oncogene/tumor suppressor gene expression; interacting with transcription factors to repress/activate their transcriptional programs; acting as anti-sense molecules for tumor suppressor mRNAs, among other mechanisms [2,18]. Importantly, regulation of lncRNA expression and function in cancer follows similar principles to that of known oncogenes and tumor suppressors, as it can be mediated by DNA methylation [19], amplification or deletion [20], and mutation or SNPs of DNA sequences [21,22]. In this section, we provide examples of well-studied lncRNAs that are reported to have oncogenic or tumor-suppressive roles in lung cancer, as well as those with controversial functions.

MALAT1 (Metastasis-Associated Lung Adenocarcinoma Transcript 1) was one of the first lncRNAs described to be associated with cancer. In 2003, Ji et al. analyzed the gene expression profile in human primary lung cancer tumors that subsequently metastasized or those that did not metastasize and compared their transcriptional signatures. They identified a Metastasis-Associated Lung Adenocarcinoma Transcript 1, named *MALAT1*, for its higher expression in primary tumors that metastasized. They also found a significant correlation between higher level of *MALAT1* expression in stage I lung cancer and worse survival outcomes [23]. Several studies since then have described *MALAT1* function in normal physiology and cancer [24–26]. Three independent groups developed an in vivo approach to describing *MALAT1* physiological function. They found that *Malat1* is highly abundant in several mouse tissues and highly conserved across species. Genetic perturbation of the *Malat1* locus in mice (using genetic deletion [27,28] or genetic inactivation approaches [29]) did not alter animal development, nuclear speckle formation, splicing, or mRNA stability. However, they described a role for *Malat1* in controlling neighboring genes expression in a tissue-specific manner that was not consistent between the three studies, especially that of *Neat1*, another lncRNA [28,29]. In the context of lung cancer, *MALAT1*-silencing did not show effects on lung cancer cell proliferation or viability in vitro [27]. To further understand the role of *MALAT1* in lung cancer metastasis, Gutschner et al. implanted EBC-1 lung cancer cells into nude mice and treated them with subcutaneous administration of an anti-sense oligonucleotide (ASO) targeting *MALAT1*. After five weeks of treatment, all primary tumors were excised. Metastasis nodules were analyzed at 12 weeks, indicating fewer and smaller metastatic nodules in the treated group, suggesting a role for *MALAT1* in promoting metastasis. Through additional in vitro studies, they report *MALAT1* inhibition results in aberrant expression of metastasis-associated genes in cell lines [30]. Although these studies support a role for *MALAT1* in promoting metastasis and regulating certain genes expression, other evidence exists for different roles. Kim et al. describe an elegant study that challenges previously reported roles for *MALAT1* in metastasis. Using the *Malat1* knock-out (KO) mouse model developed by Nakagawa et al. (LacZ and Poly-A sequences were used as transcriptional terminators inserted 69 bp downstream of the transcription start site of *Malat1* without deletion of the DNA sequence), Kim et al. crossed these mice with a breast cancer model driven by MMTV-PyMT that mimics human disease. Surprisingly, they found a 7.2-fold increase in metastatic foci and 31-fold increase in the percent of lung areas with metastatic lesions in *Malat1*-KO mice as compared to *Malat1* WT mice, suggesting a role for *Malat1* in suppressing breast cancer metastasis to the lung. This phenotype was rescued by transgenic expression of *Malat1*, suggesting that the RNA product itself diminished metastasis. Additionally, they identified *Malat1* interaction with TEAD (transcriptional enhanced associate domain) proteins in mouse-derived tumors and cell lines, which suppressed TEAD-YAP interaction and, therefore, inhibited their pro-metastatic transcriptional program [31]. This study and others highlight the importance of context (lung model vs. breast model) as well as experimental methodology to approach lncRNAs' functional characterization (such as lncRNA genomic DNA loss vs. RNA loss reviewed in detail elsewhere [24]). Of note, DNA elements themselves within a lncRNA locus may be responsible for regulatory functions that are independent from transcript function [32–35]. In summary, *MALAT1* promotes metastasis in lung cancer, but may show opposite functions in different types of cancer depending on cellular context.

The *GAS5* (growth arrest-specific 5) gene was first described as a G0-specific gene that is inhibited by serum and growth factors [36]. In vivo, *Gas5* genetic deletion ($Gas5^{+/-}$) in mice decreased bone mass and impaired bone repair, leading to osteoporosis. Mechanistically, *Gas5* positively influenced proper cell differentiation through interaction with UPF1 (a DNA/RNA helicase) to accelerate *SMAD7* mRNA decay [37]. In lung cancer, *GAS5* was found down-regulated in 72 NSCLC tumor samples as compared to their paired adjacent normal tissues, suggesting a tumor-suppressive role. Additionally, low *GAS5* expression was correlated with larger tumor size, lower differentiation levels, and higher staging of tumor–node metastasis [38]. A xenograft model of *GAS5* overexpression (OE) showed that *GAS5* OE markedly decreased tumor size compared with control [38]. Although this study did not explore a mechanism for *GAS5*-mediated tumor suppression, other studies implicate a role for *GAS5* as a miRNA sponge to negatively influence cell cycle activator genes [39] or positively influence *PTEN* levels [40]. Additionally, a recent study revealed *GAS5* is partially localized to the mitochondria where it modulates energy homeostasis by promoting the de-acetylation of malate dehydrogenase, suppressing breast cancer [41]. Taken together, a role for *GAS5* in halting the cell cycle as well as promoting cell differentiation in normal cells supports its tumor-suppressive role reported in lung cancer. Besides

GAS5, other lncRNAs have been studied for their tumor-suppressive functions, such as *MEG3* and *TUG1*, reviewed elsewhere [42].

LUCAT1 (Lung Cancer-Associated Transcript 1), first identified as smoke-induced and cancer-associated lncRNA1 (*SCAL1*) [43], has higher expression in lung cancer as compared to normal controls, and is also found to be overexpressed in several cancer types [44]. To our knowledge, a *LUCAT1*-deficient mouse model has not been reported. Additionally, patients with tumors that express high levels of *LUCAT1* showed poorer overall survival as compared to those with lower *LUCAT1* expression. Moreover, high *LUCAT1* levels were associated with late staging in tumor–lymph node metastasis and higher tumor volume. In NSCLC cell lines A549 and SPC-A1, *LUCAT1* modulates *p21* and *p57* expression by promoting loci methylation through PRC2 [45].

More recently, a role for *LUCAT1* in regulating immune responses has been described. *LUCAT1* genetic deletion in myeloid cells is found to enhance interferon-mediated gene transcription. *LUCAT1* acts as an immune suppressor by interacting with STAT1 and chromatin in the nucleus. It may also act by inhibiting NF-kB functions [46]. These findings suggest a tumor-promoting role for *LUCAT1* that is tumor-cell-intrinsic, in addition to a potential non-cell autonomous mechanism via the inhibition of immune surveillance, although this mechanism remains to be explored in the lung cancer context.

HOTAIR has been vastly studied in cancer contexts [47,48]. In NSCLC, tumor samples and cell lines expressed higher *HOTAIR* levels as compared to normal counterparts [49]. Additionally, high *HOTAIR* levels correlated with higher tumor grade and presence of lymph node metastases [49,50]. In vitro, *HOTAIR* has been reported as a direct target of the hypoxia-inducible-factor-1α (HIF-1α), therefore enhancing A549 NSCLC cells' proliferation, migration, and invasion [51]. In vivo, tail vein injections of SPC-A1 cells with or without siRNA targeting *HOTAIR* showed that the knock-down condition reduced the number of metastasis nodules found in the lungs of immunocompromised mice [49]. *HOTAIR*-silencing resulted in a decrease in matrix metalloproteinases (MMPs, which promote invasion and migration) expression and an increase in HOXA5 levels (a tumor suppressor) in cell lines, suggesting *HOTAIR* acts through the regulation of expression of cancer-related genes [49]. Although this xenograft assay does not account for all steps required for a tumor cell to achieve metastatic colonization (extravasation, survival in blood, seeding of new site, proliferation in new site), and they measured colonization of lungs using a lung cancer cell line (same tissue), it raises the possibility that *HOTAIR* may be involved in the seeding and survival of cancer cells. Additionally, the absence of a competent immune system challenges interpretation of these results. Development of a *HOTAIR* transgenic mouse model to understand in vivo implications in lung cancer initiation and progression is necessary, similar to a *HOTAIR* inducible system recently reported for breast cancer [52]. This model showed that sustained *HOTAIR* overexpression promotes breast cancer metastasis to lungs. Overall, with the data available, *HOTAIR* seems to play an oncogenic role in lung cancer; however, robust mechanisms through which this lncRNA function remain to be uncovered.

Most studies focus on the contribution of a single lncRNA to cancer phenotypes. However, whether the lncRNAs described above are expressed simultaneously in tumors with unique or redundant functions remains to be explored in depth. For example, Esposito et al. showed that at least 80 oncogenic lncRNAs are active in NSCLC through a lncRNA-focused CRISPR screen. By further dissecting the role of two candidate lncRNAs, *CHiLL1* and *GCAWKR*, they showed these have distinct cellular localization and non-overlapping targets. Importantly, ASOs targeting both these lncRNAs yielded additive effects, suggesting that they have cooperating functions in NSCLC progression [53]. LncRNAs are generally expressed at lower levels than protein-coding genes [54]. Because of this, we speculate that lncRNAs with redundant functions may be expressed simultaneously to compensate for a higher expression of their targets in disease conditions. By examining available TCGA lung adenocarcinoma datasets containing mRNA expression data, we did not find significant correlations (negative or positive) among the expression of lncRNAs described here. However, such an analysis in combination with functional studies could shed light on mutual exclusivity relationships between certain lncRNAs. Additionally, whether certain lncRNAs are predominantly expressed at different stages of tumor progression remains to be explored. A new online resource, lncRNAfunc https://ccsm.uth.edu/lncRNAfunc (accessed on 7 May 2023), provides insights on differentially expressed lncRNAs across different cancer types and stages available in TCGA, as well as functional predictions [55]. Although this analysis did not detect any correlations between the lncRNAs mentioned here and lung cancer stages, possibly due to lack of sufficient sample sizes, these lncRNAs did show correlation with stage in other cancers; for example, *LUCAT1* was correlated with cancer stage in kidney cancer.

In summary, lncRNAs have oncogenic and tumor-suppressive roles in lung cancer, illustrated in Figure 1. LncRNAs interact with protein-coding molecules, resulting in the activation or inactivation of specific signaling pathways in cancer cells. We speculate more lung-cancer-specific lncRNAs will be identified with genome-wide transcriptomic studies.

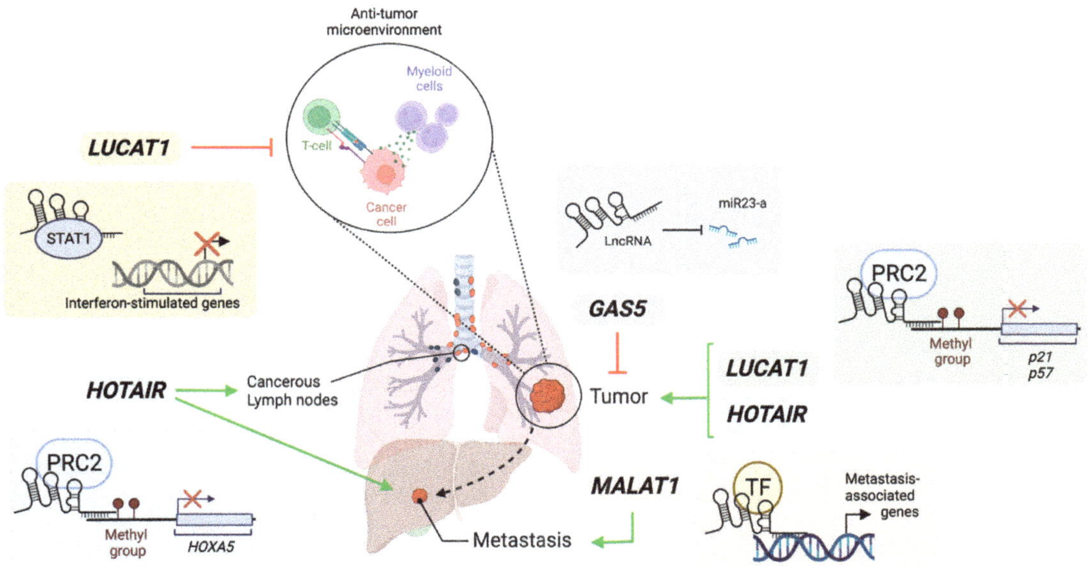

Figure 1. Illustration of roles of lncRNAs in lung cancer. *GAS5* acts as a miRNA sponge to activate their mRNA targets and inhibit cancer cell proliferation, therefore acting as a tumor suppressor. *LUCAT1* can promote cancer cell proliferation via epigenetic silencing of *p21* and *p57* loci. Additionally, *LUCAT1* may create a pro-tumorigenic microenvironment by inhibiting interferon-mediated responses through the sequestration of STAT1. *HOTAIR* can promote cancer cell proliferation and metastasis via the recruitment of PRC2 to methylate loci and repress gene expression, such as *HOXA5* which is a tumor suppressor. *MALAT1* stimulates lung cancer metastasis by potential recruitment of transcription factors (TF) to promoters of metastasis-associated genes. Green arrows represent positive influence. Red arrows represent inhibition. Created with Biorender.com (accessed on 4 June 2023).

3. LncRNAs in Lung Cancer Therapy Resistance

The role of non-coding RNAs in resistance to cancer therapies has been documented [48,56,57]. Mechanistically, lncRNAs can contribute to therapy resistance by promoting cell survival pathways (including autophagy, DNA repair), inhibiting apoptosis and cell cycle checkpoints, increasing self-renewal capacity and epithelial to mesenchymal transition (EMT), modulating the tumor microenvironment as well as the cellular xenobiotic stress response (drug efflux mechanisms), among others [48,58]. Here, we provide evidence for the roles of the lncRNAs discussed above in resistance to chemotherapy, radiotherapy, and targeted therapy in lung cancer. Additionally, we discuss new advances in understanding the role of lncRNAs in immune checkpoint therapy in lung cancer.

3.1. Role of lncRNAs in Resistance to Chemotherapy, Radiotherapy, and Targeted Therapy in Lung Cancer

In a tumor xenograft model, silencing of *MALAT1* in cisplatin resistant A549 cells led to decreased growth in nude mice (subsequently treated with cisplatin), as compared to non-targeting control. Similarly, overexpression of *MALAT1* in cisplatin-sensitive A549 lung cancer cells increased tumor volume as compared to empty vector control [59]. These data suggest a role for *MALAT1* in promoting cisplatin resistance in lung cancer. The authors suggest that modulation of STAT-3 function by *MALAT1* drives this phenotype, although a direct interaction was not confirmed [59]. While additional lung-cancer-focused studies are lacking, the *MALAT1* role in cisplatin resistance was also found in a xenograft model of oral squamous cell carcinoma [60]. In radiotherapy, *MALAT1* also promotes resistance, although this function has not been explored in lung cancer. In a xenograft model of esophageal squamous cell carcinoma, *MALAT1* levels were found to be reduced upon radiation in tumors that respond to treatment. Additionally, overexpressing *MALAT1* in xenografts did not decrease their size upon radiation exposure, while controls showed regression [61]. Furthermore, in colorectal carcinoma cell lines, *MALAT1* knockdown enhanced radiosensitivity [62]. Therefore, *MALAT1* can impact sensitivity to radiation therapy in cancer. To our knowledge, there are no robust in vivo studies addressing the role of *MALAT1* in lung cancer targeted therapy resistance (TKIs), only those in cell lines. Cheng et al. characterized differentially expressed lncRNAs in gefitinib (EGFR TKI)-sensitive PC9 cells and gefitinib-resistant PC9 cells. They found *H19* and *BC200* lncRNAs to be up-regulated in resistant cells vs. sensitive ones, while *MALAT1* and *HOTAIR* were down-regulated in the resistant setting. These data suggest that *MALAT1* may promote sensitivity to targeted therapy in lung cancer cells [63]. However, opposite roles for *MALAT1* in targeted therapy resistance in other cancers have been described. For example, *MALAT1* is overexpressed in Sutinib-resistant renal cell carcinoma tumors vs. sensitive tumors [64]. These findings suggest *MALAT1* plays a role in therapy resistance that can be highly context-specific in regard to the type of therapy (cisplatin vs. targeted therapy) or cancer primary site (lung vs. kidney) and support the need to study these functions and mechanisms in physiologically relevant settings.

GAS5, a tumor-suppressive lncRNA, plays a role in the sensitization of lung cancer cells to therapy. In cisplatin-resistant A549 and H1299 cells, *GAS5* overexpression reduced IC_{50} (half-maximal inhibitory concentration) values to cisplatin. In vivo, cisplatin-resistant A549 cells stably overexpressing *GAS5* yielded lower tumor volumes when injected into nude mice as compared to vector controls [65]. Additionally, a role for *GAS5* has been reported in sensitivity to targeted therapy (gefitinib, EGFR TKI). A xenograft mouse model of *GAS5* overexpression (OE), *GAS5* OE plus gefitinib, gefitinib alone, or vehicle, showed that *GAS5* OE plus gefitinib yielded the best tumor size reduction outcomes. This suggests that *GAS5* can synergize with targeted therapy to achieve better clinical outcomes [38]. Lastly, roles for *GAS5* in sensitizing lung cancer cells to radiotherapy have also been reported [66]. This evidence suggests *GAS5* as a promising target to sensitize lung tumors to cancer therapy.

LUCAT1 can contribute to cisplatin resistance in NSCLC. Shen et al. describe a role of *LUCAT1* as sponge of miR-514-3p, whose target is ULK1, a protein involved in autophagy. Therefore, *LUCAT1* promotes cisplatin resistance by modulating autophagy [67]. Although further investigation in the context of lung cancer is limited, *LUCAT1* is known to promote resistance to DNA-damaging agents in colorectal carcinomas [68]. Because *LUCAT1* has been recently reported to play a role in immune cell regulation [46,69], its role in immune checkpoint blockade therapies should be explored in detail as a possible target for combination therapy.

The role of *HOTAIR* in therapy resistance, similar to findings described for *MALAT1*, can be complex. For example, *HOTAIR* was found in higher levels in cisplatin-resistant NSCLC tumors as compared to sensitive ones [50]. In radiotherapy, *HOTAIR* can promote resistance to radiation therapy through the inhibition of *p21* in cervical cancer [70], it can modulate β-catenin signaling in Lewis lung cancer tumors [71], and it modulates Akt signaling in breast cancer cell lines [72]. In contrast, in targeted therapy, *HOTAIR* was reported down-regulated in tumors derived from acquired and primary resistant states to EGFR-TKIs as compared to treatment naïve tumors. Here, higher levels of *HOTAIR* expression were correlated with better survival outcomes [73]. Consistent with these findings, *HOTAIR* was down-regulated in gefitinib-resistant PC9 cells vs. sensitive ones [63]. However, another study reports higher *HOTAIR* expression in gefitinib-resistant PC9 cells as compared to gefitinib-sensitive cells [74]. Lastly, in vitro assays suggest that *HOTAIR* may mediate Crizotinib (ALK/ROS1 inhibitor) resistance through the up-regulation of autophagy [75]. Taken together, *HOTAIR* promotes resistance to cisplatin and radiotherapy in lung cancer. In targeted therapy, the type of drug and differences in experimental methodologies employed may account for the confounding role of *HOTAIR*.

Programmed cell death is crucial in attaining effective therapeutic responses regardless of the type of therapy employed. LncRNAs play a significant role in modulating cell death through several mechanisms, influencing therapeutic failure and resistance. For instance, certain lncRNAs can inhibit pro-apoptotic proteins such as P53, BAX and PARP-1, while others can promote anti-apoptotic proteins such as BCL-2 directly or through activating MYC [76–78]. Both mechanisms lead to the inhibition of apoptosis, contributing to therapy resistance. Ferroptosis is another type of programmed cell death that uses iron-dependent accumulation of reactive oxygen species (ROS) to induce death. LncRNAs can interfere with ferroptosis by modifying levels of key proteins, such as inhibiting ACSL4 or increasing GPX4 [79]. Pyroptosis is a type of cell death that triggers an inflammatory response and its role in cancer is controversial. LncRNAs can also exert an impact on pyroptosis. In particular, *XIST* can sequester SMAD2 in the cytoplasm, impeding the transcription of *NLRP3*, an essential mediator of pyroptosis. This finding was associated with an increased resistance to cisplatin in NSCLC [78]. In summary, understanding the role of lncRNAs in regulating programmed cell death, including apoptosis, ferroptosis and pyroptosis, is key for shedding light on their contribution to therapy resistance and developing strategies to overcome it.

3.2. Role of lncRNAs in Immunotherapy Responses in Lung Cancer

The role of lncRNAs in resistance to immune checkpoint inhibitors (ICI) is emerging as a field of study in many cancers [80]. ICIs target immune inhibitory molecules such as PD-1, PD-L1, and CTLA-4 with the goal of re-activating immune surveillance and tumor-cell killing [81]. Even though these therapies have favorable outcomes in certain tumor types, an effective and durable response in lung cancer is achieved only in ~25% of cases [82]. Therefore, understanding the underlying molecular mechanisms of response or resistance to ICIs is critical to improve lung cancer outcomes.

MALAT1 may be involved in regulating responses to immunotherapy. In a study of 113 NSCLC tumor samples, *MALAT1* expression was positively correlated with *PD-L1* mRNA as well as PD-L1 protein levels [83]. Here, the authors propose *MALAT1* acts as a sponge of *miR-200a-3p*, whose target is *PD-L1* [83]. Similarly, another study proposed *LINC01140* directly represses two miRNAs (*miR-377* and *miR-155-5p*) whose target is *PD-L1*. Therefore, *LINC01140* expression promotes *PD-L1* expression and a potential pro-tumorigenic microenvironment [84]. In a co-culture assay, *LINC01140-silencing* in lung cancer cells promoted higher IFN-γ secretion from cytokine-induced killer cells, as compared to non-targeting control. In a xenograft model, lung cancer cells with knock-down of *LINC01140* were injected into immunocompromised mice and received peritumoral administration of cytokine-induced killer cells upon tumor establishment. Further tumor growth was inhibited in the knock-down condition as compared to non-targeting controls. Importantly, higher levels of pro-inflammatory cytokines were found in the circulation of mice injected with *LINC01140* knock-down tumor cells compared to controls [84]. Moreover, a recent study reports an unbiased approach to understanding lncRNAs' relationship to the tumor immune microenvironment and prediction of response to immune checkpoint therapy in NSCLC [85]. Based on lncRNAs that were correlated with immune-checkpoint expression, and taking into account overall survival data, Zhang et al. identified a signature of ten lncRNAs that they used to separate patients into "low"- and "high"-risk groups. They analyzed immune infiltrates in tumor samples and found a significantly higher density of T-cells ($CD4^+$ and $CD8^+$) and dendritic cells in the low-risk group (suggesting responsiveness to immunotherapy), while macrophages were higher in the high-risk group tumors (suggesting unresponsiveness to immunotherapy) [85]. Although the gold-standards for prediction of immunotherapy response are still levels of immune-checkpoint molecules, the evidence discussed here suggests there is potential for lncRNAs to function as biomarkers to predict immunotherapy response in lung cancer, as well as to be therapeutic targets in combination with immune checkpoint inhibitors.

More recently, the cGAS/STING pathway has gained importance in modulating cancer immunotherapy responses. This pathway provides a defense against microbial pathogens and malignant cells [86]. Therefore, approaches to activate it have recently emerged, such as STING agonists. These approaches have shown to synergize with immunotherapy (anti PD-L1) and achieve better responses in cancer [87]. Importantly, lncRNAs can also regulate the cGAS/STING pathway. In NSCLC, the lncRNA *PCAT1* was reported to activate the transcription of SOX2, therefore inhibiting cGAS/STING-dependent interferon responses and causing immunosuppression [88]. In non-cancer contexts, *MALAT1* has been reported to activate cGAS/STING through CREB, therefore promoting inflammatory lung conditions [89]. It is imperative to continue researching the role of lncRNAs in modulating cGAS/STING as their de-regulation may impact cancer patient responses to immunotherapy.

In summary, the lncRNAs discussed above have all been reported to modulate resistance to cancer therapies (illustrated in Figure 2), although evidence for some therapy types is limited to other cancer settings. A caveat in the data presented is the focus on association studies, in vitro and xenograft assays (immunocompromised mice) without strong mechanistic insights.

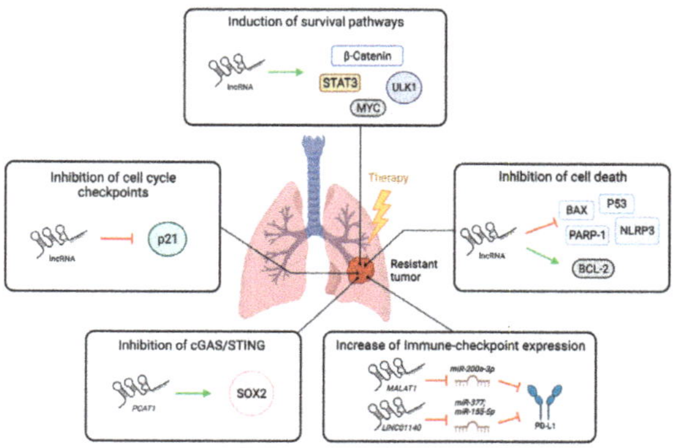

Figure 2. Illustration of lncRNAs contribution to therapy resistance. LncRNAs can induce survival pathways through activation of several proteins such as β-catenin, STAT3, MYC, or ULK1. LncRNAs inhibit cell death by inducing anti-apoptotic proteins such as BCL-2 or inhibiting pro-apoptotic proteins such as P53, BAX, PARP-1, and NLRP3. LncRNAs promote increase in PD-L1 levels by inhibiting miRNAs that target PD-L1. *PCAT1* induces *SOX2* to inhibit cGAS/STING pathway and

diminish interferon responses, thereby modulating immune-therapy efficacy. LncRNAs may inhibit cell cycle checkpoints by modulating *p21*, although this has not been explored in the context of lung cancer. Green arrows represent induction. red arrows represent inhibition. Created with Biorender.com (accessed on 4 June 2023).

4. LncRNAs as Biomarkers in NSCLC

LncRNAs are highly stable molecules that can be found in the systemic circulation. They are resistant to degradation due in part to their secondary structures, transport by exosomes, and stabilizing post-translational modifications [90,91]. Therefore, the study of lncRNAs in circulation is a plausible non-invasive method of detecting and following cancer progression. Currently, the most commonly used biomarkers for NSCLC diagnosis from circulation are carcinoembryonic antigen (CEA), cytokeratin-19 fragment (CYFRA21-1), squamous cell carcinoma antigen (SCCA), prolactin (PRL), and carbohydrate antigen 125 (CA125), which can be used individually or combined as a signature [92,93]. Additionally, lncRNAs can also be stably found in urine [94] and even in nasal mucosa [95], although the latter has not been explored in the context of cancer. In this section, we review evidence for the potential of lncRNAs as disease biomarkers in lung cancer.

Exosomes are small vesicles that facilitate the transfer of cargo from one cell to another. One of the first lncRNAs found in exosomes was *PARTICLE*. It was observed not only in exosomes isolated from the MDA-MB-361 metastatic breast adenocarcinoma cell line but also in plasma samples obtained from patients exposed to radiation [2]. In lung cancer, exosomal lncRNAs have been found to play a significant role in cancer development and metastasis. For example, the exosomal lncRNA *UFC1* was demonstrated to enhance lung cancer cell proliferation and metastasis by interacting with EZH2 and subsequently reducing *PTEN* expression [96]. Due to their packaging into exosomes, lncRNAs exhibit increased stability and can travel to distant sites, having significant potential effects in biological processes of both cancer and non-cancer cells. For instance, as reported for miRNAs, exosomal lncRNAs may sustain proliferation of cancer cells at the same time that they modify pre-metastatic niches by remodeling the microenvironment towards tumor-promoting immune cells or fibroblasts, as well as increasing vessel leakiness and angiogenesis [97]. A specific example is exosomal *HOTAIRM1*, which has been observed to modulate the expression of SPON2, an extracellular matrix protein, in cancer-associated fibroblasts. This modulation promotes cancer cell migration and invasion [98]. Additionally, *PCAT6* derived from NSCLC exosomes influences macrophages polarization, promoting the shift towards M2 phenotypes that support tumor growth [99]. These findings highlight the pivotal role of exosomal lncRNAs in intercellular communication and their implications in cancer progression.

Tang et al. analyzed lncRNAs expression in the blood from 232 patients diagnosed with NSCLC as compared to healthy controls. Expression levels of three lncRNAs (*RP11-397D12.4*, *AC007403.1*, *ERICH1-AS1*) were higher in disease versus health states. Importantly, this expression pattern was stable even after freeze–thaw cycles [100]. In a similar study, the lncRNAs *SPRY4-IT1*, *ANRIL*, and *NEAT1* were found overexpressed in NSCLC versus healthy controls (N = 50/group), and were detected stably after several freeze–thaw cycles and after the samples' exposure to room temperature for up to 24 h [101]. A more recent study analyzed exosomal lncRNAs from the blood of NSCLC patients vs. tuberculosis patients and healthy controls. They report higher levels of *RP5-977B1* in NSCLC compared to the two non-cancer groups. The diagnostic power of this lncRNA was greater than that of conventional biomarkers such as CEA and CYFRA21-1, and additionally worked for early-stage NSCLC, speaking to the promise of lncRNAs to detect early disease [102]. *HOTAIR* has also been evaluated for its diagnostic value for pathological staging of NSCLC. It was determined to have a power similar to that of biomarkers CEA and CYFRA21-1 [103]. *ESCCAL-1* is an oncogenic lncRNA, initially identified in esophageal cancer [104]. The results from a large cohort lung cancer screen show that *ESCCAL-1* has increased expression in serum samples from lung cancer patients as compared to serum from patients with benign nodules or healthy individuals. These studies support the use of elevated lncRNAs expression individually or as signatures as biomarkers to predict disease or as staging tools in NSCLC. Clinical trials that evaluate the use of lncRNAs as biomarkers are currently ongoing for other cancers, according to the ClinicalTrials.gov database (accessed on 12 April 2023).

Comparable to tumor-promoting lncRNAs, tumor-suppressive lncRNAs such as *GAS5* also show potential as biomarkers in NSCLC. Liang et al. compared *GAS5* levels in the plasma of 90 NSCLC patients vs. 33 healthy controls. They detected significantly lower levels of *GAS5* in plasma derived from cancer patients. Additionally, they measured the dynamics of *GAS5* before and after surgery, detecting an increase in *GAS5* levels seven days after the patients had surgery [105]. As discussed above, *GAS5* has a clear tumor-suppressive role in lung cancer; therefore, these studies are consistent and support the use of *GAS5* as biomarker for diagnosis and for responses to clinical intervention.

5. LncRNAs as Therapeutic Targets

The main approaches to targeting lncRNAs are similar to protein-coding genes: inhibiting oncogenic lncRNAs or restoring the function of tumor-suppressive lncRNAs. Here, we briefly describe advances in lncRNA therapeutics as well as some challenges.

Therapeutic approaches to target lncRNAs with oncogenic functions mainly use ASOs (anti-sense oligonucleotides). The core mechanisms of action of ASOs involve promoting the cleavage of their RNA targets, impeding their translation, or modulating splicing [106]. There are several types of ASOs depending on their chemical modifications, such as locked nucleic acids and morpholinos, that may make them more resistant to degradation, provide better cellular availability, and lower off-target events [106]. As mentioned in the sections above, ASOs are a strategy widely used in cell line and animal-based assays to elucidate lncRNAs role in cancer [30]. Although ASOs that target miRNAs have entered clinical trials [7,107], to our knowledge, there are no therapies in clinical trials targeting lncRNAs directly. Small molecules are another strategy to interfere with oncogenic lncRNAs function [108]. For example, quercetin, a recently developed small molecule, binds to a *MALAT1* triplex and modulates its transcript levels and functions in vitro [3]. Another small molecule recently identified, AC1NOD4Q, blocks the interaction between *HOTAIR* and EZH2 (a PRC2 subunit), impeding methylation of downstream targets [109]. Moreover, emerging strategies to silence lncRNAs in pre-clinical models use CRISPR-based approaches, although these have not reached the clinical setting [110].

Restoring the function of tumor-suppressive lncRNAs can be achieved via gene therapy or administration of synthetic RNA molecules approaches. Although gene therapy has not reached the clinic for lncRNAs as targets, there are pre-clinical models exploring it in non-cancer contexts. For example, the lncRNA *LeXis* has been explored for exogenous administration using an adeno-associated virus vector in a murine model of familial hypercholesterolemia [111].

A main limitation in translating lncRNA targeting approaches into the clinic is their relatively poor sequence conservation between humans and other species [112]. However, it is noteworthy that *MALAT1* sequence exhibits significant conservation among various species [113], rendering it a compelling candidate for in-depth investigation into its potential as a therapeutic target. Furthermore, it is worth noting that lncRNAs are highly tissue-specific thereby minimizing the likelihood of off-target effects when employed for therapeutic purposes. Several pre-clinical models have been used to explore *MALAT1* targeting, including genetically engineered mouse models, patient-derived cell lines, organoids, and xenografts. For instance, in a MMTV-PyMT breast cancer mouse model, subcutaneous delivery of *Malat1* ASO led to metastasis reduction and higher levels of cancer cell differentiation compared to non-targeting controls. Similarly, in a 3D breast cancer organoid model, *Malat1* ASOs inhibited branching morphogenesis [114]. In lung cancer, systemic administration of *Malat1* ASO in nude mice yielded a marked reduction in the colonization of patient-derived lung cancer cells in the lungs, as compared to non-targeting controls [30].

A more recent approach to targeting lncRNAs is the use of nanoparticles containing siRNA against a specific oncogenic lncRNA or nanoparticle-conjugated tumor-suppressive lncRNAs. Nanoconjugates of the tumor suppressor *MEG3* delivered intrahepatically to animals with liver cancer showed improvement in histopathology and tumor-associated biomarkers that was superior to that achieved with unconjugated *MEG3*. Although this study did not analyze differences in blood retention or side effects, it is important to consider these factors when assessing the superiority of nanoparticles as therapeutic approaches [115]. In another study, siRNA targeting *DANCR*, a tumor-promoting lncRNA, packed into nanoparticles was delivered systemically into breast cancer-bearing nude mice. This approach was effective in suppressing tumor progression with no significant changes in animal body weight and in morphology and histology of liver, kidneys, and lungs [116]. Although this strategy has not been used in in vivo models of lung cancer, its use in cell lines shows promising results as it was effective in silencing *DANCR* and reducing migration and invasion in vitro, supporting its further exploration in animal models [117].

There are several other challenges for RNA-based therapies reported from the field of miRNAs. For example, in the phase-I clinical trial for MRX34, a double-stranded *miR-34a* mimic encapsulated in liposomal nanoparticles, serious adverse events were immune-mediated toxicities, such as cytokine release syndrome, which resulted in four patient deaths and caused the trial to terminate early [118]. The delivery of these therapies is another challenge, as they may accumulate in detoxifying organs such as kidneys and liver, causing associated toxicities [119,120]. Additionally, off-target events and on-target events in non-tumor tissues can account for toxicities. Therefore, there is a need to detect immune-related toxicities in pre-clinical studies and design strategies to reduce their prevalence in order to engineer delivery strategies that target organs of interest with lower off-site effects, as well as to increase target molecule specificity.

6. Conclusions/Perspectives

In summary, lncRNAs regulate key biological pathways of lung cancer such as tumor development, metastasis, and resistance to current therapies, summarized in Figure 1 and Table 1. Although the lncRNAs described in this review have been studied for at least 10 years, several aspects (such as roles in immunotherapy responses) remain to be explored in the context of lung cancer (Table 1). Because of the ability of lncRNAs to modulate function of other biomolecules, a comprehensive approach to studying their role in signaling pathways is necessary and should take into account the interactome between coding and non-coding molecules. Such approach has been recently reported for neural cell differentiation processes using CRISPRi and single-cell RNA-seq approaches [121]. Additionally, with the increasing access to single-cell sequencing technologies, it would be important to question cell of origin of lncRNAs expression within a tumor, such as for *LUCAT1*, reported to modulate immune responses.

LncRNAs have highly context-dependent roles. Therefore, robust in vivo studies to dissect mechanisms that account for physiological functions of lncRNAs are essential. Additionally, studies should be conducted in the presence of a competent immune system. Assays relying on xenografts (using immunocompromised mice) have two important caveats: (1) they do not account for potential physical interactions between tumor cells and microenvironment that may alter outcomes; and (2) they fail to predict potential side effects of lncRNA-targeting therapies that are immune-system-dependent. This gap in knowledge presents an opportunity for the field to develop such approaches.

Table 1. Summary of roles of lncRNAs *GAS5*, *MALAT1*, *LUCAT1*, and *HOTAIR* in normal physiology and in lung cancer.

LncRNA	Normal Function	Role in Lung Cancer	Role in Chemotherapy	Role in Radiotherapy	Role in Targeted Therapy	Role in Immunotherapy	Potential as Biomarker
			In Lung Cancer				
GAS5	Cell cycle inhibition [36], cell differentiation [37]	Tumor suppressor [38]	Promotes sensitivity to cisplatin [65]	Promotes sensitivity [66]	Promotes sensitivity to EGFR TKI [38]	Not explored	Yes—lower levels in NSCLC [105]
MALAT1	Regulation of neighboring genes expression [28,29]	Oncogene [23,30]	Promotes resistance to cisplatin [59]	Not explored	May promote sensitivity: Down-regulated in EGFR-TKI resistant PC9 cells [63]	May be associated with therapeutic failure: correlated with PD-L1 expression [83]	Yes—higher levels in NSCLC [122]
LUCAT1	Inhibition of immune responses [46,69]	Oncogene [43,45]	Promotes resistance to cisplatin [67]	Not explored	Not explored	Not explored	Yes—higher levels in LUAD [123]
HOTAIR	Regulation of *HOX* genes expression [12–14] by recruitment of histone-modifier enzymes [124]	Oncogene [49,50,72]	Promotes resistance to cisplatin [50]	Promotes resistance [71]	Controversial roles: • Down-regulated in EGFR-TKI resistant tumors [73]; • Down-regulated in EGFR-TKI resistant PC9 cells [63]; • Up-regulated in EGFR-TKI resistant PC9 cells [74]; • Promotes resistance to Crizotinib (ALK/ROS1 inhibitor) [75].	Not explored	Yes—higher levels in NSCLC [103]

TKI = tyrosine-kinase inhibitor, NSCLC = non-small cell lung cancer, LUAD = lung adenocarcinoma.

Lastly, even though the role of lncRNAs in therapy resistance is documented, a role in the persister cell state (minimal residual disease) has not been explored. Such studies could shed light on important survival mechanisms that drive therapeutic failure and disease recurrence. LncRNAs show promise as disease biomarkers based on their highly stable nature in circulation. An important aspect to explore deeper is their potential to detect disease in early stages and temporal circular detection of lncRNAs to monitor the therapeutic response. Such an approach could be implemented for routine surveillance in advanced cancer patient groups.

Author Contributions: Conceptualization, J.G.-A., W.W. and T.G.B.; investigation, J.G.-A.; data curation, J.G.-A.; writing—original draft preparation, J.G.-A.; writing—review and editing, W.W. and T.G.B.; supervision, W.W. and T.G.B. All authors have read and agreed to the published version of the manuscript.

Funding: T.G.B. is supported by funding from NIH/NCI and the Chan-Zuckerberg Biohub. J.G.-A. is a recipient of a University of California President's Post-Doctoral Fellowship.

Acknowledgments: The authors acknowledge the NIH/NCI and Chan-Zuckerberg Biohub for research support.

Conflicts of Interest: T.G.B. received funding support from Revolution Medicines, Kinnate, and Verastem and is an advisor to Revolution Medicines, Relay, Rain, Scorpion, Engine, Deciphera.

References

1. Leisegang, M.S.; Bains, J.K.; Seredinski, S.; Oo, J.A.; Krause, N.M.; Kuo, C.C.; Gunther, S.; Sentruk Cetin, N.; Warwick, T.; Cao, C.; et al. HIF1alpha-AS1 is a DNA:DNA:RNA triplex-forming lncRNA interacting with the HUSH complex. *Nat. Commun.* **2022**, *13*, 6563. [CrossRef]
2. O'Leary, V.B.; Ovsepian, S.V.; Carrascosa, L.G.; Buske, F.A.; Radulovic, V.; Niyazi, M.; Moertl, S.; Trau, M.; Atkinson, M.J.; Anastasov, N. PARTICLE, a Triplex-Forming Long ncRNA, Regulates Locus-Specific Methylation in Response to Low-Dose Irradiation. *Cell Rep.* **2015**, *11*, 474–485. [CrossRef]
3. Rakheja, I.; Ansari, A.H.; Ray, A.; Chandra Joshi, D.; Maiti, S. Small molecule quercetin binds MALAT1 triplex and modulates its cellular function. *Mol. Ther. Nucleic Acids* **2022**, *30*, 241–256. [CrossRef]
4. Kung, J.T.; Colognori, D.; Lee, J. Long noncoding RNAs: Past, present, and future. *Genetics* **2013**, *193*, 651–669. [CrossRef]
5. Yao, R.W.; Wang, Y.; Chen, L.L. Cellular functions of long noncoding RNAs. *Nat. Cell Biol.* **2019**, *21*, 542–551. [CrossRef] [PubMed]
6. Guo, C.J.; Xu, G.; Chen, L.L. Mechanisms of Long Noncoding RNA Nuclear Retention. *Trends Biochem. Sci.* **2020**, *45*, 947–960. [CrossRef]
7. Chen, B.; Dragomir, M.P.; Yang, C.; Li, Q.; Horst, D.; Calin, G.A. Targeting non-coding RNAs to overcome cancer therapy resistance. *Signal Transduct. Target. Ther.* **2022**, *7*, 121. [CrossRef] [PubMed]
8. Long, Y.; Wang, X.; Youmans, D.T.; Cech, T.R. How do lncRNAs regulate transcription? *Sci. Adv.* **2017**, *3*, eaao2110. [CrossRef] [PubMed]
9. Wang, W.; Min, L.; Qiu, X.; Wu, X.; Liu, C.; Ma, J.; Zhang, D.; Zhu, L. Biological Function of Long Non-coding RNA (LncRNA) Xist. *Front. Cell Dev. Biol.* **2021**, *9*, 645647. [CrossRef]
10. Monnier, P.; Martinet, C.; Pontis, J.; Stancheva, I.; Ait-Si-Ali, S.; Dandolo, L. H19 lncRNA controls gene expression of the Imprinted Gene Network by recruiting MBD. *Proc. Natl. Acad. Sci. USA* **2013**, *110*, 20693–20698. [CrossRef]
11. Ripoche, M.A.; Kress, C.; Poirier, F.; Dandolo, L. Deletion of the H19 transcription unit reveals the existence of a putative imprinting control element. *Genes Dev.* **1997**, *11*, 1596–1604. [CrossRef] [PubMed]
12. Rinn, J.L.; Kertesz, M.; Wang, J.K.; Squazzo, S.L.; Xu, X.; Brugmann, S.A.; Goodnough, L.H.; Helms, J.A.; Farnham, P.J.; Segal, E.; et al. Functional demarcation of active and silent chromatin domains in human HOX loci by noncoding RNAs. *Cell* **2007**, *129*, 1311–1323. [CrossRef]
13. Li, L.; Liu, B.; Wapinski, O.L.; Tsai, M.C.; Qu, K.; Zhang, J.; Carlson, J.C.; Lin, M.; Fang, F.; Gupta, R.A.; et al. Targeted disruption of Hotair leads to homeotic transformation and gene derepression. *Cell Rep.* **2013**, *5*, 3–12. [CrossRef]
14. Amandio, A.R.; Necsulea, A.; Joye, E.; Mascrez, B.; Duboule, D. Hotair Is Dispensible for Mouse Development. *PLoS Genet.* **2016**, *12*, e1006232. [CrossRef]
15. Sauvageau, M.; Goff, L.A.; Lodato, S.; Bonev, B.; Groff, A.F.; Gerhardinger, C.; Sanchez-Gomez, D.B.; Hacisuleyman, E.; Li, E.; Spence, M.; et al. Multiple knockout mouse models reveal lincRNAs are required for life and brain development. *Elife* **2013**, *2*, e01749. [CrossRef]
16. Cronin, K.A.; Lake, A.J.; Scott, S.; Sherman, R.L.; Noone, A.M.; Howlader, N.; Henley, S.J.; Anderson, R.N.; Firth, A.U.; Ma, J.; et al. Annual Report to the Nation on the Status of Cancer, part I: National cancer statistics. *Cancer* **2018**, *124*, 2785–2800. [CrossRef] [PubMed]
17. Sabnis, A.J.; Bivona, T.G. Principles of Resistance to Targeted Cancer Therapy: Lessons from Basic and Translational Cancer Biology. *Trends Mol. Med.* **2019**, *25*, 185–197. [CrossRef]

18. Zhang, X.Z.; Liu, H.; Chen, S.R. Mechanisms of Long Non-Coding RNAs in Cancers and Their Dynamic Regulations. *Cancers* **2020**, *12*, 1245. [CrossRef] [PubMed]
19. Shen, S.; Chen, J.; Li, H.; Jiang, Y.; Wei, Y.; Zhang, R.; Zhao, Y.; Chen, F. Large-scale integration of the non-coding RNAs with DNA methylation in human cancers. *Cell Rep.* **2023**, *42*, 112261. [CrossRef]
20. Aprile, M.; Katopodi, V.; Leucci, E.; Costa, V. LncRNAs in Cancer: From garbage to Junk. *Cancers* **2020**, *12*, 3220. [CrossRef]
21. Tong, G.; Tong, W.; He, R.; Cui, Z.; Li, S.; Zhou, B.; Yin, Z. MALAT1 Polymorphisms and Lung Cancer Susceptibility in a Chinese Northeast Han Population. *Int. J. Med. Sci.* **2022**, *19*, 1300–1306. [CrossRef]
22. Ren, M.M.; Xu, S.; Wei, Y.B.; Yang, J.J.; Yang, Y.N.; Sun, S.S.; Li, Y.J.; Wang, P.Y.; Xie, S.Y. Roles of HOTAIR in lung cancer susceptibility and prognosis. *Mol. Genet. Genom. Med.* **2020**, *8*, e1299. [CrossRef]
23. Ji, P.; Diederichs, S.; Wang, W.; Boing, S.; Metzger, R.; Schneider, P.M.; Tidow, N.; Brandt, B.; Buerger, H.; Bulk, E.; et al. MALAT-1, a novel noncoding RNA, and thymosin beta4 predict metastasis and survival in early-stage non-small cell lung cancer. *Oncogene* **2003**, *22*, 8031–8041. [CrossRef]
24. Sun, Y.; Ma, L. New Insights into Long Non-Coding RNA MALAT1 in Cancer and Metastasis. *Cancers* **2019**, *11*, 216. [CrossRef] [PubMed]
25. Hou, J.; Zhang, G.; Wang, X.; Wang, Y.; Wang, K. Functions and mechanisms of lncRNA MALAT1 in cancer chemotherapy resistance. *Biomark. Res.* **2023**, *11*, 23. [CrossRef] [PubMed]
26. Jiang, L.; Li, Z.; Wang, R. Long non-coding RNAs in lung cancer: Regulation patterns, biologic function and diagnosis implications (Review). *Int. J. Oncol.* **2019**, *55*, 585–596. [CrossRef]
27. Eissmann, M.; Gutschner, T.; Hammerle, M.; Gunther, S.; Caudron-Herger, M.; Gross, M.; Schirmacher, P.; Rippe, K.; Braun, T.; Zornig, M.; et al. Loss of the abundant nuclear non-coding RNA MALAT1 is compatible with life and development. *RNA Biol.* **2012**, *9*, 1076–1087. [CrossRef] [PubMed]
28. Zhang, B.; Arun, G.; Mao, Y.S.; Lazar, Z.; Hung, G.; Bhattacharjee, G.; Xiao, X.; Booth, C.J.; Wu, J.; Zhang, C.; et al. The lncRNA Malat1 is dispensable for mouse development but its transcription plays a cis-regulatory role in the adult. *Cell Rep.* **2012**, *2*, 111–123. [CrossRef]
29. Nakagawa, S.; Ip, J.Y.; Shioi, G.; Tripathi, V.; Zong, X.; Hirose, T.; Prasanth, K.V. Malat1 is not an essential component of nuclear speckles in mice. *RNA* **2012**, *18*, 1487–1499. [CrossRef]
30. Gutschner, T.; Hammerle, M.; Eissmann, M.; Hsu, J.; Kim, Y.; Hung, G.; Revenko, A.; Arun, G.; Stentrup, M.; Gross, M.; et al. The noncoding RNA MALAT1 is a critical regulator of the metastasis phenotype of lung cancer cells. *Cancer Res.* **2013**, *73*, 1180–1189. [CrossRef]
31. Kim, J.; Piao, H.L.; Kim, B.J.; Yao, F.; Han, Z.; Wang, Y.; Xiao, Z.; Siverly, A.N.; Lawhon, S.E.; Ton, B.N.; et al. Long noncoding RNA MALAT1 suppresses breast cancer metastasis. *Nat. Genet.* **2018**, *50*, 1705–1715. [CrossRef] [PubMed]
32. Anderson, K.M.; Anderson, D.M.; McAnally, J.R.; Shelton, J.M.; Bassel-Duby, R.; Olson, E.N. Transcription of the non-coding RNA upperhand controls Hand2 expression and heart development. *Nature* **2016**, *539*, 433–436. [CrossRef]
33. Cho, S.W.; Xu, J.; Sun, R.; Mumbach, M.R.; Carter, A.C.; Chen, Y.G.; Yost, K.E.; Kim, J.; He, J.; Nevins, S.A.; et al. Promoter of lncRNA Gene PVT1 Is a Tumor-Suppressor DNA Boundary Element. *Cell* **2018**, *173*, 1398–1412.e22. [CrossRef]
34. Engreitz, J.M.; Haines, J.E.; Perez, E.M.; Munson, G.; Chen, J.; Kane, M.; McDonel, P.E.; Guttman, M.; Lander, E.S. Local regulation of gene expression by lncRNA promoters, transcription and splicing. *Nature* **2016**, *539*, 452–455. [CrossRef]
35. Nunez-Martinez, H.N.; Recillas-Targa, F. Emerging Functions of lncRNA Loci beyond the Transcript Itself. *Int. J. Mol. Sci.* **2022**, *23*, 6258. [CrossRef] [PubMed]
36. Schneider, C.; King, R.M.; Philipson, L. Genes specifically expressed at growth arrest of mammalian cells. *Cell* **1988**, *54*, 787–793. [CrossRef]
37. Li, M.; Xie, Z.; Li, J.; Lin, J.; Zheng, G.; Liu, W.; Tang, S.; Cen, S.; Ye, G.; Li, Z.; et al. GAS5 protects against osteoporosis by targeting UPF1/SMAD7 axis in osteoblast differentiation. *Elife* **2020**, *9*, e59079. [CrossRef] [PubMed]
38. Dong, S.; Qu, X.; Li, W.; Zhong, X.; Li, P.; Yang, S.; Chen, X.; Shao, M.; Zhang, L. The long non-coding RNA, GAS5, enhances gefitinib-induced cell death in innate EGFR tyrosine kinase inhibitor-resistant lung adenocarcinoma cells with wide-type EGFR via downregulation of the IGF-1R expression. *J. Hematol. Oncol.* **2015**, *8*, 43. [CrossRef]
39. Mei, Y.; Si, J.; Wang, Y.; Huang, Z.; Zhu, H.; Feng, S.; Wu, X.; Wu, L. Long Noncoding RNA GAS5 Suppresses Tumorigenesis by Inhibiting miR-23a Expression in Non-Small Cell Lung Cancer. *Oncol. Res.* **2017**, *25*, 1027–1037. [CrossRef]
40. Guo, C.; Song, W.Q.; Sun, P.; Jin, L.; Dai, H.Y. LncRNA-GAS5 induces PTEN expression through inhibiting miR-103 in endometrial cancer cells. *J. Biomed. Sci.* **2015**, *22*, 100. [CrossRef]
41. Sang, L.; Ju, H.Q.; Yang, Z.; Ge, Q.; Zhang, Z.; Liu, F.; Yang, L.; Gong, H.; Shi, C.; Qu, L.; et al. Mitochondrial long non-coding RNA GAS5 tunes TCA metabolism in response to nutrient stress. *Nat. Metab.* **2021**, *3*, 90–106. [CrossRef] [PubMed]
42. Ye, R.; Tang, R.; Gan, S.; Li, R.; Cheng, Y.; Guo, L.; Zeng, C.; Sun, Y. New insights into long non-coding RNAs in non-small cell lung cancer. *Biomed. Pharmacother.* **2020**, *131*, 110775. [CrossRef]
43. Thai, P.; Statt, S.; Chen, C.; Liang, E.; Campbell, C.; Wu, R. Characterization of a novel long noncoding RNA, SCAL1, induced by cigarette smoke and elevated in lung cancer cell lines. *Am. J. Respir. Cell Mol. Biol.* **2013**, *49*, 204–211. [CrossRef]
44. Xing, C.; Sun, S.G.; Yue, Z.Q.; Bai, F. Role of lncRNA LUCAT1 in cancer. *Biomed. Pharmacother.* **2021**, *134*, 111158. [CrossRef] [PubMed]

45. Sun, Y.; Jin, S.D.; Zhu, Q.; Han, L.; Feng, J.; Lu, X.Y.; Wang, W.; Wang, F.; Guo, R.H. Long non-coding RNA LUCAT1 is associated with poor prognosis in human non-small lung cancer and regulates cell proliferation via epigenetically repressing p21 and p57 expression. *Oncotarget* **2017**, *8*, 28297–28311. [CrossRef]
46. Agarwal, S.; Vierbuchen, T.; Ghosh, S.; Chan, J.; Jiang, Z.; Kandasamy, R.K.; Ricci, E.; Fitzgerald, K.A. The long non-coding RNA LUCAT1 is a negative feedback regulator of interferon responses in humans. *Nat. Commun.* **2020**, *11*, 6348. [CrossRef] [PubMed]
47. Mahpour, A.; Mullen, A.C. Our emerging understanding of the roles of long non-coding RNAs in normal liver function, disease, and malignancy. *JHEP Rep.* **2021**, *3*, 100177. [CrossRef] [PubMed]
48. Zhu, C.; Wang, X.; Wang, Y.; Wang, K. Functions and underlying mechanisms of lncRNA HOTAIR in cancer chemotherapy resistance. *Cell Death Discov.* **2022**, *8*, 383. [CrossRef]
49. Liu, X.H.; Liu, Z.L.; Sun, M.; Liu, J.; Wang, Z.X.; De, W. The long non-coding RNA HOTAIR indicates a poor prognosis and promotes metastasis in non-small cell lung cancer. *BMC Cancer* **2013**, *13*, 464. [CrossRef]
50. Liu, M.Y.; Li, X.Q.; Gao, T.H.; Cui, Y.; Ma, N.; Zhou, Y.; Zhang, G.J. Elevated HOTAIR expression associated with cisplatin resistance in non-small cell lung cancer patients. *J. Thorac. Dis.* **2016**, *8*, 3314–3322. [CrossRef]
51. Zhou, C.; Ye, L.; Jiang, C.; Bai, J.; Chi, Y.; Zhang, H. Long noncoding RNA HOTAIR, a hypoxia-inducible factor-1alpha activated driver of malignancy, enhances hypoxic cancer cell proliferation, migration, and invasion in non-small cell lung cancer. *Tumour Biol.* **2015**, *36*, 9179–9188. [CrossRef]
52. Ma, Q.; Yang, L.; Tolentino, K.; Wang, G.; Zhao, Y.; Litzenburger, U.M.; Shi, Q.; Zhu, L.; Yang, C.; Jiao, H.; et al. Inducible lncRNA transgenic mice reveal continual role of HOTAIR in promoting breast cancer metastasis. *Elife* **2022**, *11*, e79126. [CrossRef]
53. Esposito, R.; Polidori, T.; Meise, D.F.; Pulido-Quetglas, C.; Chouvardas, P.; Forster, S.; Schaerer, P.; Kobel, A.; Schlatter, J.; Kerkhof, E.; et al. Multi-hallmark long noncoding RNA maps reveal non-small cell lung cancer vulnerabilities. *Cell Genom.* **2022**, *2*, 100171. [CrossRef]
54. Pacholewska, A.; Sung, M.H. lncRNA expression predicts mRNA abundance. *Epigenomics* **2019**, *11*, 1121–1128 [CrossRef] [PubMed]
55. Yang, M.; Lu, H.; Liu, J.; Wu, S.; Kim, P.; Zhou, X. lncRNAfunc: A knowledgebase of lncRNA function in human cancer. *Nucleic Acids Res.* **2022**, *50*, D1295–D1306. [CrossRef]
56. Mondal, P.; Meeran, S.M. Emerging role of non-coding RNAs in resistance to platinum-based anti-cancer agents in lung cancer. *Front. Pharmacol.* **2023**, *14*, 1105484. [CrossRef]
57. Wang, W.T.; Han, C.; Sun, Y.M.; Chen, T.Q.; Chen, Y.Q. Noncoding RNAs in cancer therapy resistance and targeted drug development. *J. Hematol. Oncol.* **2019**, *12*, 55. [CrossRef] [PubMed]
58. Xie, Y.; Han, J.; Xie, K.; Gou, Q. LncRNAs as biomarkers for predicting radioresistance and survival in cancer: A meta-analysis. *Sci. Rep.* **2022**, *12*, 18494. [CrossRef]
59. Fang, Z.; Chen, W.; Yuan, Z.; Liu, X.; Jiang, H. LncRNA-MALAT1 contributes to the cisplatin-resistance of lung cancer by upregulating MRP1 and MDR1 via STAT3 activation. *Biomed. Pharmacother.* **2018**, *101*, 536–542. [CrossRef]
60. Wang, R.; Lu, X.; Yu, R. lncRNA MALAT1 Promotes EMT Process and Cisplatin Resistance of Oral Squamous Cell Carcinoma via PI3K/AKT/m-TOR Signal Pathway. *OncoTargets Ther.* **2020**, *13*, 4049–4061. [CrossRef] [PubMed]
61. Li, Z.; Zhou, Y.; Tu, B.; Bu, Y.; Liu, A.; Kong, J. Long noncoding RNA MALAT1 affects the efficacy of radiotherapy for esophageal squamous cell carcinoma by regulating Cks1 expression. *J. Oral Pathol. Med.* **2017**, *46*, 583–590. [CrossRef]
62. Yao, P.A.; Wu, Y.; Zhao, K.; Li, Y.; Cao, J.; Xing, C. The feedback loop of ANKHD1/lncRNA MALAT1/YAP1 strengthens the radioresistance of CRC by activating YAP1/AKT signaling. *Cell Death Dis.* **2022**, *13*, 103. [CrossRef]
63. Cheng, N.; Li, X.; Zhao, C.; Ren, S.; Chen, X.; Cai, W.; Zhao, M.; Zhang, Y.; Li, J.; Wang, Q.; et al. Microarray expression profile of long non-coding RNAs in EGFR-TKIs resistance of human non-small cell lung cancer. *Oncol. Rep.* **2015**, *33*, 833–839. [CrossRef] [PubMed]
64. Wang, Z.; Chang, X.; Zhu, G.; Gao, X.; Chang, L. Depletion of lncRNA MALAT1 inhibited sunitinib resistance through regulating miR-362-3p-mediated G3BP1 in renal cell carcinoma. *Cell Cycle* **2020**, *19*, 2054–2062. [CrossRef]
65. Yang, X.; Meng, L.; Zhong, Y.; Hu, F.; Wang, L.; Wang, M. The long intergenic noncoding RNA GAS5 reduces cisplatin-resistance in non-small cell lung cancer through the miR-217/LHPP axis. *Aging* **2021**, *13*, 2864–2884. [CrossRef] [PubMed]
66. Lambrou, G.I.; Hatziagapiou, K.; Zaravinos, A. The Non-Coding RNA GAS5 and Its Role in Tumor Therapy-Induced Resistance. *Int. J. Mol. Sci.* **2020**, *21*, 7633. [CrossRef]
67. Shen, Q.; Xu, Z.; Xu, S. Long non-coding RNA LUCAT1 contributes to cisplatin resistance by regulating the miR-514a-3p/ULK1 axis in human non-small cell lung cancer. *Int. J. Oncol.* **2020**, *57*, 967–979. [CrossRef] [PubMed]
68. Huan, L.; Guo, T.; Wu, Y.; Xu, L.; Huang, S.; Xu, Y.; Liang, L.; He, X. Hypoxia induced LUCAT1/PTBP1 axis modulates cancer cell viability and chemotherapy response. *Mol. Cancer* **2020**, *19*, 11. [CrossRef]
69. Vierbuchen, T.; Agarwal, S.; Johnson, J.; Galia, L.; Lei, X.; Stein, K.; Olagnier, D.; Gaede, K.I.; Herzmann, C.; Holm, C.K.; et al. The lncRNA LUCAT1 is elevated in inflammatory disease and restrains inflammation by regulating the splicing and stability of NR4A. *Proc. Natl. Acad. Sci. USA* **2023**, *120*, e2213715120. [CrossRef]
70. Jing, L.; Yuan, W.; Ruofan, D.; Jinjin, Y.; Haifeng, Q. HOTAIR enhanced aggressive biological behaviors and induced radio-resistance via inhibiting p21 in cervical cancer. *Tumour Biol.* **2015**, *36*, 3611–3619. [CrossRef]
71. Chen, J.; Shen, Z.; Zheng, Y.; Wang, S.; Mao, W. Radiotherapy induced Lewis lung cancer cell apoptosis via inactivating beta-catenin mediated by upregulated HOTAIR. *Int. J. Clin. Exp. Pathol.* **2015**, *8*, 7878–7886.

72. Zhou, Y.; Wang, C.; Liu, X.; Wu, C.; Yin, H. Long non-coding RNA HOTAIR enhances radioresistance in MDA-MB231 breast cancer cells. *Oncol. Lett.* **2017**, *13*, 1143–1148. [CrossRef]
73. Wang, Q.; Li, X.; Ren, S.; Su, C.; Li, C.; Li, W.; Yu, J.; Cheng, N.; Zhou, C. HOTAIR induces EGFR-TKIs resistance in non-small cell lung cancer through epithelial-mesenchymal transition. *Lung Cancer* **2020**, *147*, 99–105. [CrossRef]
74. Liu, Y.; Jiang, H.; Zhou, H.; Ying, X.; Wang, Z.; Yang, Y.; Xu, W.; He, X.; Li, Y. Lentivirus-mediated silencing of HOTAIR lncRNA restores gefitinib sensitivity by activating Bax/Caspase-3 and suppressing TGF-alpha/EGFR signaling in lung adenocarcinoma. *Oncol. Lett.* **2018**, *15*, 2829–2838.
75. Yang, Y.; Jiang, C.; Yang, Y.; Guo, L.; Huang, J.; Liu, X.; Wu, C.; Zou, J. Silencing of LncRNA-HOTAIR decreases drug resistance of Non-Small Cell Lung Cancer cells by inactivating autophagy via suppressing the phosphorylation of ULK1. *Biochem. Biophys. Res. Commun.* **2018**, *497*, 1003–1010. [CrossRef]
76. Tang, X.D.; Zhang, D.D.; Jia, L.; Ji, W.; Zhao, Y.S. lncRNA AFAP1-AS1 Promotes Migration and Invasion of Non-Small Cell Lung Cancer via Up-Regulating IRF7 and the RIG-I-Like Receptor Signaling Pathway. *Cell Physiol. Biochem.* **2018**, *50*, 179–195. [CrossRef]
77. Jing, C.; Wang, Z.; Lou, R.; Wu, J.; Shi, C.; Chen, D.; Ma, R.; Liu, S.; Cao, H.; Feng, J. Nedaplatin reduces multidrug resistance of non-small cell lung cancer by downregulating the expression of long non-coding RNA MVIH. *J. Cancer* **2020**, *11*, 559–569. [CrossRef] [PubMed]
78. Xu, X.; Zhou, X.; Chen, Z.; Gao, C.; Zhao, L.; Cui, Y. Silencing of lncRNA XIST inhibits non-small cell lung cancer growth and promotes chemosensitivity to cisplatin. *Aging* **2020**, *12*, 4711–4726. [CrossRef] [PubMed]
79. Wu, H.; Liu, A. Long non-coding RNA NEAT1 regulates ferroptosis sensitivity in non-small-cell lung cancer. *J. Int. Med. Res.* **2021**, *49*, 300060521996183. [CrossRef] [PubMed]
80. Pan, X.; Li, C.; Feng, J. The role of LncRNAs in tumor immunotherapy. *Cancer Cell Int.* **2023**, *23*, 30. [CrossRef]
81. Sharma, P.; Allison, J.P. Dissecting the mechanisms of immune checkpoint therapy. *Nat. Rev. Immunol.* **2020**, *20*, 75–76. [CrossRef] [PubMed]
82. Hong, L.; Aminu, M.; Li, S.; Lu, X.; Petranovic, M.; Saad, M.B.; Chen, P.; Qin, K.; Varghese, S.; Rinsurongkawong, W.; et al. Efficacy and clinicogenomic correlates of response to immune checkpoint inhibitors alone or with chemotherapy in non-small cell lung cancer. *Nat. Commun.* **2023**, *14*, 695. [CrossRef] [PubMed]
83. Wei, S.; Wang, K.; Huang, X.; Zhao, Z.; Zhao, Z. LncRNA MALAT1 contributes to non-small cell lung cancer progression via modulating miR-200a-3p/programmed death-ligand 1 axis. *Int. J. Immunopathol. Pharmacol.* **2019**, *33*, 2058738419859699. [CrossRef] [PubMed]
84. Xia, R.; Geng, G.; Yu, X.; Xu, Z.; Guo, J.; Liu, H.; Li, N.; Li, Z.; Li, Y.; Dai, X.; et al. LINC01140 promotes the progression and tumor immune escape in lung cancer by sponging multiple microRNAs. *J. Immunother. Cancer* **2021**, *9*, e002746. [CrossRef]
85. Zhang, H.; Liu, M.; Du, G.; Yu, B.; Ma, X.; Gui, Y.; Cao, L.; Li, X.; Tan, B. Immune checkpoints related-LncRNAs can identify different subtypes of lung cancer and predict immunotherapy and prognosis. *J. Cancer Res. Clin. Oncol.* **2022**, *148*, 1597–1612. [CrossRef] [PubMed]
86. Woo, S.R.; Fuertes, M.B.; Corrales, L.; Spranger, S.; Furdyna, M.J.; Leung, M.; Duggan, R.; Wang, Y.; Barber, G.N.; Fitzgerald, K.A.; et al. STING-dependent cytosolic DNA sensing mediates innate immune recognition of immunogenic tumors. *Immunity* **2014**, *41*, 830–842. [CrossRef]
87. Wu, Y.T.; Fang, Y.; Wei, Q.; Shi, H.; Tan, H.; Deng, Y.; Zeng, Z.; Qiu, J.; Chen, C.; Sun, L.; et al. Tumor-targeted delivery of a STING agonist improvescancer immunotherapy. *Proc. Natl. Acad. Sci. USA* **2022**, *119*, e2214278119. [CrossRef]
88. Gao, Y.; Zhang, N.; Zeng, Z.; Wu, Q.; Jiang, X.; Li, S.; Sun, W.; Zhang, J.; Li, Y.; Li, J.; et al. LncRNA PCAT1 activates SOX2 and suppresses radioimmune responses via regulating cGAS/STING signalling in non-small cell lung cancer. *Clin. Transl. Med.* **2022**, *12*, e792. [CrossRef]
89. Chen, J.H.; Feng, D.D.; Chen, Y.F.; Yang, C.X.; Juan, C.X.; Cao, Q.; Chen, X.; Liu, S.; Zhou, G.P. Long non-coding RNA MALAT1 targeting STING transcription promotes bronchopulmonary dysplasia through regulation of CREB. *J. Cell Mol. Med.* **2020**, *24*, 10478–10492. [CrossRef]
90. Badowski, C.; He, B.; Garmire, L.X. Blood-derived lncRNAs as biomarkers for cancer diagnosis: The Good, the Bad and the Beauty. *NPJ Precis. Oncol.* **2022**, *6*, 40. [CrossRef]
91. Fan, T.; Sun, N.; He, J. Exosome-Derived LncRNAs in Lung Cancer. *Front. Oncol.* **2020**, *10*, 1728. [CrossRef] [PubMed]
92. Fang, R.; Zhu, Y.; Khadka, V.S.; Zhang, F.; Jiang, B.; Deng, Y. The Evaluation of Serum Biomarkers for Non-small Cell Lung Cancer (NSCLC) Diagnosis. *Front. Physiol.* **2018**, *9*, 1710. [CrossRef]
93. Sutic, M.; Vukic, A.; Baranasic, J.; Forsti, A.; Dzubur, F.; Samarzija, M.; Jakopovic, M.; Brcic, L.; Knezevic, J. Diagnostic, Predictive, and Prognostic Biomarkers in Non-Small Cell Lung Cancer (NSCLC) Management. *J. Pers. Med.* **2021**, *11*, 1102. [CrossRef] [PubMed]
94. Du, L.; Duan, W.; Jiang, X.; Zhao, L.; Li, J.; Wang, R.; Yan, S.; Xie, Y.; Yan, K.; Wang, Q.; et al. Cell-free lncRNA expression signatures in urine serve as novel non-invasive biomarkers for diagnosis and recurrence prediction of bladder cancer. *J. Cell Mol. Med.* **2018**, *22*, 2838–2845. [CrossRef] [PubMed]
95. Wei, X.; Xu, M.; Wang, C.; Fang, S.; Zhang, Y.; Wang, W. Genome-wide analysis of long noncoding RNA expression profile in nasal mucosa with allergic rhinitis. *BMC Med. Genom.* **2021**, *14*, 100. [CrossRef]

96. Entezari, M.; Ghanbarirad, M.; Taheriazam, A.; Sadrkhanloo, M.; Zabolian, A.; Goharrizi, M.; Hushmandi, K.; Aref, A.R.; Ashrafizadeh, M.; Zarrabi, A.; et al. Long non-coding RNAs and exosomal lncRNAs: Potential functions in lung cancer progression, drug resistance and tumor microenvironment remodeling. *Biomed. Pharmacother.* **2022**, *150*, 112963. [CrossRef]
97. Kanada, M.; Bachmann, M.H.; Contag, C.H. Signaling by Extracellular Vesicles Advances Cancer Hallmarks. *Trends Cancer* **2016**, *2*, 84–94. [CrossRef]
98. Chen, Z.; Bian, C.; Huang, J.; Li, X.; Chen, L.; Xie, X.; Xia, Y.; Yin, R.; Wang, J. Tumor-derived exosomal HOTAIRM1 regulates SPON2 in CAFs to promote progression of lung adenocarcinoma. *Discov. Oncol.* **2022**, *13*, 92. [CrossRef]
99. Chen, Y.; Hong, C.; Qu, J.; Chen, J.; Qin, Z. Knockdown of lncRNA PCAT6 suppresses the growth of non-small cell lung cancer cells by inhibiting macrophages M2 polarization via miR-326/KLF1 axis. *Bioengineered* **2022**, *13*, 12834–12846. [CrossRef]
100. Tang, Q.; Ni, Z.; Cheng, Z.; Xu, J.; Yu, H.; Yin, P. Three circulating long non-coding RNAs act as biomarkers for predicting NSCLC. *Cell Physiol. Biochem.* **2015**, *37*, 1002–1009. [CrossRef]
101. Hu, X.; Bao, J.; Wang, Z.; Zhang, Z.; Gu, P.; Tao, F.; Cui, D.; Jiang, W. The plasma lncRNA acting as fingerprint in non-small-cell lung cancer. *Tumour Biol.* **2016**, *37*, 3497–3504. [CrossRef] [PubMed]
102. Min, L.; Zhu, T.; Lv, B.; An, T.; Zhang, Q.; Shang, Y.; Yu, Z.; Zheng, L.; Wang, Q. Exosomal LncRNA RP5-977B1 as a novel minimally invasive biomarker for diagnosis and prognosis in non-small cell lung cancer. *Int. J. Clin. Oncol.* **2022**, *27*, 1013–1024. [CrossRef] [PubMed]
103. Yao, X.; Wang, T.; Sun, M.Y.; Yuming, Y.; Guixin, D.; Liu, J. Diagnostic value of lncRNA HOTAIR as a biomarker for detecting and staging of non-small cell lung cancer. *Biomarkers* **2022**, *27*, 526–533. [CrossRef] [PubMed]
104. Cao, W.; Lee, H.; Wu, W.; Zaman, A.; McCorkle, S.; Yan, M.; Chen, J.; Xing, Q.; Sinnott-Armstrong, N.; Xu, H.; et al. Multi-faceted epigenetic dysregulation of gene expression promotes esophageal squamous cell carcinoma. *Nat. Commun.* **2020**, *11*, 3675. [CrossRef] [PubMed]
105. Liang, W.; Lv, T.; Shi, X.; Liu, H.; Zhu, Q.; Zeng, J.; Yang, W.; Yin, J.; Song, Y. Circulating long noncoding RNA GAS5 is a novel biomarker for the diagnosis of nonsmall cell lung cancer. *Medicine* **2016**, *95*, e4608. [CrossRef]
106. Dhuri, K.; Bechtold, C.; Quijano, E.; Pham, H.; Gupta, A.; Vikram, A.; Bahal, R. Antisense Oligonucleotides: An Emerging Area in Drug Discovery and Development. *J. Clin. Med.* **2020**, *9*, 2004. [CrossRef]
107. Winkle, M.; El-Daly, S.M.; Fabbri, M.; Calin, G.A. Noncoding RNA therapeutics—Challenges and potential solutions. *Nat. Rev. Drug Discov.* **2021**, *20*, 629–651. [CrossRef]
108. Pedram Fatemi, R.; Salah-Uddin, S.; Modarresi, F.; Khoury, N.; Wahlestedt, C.; Faghihi, M.A. Screening for Small-Molecule Modulators of Long Noncoding RNA-Protein Interactions Using AlphaScreen. *J. Biomol. Screen* **2015**, *20*, 1132–1141. [CrossRef]
109. Ren, Y.; Wang, Y.; Zhang, J.; Wang, Q.X.; Han, L.; Mei, M.; Kang, C.S. Targeted design and identification of AC1NOD4Q to block activity of HOTAIR by abrogating the scaffold interaction with EZH2. *Clin. Epigenetics* **2019**, *11*, 29. [CrossRef]
110. Zhen, S.; Li, X. Application of CRISPR-Cas9 for Long Noncoding RNA Genes in Cancer Research. *Hum. Gene Ther.* **2019**, *30*, 3–9. [CrossRef]
111. Tontonoz, P.; Wu, X.; Jones, M.; Zhang, Z.; Salisbury, D.; Sallam, T. Long Noncoding RNA Facilitated Gene Therapy Reduces Atherosclerosis in a Murine Model of Familial Hypercholesterolemia. *Circulation* **2017**, *136*, 776–778. [CrossRef]
112. Huang, C.K.; Kafert-Kasting, S.; Thum, T. Preclinical and Clinical Development of Noncoding RNA Therapeutics for Cardiovascular Disease. *Circ. Res.* **2020**, *126*, 663–678. [CrossRef] [PubMed]
113. Ma, X.Y.; Wang, J.H.; Wang, J.L.; Ma, C.X.; Wang, X.C.; Liu, F.S. *Malat1* as an evolutionarily conserved lncRNA, plays a positive role in regulating proliferation and maintaining undifferentiated status of early-stage hematopoietic cells. *BMC Genom.* **2015**, *16*, 676. [CrossRef] [PubMed]
114. Arun, G.; Diermeier, S.; Akerman, M.; Chang, K.C.; Wilkinson, J.E.; Hearn, S.; Kim, Y.; MacLeod, A.R.; Krainer, A.R.; Norton, L.; et al. Differentiation of mammary tumors and reduction in metastasis upon *Malat1* lncRNA loss. *Genes Dev.* **2016**, *30*, 34–51. [CrossRef] [PubMed]
115. Elzallat, M.; Hassan, M.; Elkramani, N.; Aboushousha, T.; AbdelLatif, A.; Helal, N.; Abu-Taleb, H.; El-Ahwany, E. Nanoconjugated long non-coding RNA MEG3 as a new therapeutic approach for Hepatocellular carcinoma. *Heliyon* **2023**, *9*, e15288. [CrossRef]
116. Vaidya, A.M.; Sun, Z.; Ayat, N.; Schilb, A.; Liu, X.; Jiang, H.; Sun, D.; Scheidt, J.; Qian, V.; He, S.; et al. Systemic Delivery of Tumor-Targeting siRNA Nanoparticles against an Oncogenic LncRNA Facilitates Effective Triple-Negative Breast Cancer Therapy. *Bioconjug Chem.* **2019**, *30*, 907–919. [CrossRef]
117. Nicolescu, C.; Vaidya, A.; Schilb, A.; Lu, Z.R. Regulating Oncogenic LncRNA DANCR with Targeted ECO/siRNA Nanoparticles for Non-Small Cell Lung Cancer Therapy. *ACS Omega* **2022**, *7*, 22743–22753. [CrossRef]
118. Hong, D.S.; Kang, Y.K.; Borad, M.; Sachdev, J.; Ejadi, S.; Lim, H.Y.; Brenner, A.J.; Park, K.; Lee, J.L.; Kim, T.Y.; et al. Phase 1 study of MRX34, a liposomal miR-34a mimic, in patients with advanced solid tumours. *Br. J. Cancer* **2020**, *122*, 1630–1637. [CrossRef]
119. Reda El Sayed, S.; Cristante, J.; Guyon, L.; Denis, J.; Chabre, O.; Cherradi, N. MicroRNA Therapeutics in Cancer: Current Advances and Challenges. *Cancers* **2021**, *13*, 2680. [CrossRef]
120. Segal, M.; Biscans, A.; Gilles, M.E.; Anastasiadou, E.; De Luca, R.; Lim, J.; Khvorova, A.; Slack, F.J. Hydrophobically Modified let-7b miRNA Enhances Biodistribution to NSCLC and Downregulates HMGA2 In Vivo. *Mol. Ther. Nucleic Acids* **2020**, *19*, 267–277. [CrossRef]

121. Wu, D.; Poddar, A.; Ninou, E.; Hwang, E.; Cole, M.A.; Liu, S.J.; Horlbeck, M.A.; Chen, J.; Replogle, J.M.; Carosso, G.A.; et al. Dual genome-wide coding and lncRNA screens in neural induction of induced pluripotent stem cells. *Cell Genom.* **2022**, *2*, 100177. [CrossRef] [PubMed]
122. Pan, J.; Bian, Y.; Cao, Z.; Lei, L.; Pan, J.; Huang, J.; Cai, X.; Lan, X.; Zheng, H. Long noncoding RNA MALAT1 as a candidate serological biomarker for the diagnosis of non-small cell lung cancer: A meta-analysis. *Thorac. Cancer* **2020**, *11*, 329–335. [CrossRef] [PubMed]
123. Wang, L.; Xie, Y.; Wang, J.K.; Zhang, Y.; Liu, S.; Zhan, Y.; Zhao, Y.; Li, J.; Li, P.; Wang, C. Characterization of a Novel LUCAT1/miR-4316/VEGF-A Axis in Metastasis and Glycolysis of Lung Adenocarcinoma. *Front. Cell Dev. Biol.* **2022**, *10*, 833579. [CrossRef] [PubMed]
124. Tsai, M.C.; Manor, O.; Wan, Y.; Mosammaparast, N.; Wang, J.; Lan, F.; Shi, Y.; Segal, E.; Chang, H.Y. Long noncoding RNA as modular scaffold of histone modification complexes. *Science* **2010**, *329*, 689–693. [CrossRef] [PubMed]

cancers

MDPI

Review

Multi-Omics Approaches in Colorectal Cancer Screening and Diagnosis, Recent Updates and Future Perspectives

Ihsan Ullah [1], Le Yang [2], Feng-Ting Yin [1], Ye Sun [2], Xing-Hua Li [1], Jing Li [1] and Xi-Jun Wang [1,2,*]

1 National Chinmedomics Research Center, National TCM Key Laboratory of Serum Pharmacochemistry, Heilongjiang University of Chinese Medicine, Harbin 150040, China
2 State Key Laboratory of Dampness Syndrome, The Second Affiliated Hospital of Guangzhou University of Chinese Medicine, Dade Road 111, Guangzhou 510260, China
* Correspondence: xijunw@sina.com; Tel.: +86-13339512688

Simple Summary: Colorectal cancer (CRC) is one of the most prevalent cancers worldwide. Due to the absence of specific early symptoms, most of CRC patients are often diagnosed at late stages. Different screening and diagnostic biomarkers are currently used for risk stratification and early detection of CRC, which might prolong the overall survival. High-throughput technologies have witnessed rapid advancements in the last decade. Consequently, the development of multiple omics technologies, such as genomics, transcriptomics, proteomics, metabolomics, microbiomics, and lipidomics, has been widely applied to develop novel biomarkers that could contribute to the clinical management of CRC. In this paper, we aim to summarize the recent advances and future perspectives in using multi-omics technologies in CRC research, and reveal the potential implications of multi-omics for discovering novel biomarkers and enhancing clinical evaluations.

Abstract: Colorectal cancer (CRC) is common Cancer as well as the third leading cause of mortality around the world; its exact molecular mechanism remains elusive. Although CRC risk is significantly correlated with genetic factors, the pathophysiology of CRC is also influenced by external and internal exposures and their interactions with genetic factors. The field of CRC research has recently benefited from significant advances through Omics technologies for screening biomarkers, including genes, transcripts, proteins, metabolites, microbiome, and lipidome unbiasedly. A promising application of omics technologies could enable new biomarkers to be found for the screening and diagnosis of CRC. Single-omics technologies cannot fully understand the molecular mechanisms of CRC. Therefore, this review article aims to summarize the multi-omics studies of Colorectal cancer, including genomics, transcriptomics, proteomics, microbiomics, metabolomics, and lipidomics that may shed new light on the discovery of novel biomarkers. It can contribute to identifying and validating new CRC biomarkers and better understanding colorectal carcinogenesis. Discovering biomarkers through multi-omics technologies could be difficult but valuable for disease genotyping and phenotyping. That can provide a better knowledge of CRC prognosis, diagnosis, and treatments.

Keywords: colorectal cancer; multi-omics; biomarkers

Citation: Ullah, I.; Yang, L.; Yin, F.-T.; Sun, Y.; Li, X.-H.; Li, J.; Wang, X.-J. Multi-Omics Approaches in Colorectal Cancer Screening and Diagnosis, Recent Updates and Future Perspectives. *Cancers* **2022**, *14*, 5545. https://doi.org/10.3390/cancers14225545

Academic Editor: Wei Wu

Received: 15 October 2022
Accepted: 9 November 2022
Published: 11 November 2022

1. Introduction

Colorectal cancer (CRC) is the third most common cancer accounting for 10.2% of new cases and 9.2% of Cancer-related mortality, thus accounting for the second most deadly cancer globally [1]. It has been reported that the overall survival rate of metastatic CRC (mCRC) at 5 years over the first examination lowers from 87–90% in stages I–II, and 68–72% in stages III; in stage IV, the rate drops to 11–14% [2]. Most CRC treatment options currently rely on cancer staging, patient performance status, RAS, BRAF, ERBB2, and mismatch repair (MMR) status assessments using tumour samples taken during surgery or core biopsy [3,4]. At present, for patients with mCRC, it is recommended to determine

KRAS/NRAS and BRAF mutation status, as well as HER2 amplification and microsatellite instability high (MSI)/mismatch repair (MMR) status (if not performed already) [4]. Recent studies have demonstrated that immune checkpoint inhibitor therapy is effective in treating dMMR/MSI-H mCRC tumours at advanced stages of the disease [5]. The discovery of new molecular biomarkers in CRC and other cancers has begun to follow the approval of tumour-agnostic drugs, including NTRK1-3 translocations and high tumour mutational burdens (TMBs) [6,7]. As opposed to metastatic cancer, there are still no validated biomarkers indicating which patients are more likely to benefit from adjuvant cytotoxic therapy in stage II or III CRC, except for microsatellite instability (MSI) [8]. Additionally, postoperative treatments are often administered following metastatic resections, despite the absence of predictive biomarkers [3,4]. Currently, Colonoscopy, tissue biopsy, and fecal occult blood test (FOBT) are the major techniques used in CRC screening and detection. However, in the case of Colonoscopy or biopsy, these techniques are invasive, causing discomfort for the patient, or in the case of FOBT, they may also have low sensitivity [9–11]. Therefore, it is demonstrated that a less invasive test with higher sensitivity is needed in clinical practice.

In particular, high throughput "multi-omics" technologies, including genomics, transcriptomics, proteomics, microbiomics, and metabolomics, provide less or noninvasive approaches for diagnosing CRC. Each method offers a unique advantage for the discovery of novel diagnostic cancer biomarkers, such as Genomics, which is incredibly efficient for evaluating CRC vulnerability and the disease's genetic risk. However, it has little diagnostic potential since DNA sequences seldom translate directly to phenotype due to epigenetic, post-transcriptional, and post-translational alterations [12]. Transcriptomics and proteomics have great therapeutic potential as they are more closely tied to organisms' physiological states. Still, their diagnostic power is not as good as that of metabolomics, which enables quick and precise phenotypic characterization of the organism and its metabolic pathways as well as the potential to evaluate how host and gut bacterial metabolites interact, which is a crucial step in the CRC progression [13]. Additionally, a number of recent research have shown that the gut microbial community and microbial metabolites play a crucial role in the emergence of CRC [14,15]. Recent years have seen the emergence of lipidomics as a research tool and a multi-omics technology that holds great promise. As a result, this tool has been demonstrated to be useful for both the quantification of cellular lipids and their characterization. This is not only for disease diagnosis but also for other mechanistic studies [16,17]. The mechanism of CRC initiation and progression has remained largely enigmatic despite the discovery of more diagnostic methods and potential therapies; many challenges remain unresolved due to the lack of new biomarkers and the heterogeneity of tumours. After the completion of the human genome project, omics science has revolutionized CRC research [18]. In order to enable personalized medicine and to define CRC treatment, the identification of novel biomarkers has become an essential part of molecular diagnosis and treatment [19]. The use of new biomarkers in clinical practice is still challenging despite developments in the molecular analysis [20]. However, genomic advances have made significant contributions to understanding cancer biology over the past few years [21]. In oncology, the structure and functions of the genome, as well as mechanisms governing genes' expression, have been extensively investigated since the completion of the Human Genome Project and the development of next-generation sequencing (NGS) techniques [21,22]. A constantly expanding understanding of genomic hallmarks of malignant transformation provides a new perspective on pathogenesis and targeted treatment of particular tumours [23,24]. Genomics, transcriptomics, proteomics, metabolomics, microbiomics, and lipidomics, make a significant contribution to a fundamental change toward a multiparametric, innovative, immunological, and stromal model, which helps us to understand how CRC develops and categorizes it into various molecular subtypes for clinical diagnosis as well as the emergence of new biomarkers and therapeutic strategies [25], (Figure 1).

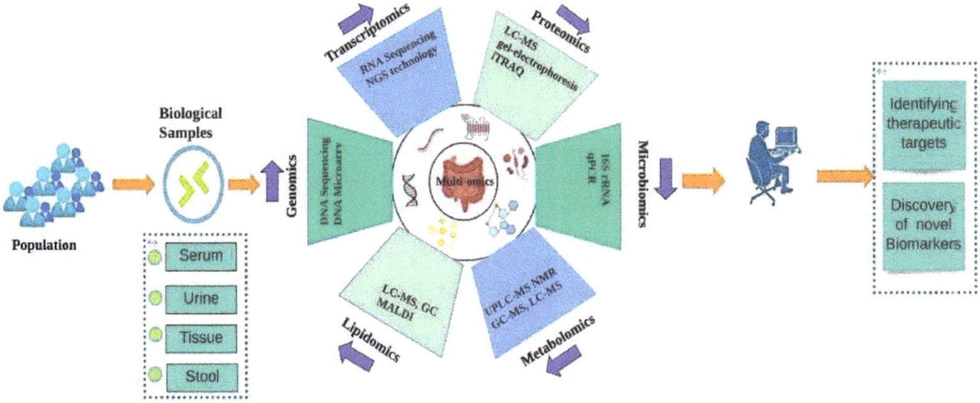

Figure 1. Graphical representation of different multi-omics-based approaches in discovering novel CRC biomarkers and therapeutic targets.

The aim of this review is to summarize the recent developments in multiple multi-omics technologies in the exploration of CRC biomarkers signatures via genomics, transcriptomics, proteomics, microbiomics, metabolomics, and lipidomics. These promising multi-omics base CRC biomarkers could be useful for clinical research.

2. Genomics of CRC

Genomic science comprises the study of an individual's entire set of DNA (including all of their genes) [26]. An individual's genomes are a comprehensive collection of information that enables them to grow and develop [27]. Using genomic analysis, researchers may better understand gene interactions, environmental effects, and how several conditions, such as cancer and diabetes, develop [28]. The development of these new approaches may facilitate disease diagnosis, treatment, and prevention [29]. During carcinogenesis, genetic and epigenetic changes occur that contribute to the identification of ideal biomarkers of CRC [30]. There is growing evidence that genetic changes play a key role in tumorigenesis. Due to this, genomics is becoming a powerful tool for finding genetic markers that can be used to diagnose and prognosis cancer, as well as improving our understanding of the disease. High-throughput next-generation sequencing is a genomic technique for sequencing an organism's DNA [31]. Multiple- biomarker panels are usually more sensitive than single biomarkers, as demonstrated by many research studies over the last few years [32]. For illustration, Ghatak S et al. used differential gene expression analysis in five independent in silico CRC cohorts and immunohistochemistry in one clinical cohort to validate their results. The authors developed a novel biomarker for early diagnosis and prognosis of cancer based on a five-panel gene signature [33]. All five in silico datasets showed that four genes (PTGS2, BDNF, CTNNB1, and GSK3B) were highly upregulated. One gene (HPGD) was substantially downregulated in primary tumour tissues compared to neighbouring normal tissues. Based on independent clinical validation cohorts, this five-gene signature was significantly associated with poor overall survival (AUC = 0.82) among colon cancer patients.

An epigenetic change happens when modified nucleotide sequences in the genome appear to be altered beyond their original form [34]. Gene expression is regulated by epigenetic mechanisms such as DNA methylation, histone modification, and nucleosome positioning Inhibitions in these regulatory processes promote malignant transformation by impairing gene function [35,36]. There is abnormal methylation of the CpG promoter during CRC, which leads to promoter hypermethylation in the promoters of tumour suppressor genes and the silencing of the transcriptional activity of DNA repair genes, which is accompanied by a loss of methylation (hypomethylation) that contributes to oncogene activation, chromosome instability, and microsatellite instability [37,38]. There is evidence that CRC biomarkers such as methylation in cfDNA and CTCs may be useful for the noninvasive diagnosis of CRC [39,40]. In addition, the epigenetic modification of 5-methylcytosine (5mC) has been associated with the emergence of several disorders, including CRC. An increasing number of studies suggest that 5mC can be used in diagnoses and prognosis of colorectal Cancer [41–47]. Furthermore, members of the ten-eleven translocation family catalyze the production of 5-hydroxymethylcytosine (5hmC), a persistent byproduct of DNA epigenetic regulation. The change of 5hmC, a new epigenetic biomarker, is linked to several disorders, particularly Cancer [48–53]. There is evidence that 5hmC plays an important role in the progression of CRC [47,53]. However, it has rarely been studied as a potential diagnostic marker for the early detection of CRC. The potential genomics biomarkers are shown in Table 1.

Table 1. Potential multi-omics base Genomics biomarkers in CRC.

Biomarker	Sample Type	Change	Application	References
CBX8, CD96	datasets	downregulated	diagnostic	[54]
MTUS1	tissue	downregulated	diagnostic and prognostic	[55]
SDC2, NDRG4	stool	upregulated	Screening	[56]
SOX21	stool	upregulated	diagnostic	[57]
BDNF, PTGS2, GSK3B and CTNNB1	tissue	upregulated	prognostic and diagnostic	[33]
HPGD	tissue	downregulated	prognostic and diagnostic	[33]
YWHAB, MCM4, and FBXO46	datasets	overexpress	prognostic	[58]
DPP72	datasets	lower expression	prognostic	[58]
SDC2, TFPI2	stool	hypermethylated	screening	[59]
SNORD15B, SNORA5C	tissue	upregulated	diagnostic and prognostic	[60]
GALR1	tissue	hypermethylation	screening	[61]
LRRC19	datasets	downregulated	prognosis	[62]
KRAS, BRAF, PIK3CA	tissue	mutation	detection	[63]

3. Transcriptomics of CRC

A transcriptomic study analyzes an organism's entire RNA content. Transcription represents an overview of the cell's activity at a particular moment due to the information in DNA [64]. In recent years, transcriptomics has made unprecedented progress in molecular genetics [21,65]. At certain developmental stages and under certain physiological or pathological conditions, transcriptomes represent all RNA molecules produced in a cell from the genome [66,67]. It consists of protein-coding RNAs (pcRNAs), also known as messenger RNAs (mRNAs), and non-coding RNAs (ncRNAs), of which each molecule exhibits a wide range of cellular functions and responses to external stimuli [68–71]. As a result of epigenetic changes and genomic instability, transcriptome changes may occur in CRC. In CRC, ncRNAs play an important role in angiogenesis, migration, differentiation, and apoptosis. Therefore, the study of ncRNAs is one of the most prominent areas of RNA research. Numerous studies have provided evidence that ncRNA expression is abnormal in CRCs. A study of ncRNA stability in stool, plasma, and serum may provide new possibilities for developing new methods of detecting ncRNAs, and it has been demonstrated that among ncRNAs, microRNAs have significant impacts on CRC [72,73]. By using next-generation sequencing, deep sequencing of CRC tumours was performed to examine the miRNA transcriptome results demonstrating that CRC patients had increased levels of miRNA-615-3p and miRNA-10b-5p expression in both the right and left side of the colons correspondingly. Additionally, five miRNAs were found to be significantly elevated in CRC patients in the study, including miR-143-3p, miR-22-3p, miR-192-5p, miR-

21-5p and miR-10a-5p [74]. Several studies have been conducted to identify novel miRNA as biomarkers, and several studies have demonstrated an important role for miR-92a and miR-429 in CRC pathogenesis [75,76]. In contrast, several miRNA molecules have been demonstrated to have significant diagnostic value for advanced neoplasia, including miR-17-92, miR-135, miR-143, and miR-145.

Moreover, a recent study improved and facilitated exosome-miRNA identification in blood using SHERLOCK-based miRNA detection. It revealed that miR-23a, miR 126, miR-940 and miR-1290, are the best good prognostic indicators for the initial stages of CRC [77]. Several miRNAs including MiR-192a, miR-29a, miR-19a-3p, miR-92a-3p, miR-125b, miR-422a and miR-223-3p, have been considering significant CRC marker. However, miR-21 has been studied extensively for diagnosing CRC [78]. In another study, miR-429 was found to reside at the centre of a miRNA-target gene network, indicating that it plays a critical role in cancer development. The miRNA samples from 28 patients with CRC markedly showed an increase in miR-32 levels. It has been determined that miR-32 expression and CRC lymphatic invasion and metastasis are correlated by the cancer genome atlas (TCGA), and a negative association was also observed between miR-32 and bone morphogenetic protein 5 (BMP5) [79]. By inhibiting EPST11 activation, BMP5 acts as a tumour suppressor. Alteration in BMP5 levels triggers the epithelial-mesenchymal transition, which stimulates tumorigenesis. Sporadic CRC tissues show a positive correlation between BMP5 expression and E-cadherin expression. Yamada et al. identified four lncRNAs, including CRCAL-1, CRCAL-2, CRCAL-3, and CRCAL-4, which differ in expression among normal mucosa and CRC patients through RNA sequencing [80]. The findings of this research highlight the implication of RNA-Seq for identifying new lncRNAs in colorectal cancer. CRC tissue also showed downregulation of NONHSAT074176.2 under GO and KEGG analysis, which may serve as a valuable diagnostic biomarker [81].

A fusion transcript (FT) is a chimeric RNA that comes from a single gene product or the trans-splicing of a transcript made by two gene products. FTs play an important role in the regulation of cancerous cells. It has been reported that transcripts from COMMD10-AP3S1, CTB-35F21.1-PSD2, and AKAP13-PDE8A are the most frequently reported transcripts in CRC. According to another study, higher levels of NFATC3-PLA2G15 fusion transcript were detected in 19 pairs of CRC tumours and adjacent normal tissue samples. As a result of the knockdown of NFATC3-PLA2G15, invasion and proliferation are inhibited in cancer cells, suggesting that NFATC3-PLA2G15 FTs may influence CRC progression; these impact findings show that this fusion transcript can serve as a novel biomarker for CRC [82] TGFRN-NOTCH2 fusion transcripts were the only transcripts detected in CRC and adjacent normal tissues from deep transcriptome sequencing. RT-PCR analysis confirmed the findings, suggesting that PTGFRN-NOTCH2 may be an FT gene in CRCs and may serve as a potential biomarker [83].

Furthermore, single-cell RNA sequencing (scRNA-seq) assesses the transcriptomic status of specific populations of single cells compared with RNA sequencing (RNA-seq) in which transcript levels are measured across different cell types [84,85]. In microdroplets and microwells, thousands of single cells can be simultaneously barcoded and handled at the same time [84]. Several technologies have been developed that measure mRNAs that are isolated from a single cell, including Quartz-Seq, Smart-seq, Smart-seq2, and CEL-seq [84,86,87]. These different types of mRNA sequencing technologies with distinct purposes. Smart-seq, for example, detects full-length transcripts. During Quartz-Seq, samples are analyzed and pooled according to the 30 end of transcripts and the CEL-Seq barcodes before linear amplification of mRNA [86]. In a recent study, the transcriptional profiles of 371,223 cells from colorectal cancer and neighbouring normal tissues were taken from 28 tumours with mismatch repair proficiency, and 34 tumours with mismatch repair deficient [88]. a significant finding of this study is that there is a structured arrangement of T cells within a tumour. In summary, the authors have provided a large number of individuals with colorectal cancer with datasets that contain information about cellular states, gene networks, and tumour transformations [88,89]. The results of scRNA-seq studies are promising because, for each cell type in a tumour, alterations may be associated with patient characteristics, diagnostic methods, therapeutic approaches, and prognosis. In the near future, scRNA-seq could be used clinically to develop customized treatment regimens for each patient based on their genetic information [90]. The potential transcriptomics biomarker is shown in Table 2.

Table 2. Potential multi-omics base transcriptomics biomarkers in CRC.

Biomarker	Sample Type	Change	Application	References
miR-92a, miR-21	serum	upregulated	diagnostic and prognostic	[91]
hsa_circ_0000567	CRC tissue and cell lines	downregulated	diagnostic	[92]
hsa-circ-0006282	plasma	upregulated	Diagnostic	[93]
hsa_circ_000592, hsa_circ_0001900 and hsa_circ_0001178	plasma	upregulated	diagnostic	[94]
miR-129-1-3p mmiR-566	urine	upregulated	detection	[95]
GPR55	CRC tissue and cell lines	downregulated	prognostic	[96]
miR-1290	plasma	upregulated	prognostic	[97]
miR-320d	plasma	downregulated	diagnostic	[98]
miR-103a-3p, miR-127-3p, miR-17-5p, miR151a5p, miR-181a-5p, miR-18a-5p and miR-18b-5p	plasma	upregulated	diagnostic	[99]
CCAT2, CCAT1, H19, MALAT1, MEG3, HULC, HOTAIR, PCAT1, PTENP1 and TUSC7	stool	upregulated	detection	[100]
miR-214, miR-199a-3p, miR-196a, miR-106a, miR-183, miR-134, miR-92a, miR-96, miR-20a, miR-21, miR-17, miR-7.	stool	upregulated	screening	[101]
miR-138, miR-143, miR-29b, miR-9, miR-146a, miR-127-5p, miR-938, miR-222.	stool	downregulated	screening	[101]

4. Proteomics of CRC

Proteins regulate many biological processes, and gene mutations could alter their expression. As well as serving as a source of potential biomarkers, the proteome is also the functional translation of the genome. Compared to the normal proteome, cancer proteome biomarkers are up- or downregulated; Thus, researchers have recently focused their attention on identifying differences between cancerous and normal cells in terms of their expression characteristics. To develop new classification tools for CRC, diagnostic, prognostic, and predictive biomarkers must be developed to detect proteins involved in its development and progression and observe the effects of protein perturbations and modifications.

Many proteomic techniques have been employed in order to find putative diagnostic biomarkers. According to Ghazanfar et al. [102], protein expression in fresh freeze samples of colorectal cancer tissue (12 individuals) was analyzed using gel electrophoresis in combination with mass spectrometry, demonstrating that number of proteins has been upregulated in colorectal Cancer. These include actin beta-like 2 (ACTBL2). Another study by Hao et al. [103], using high-resolution Fourier transform mass spectrometry, revealed that colorectal tumour tissue overexpressed dipeptidase 1 (DPEP1) Based on the examination of 22 pairs of normal tissues adjacent to cancerous tissue. Yamamoto and colleagues used a global proteomic approach to study formalin-fixed and paraffin-embosted (FFPE) CRC tissue with liquid chromatography (LC)/mass spectrometry (MS). They found a higher concentration of cyclophilin A, annexin A2, and aldolase A in cancerous tissues versus non-cancer tissues [104]. Similarly, in another study, fibroblasts associated with cancer progression were identified from human and mouse tissue. As a result of this study, it has been demonstrated that the proteins LTBP2, OLFML3, CDH11, CDH11, CALU, and FSTL1 play an important role in the migration and invasion of CRCs and have been implicated as stromal biomarkers [105].

Among the potential biomarkers of colorectal Cancer, blood-based markers are some of the most promising for performing early detection and surveillance of CRC because obtaining the specimens is relatively easy and noninvasive with minimal risk [106,107]. A targeted liquid chromatography-tandem mass spectrometry analysis was performed on 213 healthy subjects and 50 colorectal cancer patients by Ivancic et al. [108]. This study identified five proteins, including inter-alpha-trypsin inhibitor heavy-chain family member 4, leucine-rich alpha-2-glycoprotein 1, EGFR, hemopexin, and superoxide dismutase 3, that play a significant role in detecting CRC with 89% specificity and over 70%sensitivity. Furthermore, A protein panel for early detection of CRC was discovered by Bhardwaj et al. [109], by using liquid chromatography/multiple reaction monitoring-mass spectrometry in plasma samples from 96 CRC patients, and 94 controls, using a blood-based profile of five markers, osteopontin, serum paraoxonase lactonase 3, transferrin receptor protein 1, mannan-binding lectin serine protease 1, and amphiregulin. Demonstrated promising performance in screening for colorectal Cancer. Additionally, a number of members of the Serpin family including SERPINC1 (antithrombin-3, AT-III), SERPINA3 (alpha-1 antichymotrypsin, AACT), and SERPINA1 (alpha-1 antitrypsin, A1AT), have been identified as potential biomarkers for colorectal carcinoma and adenomatous polyps by using multiplexed quantification isobaric tags for absolute and relative quantitation (iTRAQ), [110]. The importance of CC chemokines (CCL15, CCL4 and CCL2) has also been assessed in CRC however further research is needed for their utility as diagnostic and clinical markers [111].

Numerous LC-MS-based research has been conducted demonstrating different CRC biomarkers. For instance, Quesada-Calvo et al. [112] suggested KNG1, Sec24C, and OLFM4 as diagnostic biomarkers out of 561 proteins with different expression levels. One other study demonstrated that ACTBL2, Annexin A2, Aldose A, DPEP1, and cyclophilin A could also serve as a biomarker for the early detection and treatment of CRC and provide new therapeutic targets [102–104]. As a biomarker source, circulating proteins are widely accepted as a better diagnostic tool for many diseases, particularly CRC [113]. Western blot (WB) and ELISA verification studies demonstrate that MRC1 and S100A9 are higher in CRC patients' serum compared to healthy individuals [114]. Furthermore, Ivancic et al. demonstrated that serum samples containing LRG1, EGFR, ITIH4, HPX, and SOD3 could detect CRC with 89% specificity and 70% sensitivity. According to these findings, GC, CRP, CD44, and ITIH3 proteins may be able to differentiate CRC depending on its stage [108]. Additionally, Bhardwaj and colleagues [109] showed five protein signatures, including MASP1, AREG, PON3, TR, and OPN, compared with FDA-approved biomarkers derived from blood superior diagnostic performance. CXCL-1 (C-X-C motif ligand 1) and CXCL-8 (C-X-C motif ligand 8) and their receptors have also demonstrated a potential role as biomarkers for CRC prognosis and diagnosis [115], Pczek S et al. conducted a study in which increased levels of CXCL-8 were found in CRC patients when compared to normal subjects. The findings of their research revealed enhanced versatility of CXCL-8 as compared to CEA in CRC diagnosis [116]. The potential proteomics biomarkers are shown in Table 3.

Table 3. Potential multi-omics based proteomics biomarkers in CRC.

Biomarker	Sample Type	Change	Application	References
CHD 9	tissue	upregulated	prognostic	[117]
ACTBL2	tissue	upregulated	diagnostic	[102]
CDK3, CDK5, and CDK8	tissue	upregulated	diagnostic	[118]
STK4 or MST1	serum	downregulated	detection	[119]
MRC1 and S100A90	serum	upregulated	diagnostic	[114]
CEACAM-7	tissue	downregulated	predictive	[120]
CEA	plasma	upregulated	predictive and prognostic	[121]
SPG20 and STK31	blood	upregulated	diagnostic	[122]
TPM3	tissue/plasma	upregulated	detection	[123]
FJX1	serum	upregulated	prognostic and diagnostic	[124]
NOP14	datasets	upregulated	Prognosis	[125]
SPARCL1	datasets	Downregulated	diagnosis	[126]

5. Microbiomics of CRC

Microbiomics is an emerging field of omics technologies that examines a symbiotic or pathological relationship between microbial communities [127]. Many microorganisms exist in the human microbiota (microbiome), such as bacteria, viruses, fungi, etc. [128–132]. An individual's gut microbiome is composed of microorganisms and their genetic materials. Over 3 million genes exist in the gastrointestinal tract, which is 150 times more than the human genome. In the gastrointestinal tract, 1013 to 1014 different microorganisms live, and over 30 million genes exist. Approximately 7000 different strains of bacteria comprise the gut microbiome in adults [133]. Gut microbiome signatures in CRC were studied using different approaches by various researchers. Various methods enrich 16S rRNA for variable regions in stool DNA, from amplifying and sequencing the V1, V2, and V4 regions to shot-gun metagenomic sequencing. Here are various methods to enrich 16S rRNA for variable regions in stool DNA, from amplification and sequencing of the V1, V2, and V4 regions to shot-gun metagenomic sequencing [134–136]. Various qPCR methods have been used to quantitate the abundance of target microbial genes in samples of interest [137–139].

Gut microbiomes have shown a significant role in the treatment of CRC; As an example, the gut microbiome may be able to be used for screening, diagnosis, prediction and/or predictive biomarkers. Alternatively, it might be a changeable factor affecting systemic CRC treatment efficacy or prevention [127,140]. The gut microbiota is a screening marker among asymptomatic individuals with high-risk adenomas or CRC. The Fusobacterium nucleatum bacteria, for example, can be examined in faecal samples from patients with adenomas and CRC to serve as a screening biomarker. Detecting and screening for CRC early may also be possible based on metabolic markers and genotoxic metabolites of specific strains [139].

Recent research published in a nature journal examined 33 cancer patients' blood and tissue samples. It revealed that the blood contained specific gut-derived pathogenic bacterial DNA that may be used to distinguish various types of tumours [141]. Therefore, the authors concluded that further research should be undertaken on this possible microbiome-based tumour diagnosis tool. In addition, the study of pathogenic bacteria (intestinal flora), and their metabolites have been linked to CRC, and the correlation analysis of gut-microbiome and metabolomics have shown a promising role in CRC prevention, treatment and diagnosis [142]. This common gastrointestinal malignancy has also become a hotspot of research in recent years [143,144]. Chen F et al. investigated the macro genomic and metabolomic compositions of serum collected from normal patients, colorectal adenomas, and CRC patients. A total of 885 differential metabolites were found in the serum associated with intestinal bacteria. This led to the Identification of eight serum metabolites that were reproducible and were used to develop categorical diagnostic models for healthy/colorectal adenoma (AUC = 0.84) and healthy/colorectal Cancer (AUC = 0.93) [145].

Some common metabolites of intestinal bacteria in the blood, bile acids, such as short-chain fatty acids and oxotrimethylamine, could be biomarkers for early CRC detection [146–148]. According to research by Huang Y et al. [149], Fusobacterium nucleatum plays a major role in promoting colorectal carcinogenesis by increasing tumour-associated metabolites, including 12a hydroxy3oxycholic acid and phosphorylcholine in the serum. Another area of active research is the discovery of biomarkers from microbial metabolomes, as some metabolites derived from the microbiota are associated with colorectal cancer. Microbial metabolites have been identified in several studies as potential biomarkers for CRC; for example, using GC-MS, an analysis of stool metabolites was conducted for CRC patients using a GC-MS technique with the result that there was a higher concentration of acetate and a lower concentration of butyrate and ursodeoxycholic, acid (UDCA) in their stool [150]. Another GC-MS metabolomic study was conducted in CRC tissue in which 19 differentiating metabolites were identified, along with pathway enrichment analyses that demonstrated that CRC patients exhibit a significant disruption of several metabolism pathways, including short-chain fatty acid metabolism, secondary bile acid metabolism, and carbohydrate metabolism [151]. Using NMR, a combined examination of tumour tissue and feces revealed a decrease in butyrate levels in patients with CRC; Fecal and tissue samples had AUCs of 0.692 and 0.717, respectively for diagnosing CRC from normal subjects, An AUC of 0.843 was reported for fecal acetate, which was the strongest indicator of diagnostic performance [152]. According to an MS-based metabolomic analysis in CRC cohorts, polyamine-based metabolites also showed a significant upregulation (N1-acetylspermidine, citrulline, arginine and ornithine) [153]. Integrating microbiome and metabolome data has demonstrated that fecal abundances of microbial-associated polyamines (cadaverine and putrescine) may play a role in colorectal cancer diagnosis [154]. CRC screening can benefit from metabolic markers, as demonstrated in these examples. The potential microbiomics and metabolomics biomarkers are shown in the Table 4.

Table 4. Potential multi-omics base microbiomics and metabolomics biomarkers in CRC.

Biomarker	Sample	Change	Application	References
F.nucleatum, P. anaerobius and *P. Micra*	stool	increase	detection	[155]
P. micra, Streptococcus anginosus	stool	increase	diagnosis	[156]
P. Micra F. nucleatum	stool	increase	diagnosis	[157]
norvaline and myristic acid	stool	upregulated	diagnosis	[158]
menaquinone-10	stool	upregulated	diagnosis	[159]
F. nucleatum	stool	upregulated	detection	[160]
Oleic acid	stool	Upregulated	screening	[161]
Succinate, Butyrate, Lactate, Glutamate, and Alanine.	tumour tissue/feces	Upregulated (excluding Butyrate downregulated)	detection	[152]
Biomarker	**Sample**	**Change**	**Application**	**References**
Cholesteryl esters, Sphingomyelins	stool	Upregulated	diagnosis	[134]
Fusobacterium, Parvimonas and Staphylococcus	stool	increase	diagnosis	[134]
Pyruvic acid, lysine, glycolic acid, fumaric acid, ornithine	blood	upregulated	detection	[162]

Table 4. *Cont.*

tryptophan, Palmitoleic acid, lysine, 3hydroxyisovaleric acid	blood	decrease	detection	[162]
octadecanoic acid, citric acid, hexadecanoic acid, and propanoic acid-2-methyl-1-(1,1-dimethylethyl)-2-methyl-1,3-propanediyl este	urine	downregulated	screening	[163]
Hydroxyproline dipeptide, tyrosine, tryptophan, pseudouridine, glucuronic acid, glycine, histidine, glucose, 5-oxoproline, threonic acid, and isocitric acid	urine	upregulated	screening	[163]

6. Metabolomics of CRC

Metabolomics is a new research area in the omics arena. Refers to an in-depth investigation of low molecular weight substances formed by metabolism in a biological fluid, including metabolic substrates and products, small peptides, lipids, vitamins, and other protein cofactors. In biomarker discovery, metabolomics is one of the fast-growing fields [164,165]. Furthermore, unlike genomics, transcriptomics, and proteomics, it represents the connections between genes and the environment, which allows it to be more precise in describing multifactorial diseases [164,166]. Many biological specimens can be used for metabolomics, most of which can be obtained using noninvasive techniques. Although biomarkers and metabolites can vary from study to study and even between specimens and colorectal cancer levels, they remain useful for diagnosing colorectal cancer [167]. Targeted metabolomics involves quantifying the metabolites linked with particular pathways associated with a specific state of disease [168]. In contrast, the untargeted approach was used in many samples and did not undergo any bias; it often measures as many metabolites as possible [169]. Due to its unique insight into disease origin and development processes, the metabolome remains a key component of disease research. Metabolomics may provide valuable information about the pathology of CRC, identify predictive biomarkers, and evaluate the severity of the disease by examining the metabolomic fingerprint in detail [170]. The metabolomics approach based on urine metabolites can be used to identify cancer biomarkers to distinguish patients with early-stage and advanced-stage colorectal cancer [171]. Lactosylceramide has also been identified as a key metabolite distinguishing Crohn's disease from ulcerative colitis in untargeted metabolomics [172].

Several metabolomic research studies have been carried out in a small cohort of colorectal cancer patients using several biological samples such as blood, urine, stool, and tissue [173,174]. A comparison was made between metabolic profiles of healthy individuals, as well as of individuals with benign polypoid pathology [175] employing nuclear magnetic resonance (NMR) or gas- and liquid-chromatography coupled to mass spectrometry (GC-MS, LC-MS) as analytical tools. Several studies have shown a negative correlation between stool and urine metabolites in patients with advanced colon cancer. The study's authors conducted a comparison of plasma, stool, and urine metabolic profiles [176]. There have been studies conducted that identified 154 different metabolites, including those that are produced during the tricarboxylic acid (TCA) cycle, urea cycle, polyamine pathways, glycolysis and amino acids, among others. With the progression of Cancer, the concentrations of these metabolites increased, with the greatest difference found in stage IV. Moreover, polyps and CRC samples were discriminated by metabolite analysis [177]. Ning et al. carried out a research study that revealed 11 upregulated and four downregulated metabolites in urine samples collected from CRC patients and healthy subjects, as shown in Table 4. Patients with CRC who were examined for pathways involved in these metabolites showed increased glycolysis, and amino acid metabolism while showing a decrease in lipid metabolism [163].

Another research has been conducted; they studied the relationships between metabolites and health status in healthy individuals and CRC patients using GC-MS analysis based on a metabolomics-based approach. This study identified several polyamines (putrescine, cadaverine) as potential biomarkers for cancer prognosis [154]. By observing metabolomic alterations in patients with CRC, another study utilizing gas chromatography-mass spectrometry (GC-MS) found that stool fatty acids, particularly increased oleic acid, may be used to screen CRC [161]. UHPLC-MS analysis of stool samples from CRC patients revealed different sphingolipid and cholesteryl esters levels [134]. A recent study of CRC tissues and stools conducted through the proton nuclear magnetic resonance (1H NMR) technology showed that butyrate was downregulated in CRC tissue and stools. At the same time, alanine, lactate, glutamate, and succinate were upregulated [152]. As metabolomics have been made a great contribution to drug discovery, UPLC-MS base metabolites biomarkers from natural compounds have also played an essential role in disease treatments [178], for instance our recent pharmacodynamic metabolomics base study using mice serum revealed that flavonoids and anthraquinones have a role in CRC treatment [179]. A combination of several multi-omics technologies could provide a powerful strategy for making valid conclusions about biomarker signatures for Colorectal Cancer. So far, no single Omics technology offers enough information to demonstrate the detailed molecular mechanism and validation of biomarker signatures.

7. Lipidomics of CRC

The field of lipidomics is one of the newest branches of multi-omics technologies. With the help of various analytical techniques, this technology can classify and analyze almost all cellular lipids, to understand their role and characteristics within biological systems. It has been studied on a larger scale for lipid species molecules. Several kinds of disease-specific biomarkers have been found through lipidomics, and the lipid species are linked to disease severity [180,181]. Regarding CRC, A very recent study, by Zaytseva et al. suggested that fatty acid metabolism might be used as a strong predisposition in CRC. It emphasized the significance of targeting lipid dysregulation in future therapeutic strategies [182]. Many studies have been conducted to examine the complex lipid profile of serum tissue samples; Consequently, a specific CRC lipidome has been the subject of ongoing discussions that may have implications for clinical treatment. The elevated levels of VLCFA (Very Long-chain Fatty acids) and lower levels of LCFA (long-chain fatty acids) have been observed in CRC patients' serum. It was explained that ELOVLs (Elongation of Very Long-chain Fatty acids Protein) may increase VLCFA elongation by increasing saturated or monosaturated fatty acids [183,184]. Based on LC-MS analysis, it was found that saturated triacylglycerols accounted for the majority of perturbations that occurred in CRC progression. The authors attributed these perturbations to LD (lipid droplet) accumulation [185]. According to another study, glycolipids, glycerophosphocholine, and acylcarnitines serum concentrations decreased in CRC patients [186]. In a recent study, Ecker et al. also found an independent prognostic marker in triglyceride lipidomic tissue signatures capable of discriminating against patients and can be used as a prognostic indicator. Through quantitative lipidomics analysis, the author demonstrated altered levels of glycerol, glycerophospholipids and sphingolipids in matched tumour samples. It has been shown that glycerol and sphingolipids can discriminate among patients with distinct mismatch repair proficient and deficient statuses, oncogenic mutations (KRAS/BRAF), or grading [187]. Several diseases have multidimensions and networks of lipid molecules fused with genes and proteins in the molecular mechanism. Lipidomics platforms can be used to analyze and characterize these compounds. Furthermore, common and traditional disease diagnoses can be more difficult to identify therapeutic targets. However, lipidomics technology offers the possibility of easier diagnosis for certain types of diseases. A diagnosis of the disease can be made by lipidomics based on existing biomarkers as therapeutic agents [188,189]. It is also possible to study various protein-lipid interactions, lipid-lipid interactions, and lipid-gene interactions, enabling the development of better diagnostic procedures for advanced diseases. A lipidomics approach has been considered better than traditional approaches for disease investigation because it provides an understanding of systemic metabolisms and their mechanisms and precisely identifies therapeutic targets and diagnostic biomarkers [190].

8. Future Perspectives and Conclusions

It is well acknowledged that early cancer diagnosis would enhance patient prognosis and provide a greater knowledge of the disease, decrease mortality, and increase patient satisfaction. There has been a significant advancement in identifying new biomarkers in recent years, paving the way for a more personalized approach to the clinical diagnosis and treatment of CRC [19]. Several DNA, RNA, and protein-based cancer biomarkers have been developed recently through high-throughput research in cancer biomarkers that can be discovered from readily available biological samples such as blood, serum, urine, stool, and tissues. Technological advancements have improved the sensitivity and specificity of cancer-specific biomarkers in CRCs. However, traditional biomarkers in clinical practice do not have high specificities and sensitivity. Therefore, in order to develop an accurate and clinically useful test, it is recommended to discover multiple biomarker panels instead of a single biomarker. By identifying prospective new therapeutic intervention targets that might contribute to the diagnosis of CRC, it is possible to develop an alternative to conventional methods of early detection of cancer.

With the recent developments in high-throughput sequencing technologies, Increasingly, cancer researchers are relying on "multi-omics" data sets. Multi-omics combines a range of omics data sets, including genomics, proteomics, transcriptomics, metabolomics, and microbiomes, for analysis [191,192]. By combining quantitative analyses of multi-omics data and clinical features, we can get insight into alterations at the molecular level and gain a more comprehensive, systemic comprehension of various biological pathways [192,193]. By integrating multi-omics approaches, we can simultaneously uncover how information flows between different levels of omics. It will, therefore, help us to bridge and close the gap between genotypes and phenotypes data. With the advent of this Technology, colorectal cancer can be diagnosed, prognosed, treated, and prevented with greater accuracy in the future. Due to the huge amount of data available, multi-omics and big data analytics are required to interconnect all available information. In particular, integrating patient demographic, genomic, transcriptomic, proteomic, metabolomic, lipidomic, and microbiota data could assist in developing new biomarkers discovery and clinical outcomes prediction.

Another emerging field of multi-omics technologies is microbiomics which offers a non-traditional tool with potential applications in more significant comprehension of tumour biology. Identifying microbial metabolites correlated with the development of colorectal tumours has significant implications for identifying new treatment targets and possible biomarkers for disease screening [194]. There has been significant progress in recently using intestinal bacteria and their metabolites as early detection markers for CRC. The association between CRC and gut bacteria and their metabolites has received much attention recently. Moreover, the gut microbiota's microbial metabolite composition is frequently renewed and changes depending on the diet, making it more amenable to therapeutic intervention in developing CRC. New paradigms in CRC diagnosis, prevention, and treatment will be provided by elucidating the role of microbial metabolites [15]. CRC prevention, comprehensive treatment planning, and minimizing adverse effects of treatment will be significantly impacted by the gut microbiome in the near future. Individuals' gut microbiomes vary according to their geographic location, ethnicity, dietary habits, and lifestyles. In the future, clinical research will need to include several factors that contribute to the microbiota of patients, including geography, race, sex, and diet, as well as how systemic cancer treatment affects the microbiome, especially chemotherapy and immunotherapy [195].

In recent years, lipidomics has been actively used in the research community and regarded as a cutting-edge example of multi-Omics Technology. Particularly useful in analyzing the structure and function of lipid molecules to analyze changes in their dynamic composition during certain pathological and physiological changes. The alterations in lipid metabolism have also been linked to several kinds of cancer development and progression. Despite this, there is a limited understanding of the metabolic changes of lipids in cancer due to their structural diversity and characteristics, distinct from those of other biomolecules. Several analytical tools in cancer research have been used in lipidomic analysis to determine lipid composition at a large scale. Additionally, in cohort studies, glycero-, glycerophospho-, and sphingolipid levels have been significantly changed between tumours and normal tissue. A marked difference between cancerous and non-diseased tissue in sphingomyelin and triacylglycerol (TG) species [187]. Recent research demonstrated that GZMs (granzymes) proteins have a significant role in carcinogenesis, their role as new biomarkers for CRC prognosis and diagnosis will need further exploration [196]. Furthermore, it is imperative to emphasize the importance of lipidomics and proteomics research for discovering novel biomarkers and diagnosing CRCs. Integrating lipidomics with other omics, such as metabolomics, microbiomics, proteomics, etc., would provide a powerful tool that could help researchers identify novel therapeutic targets and biomarkers.

Many studies have been conducted using different multi-omics techniques and clinical samples to discover novel biomarkers. However, clearing up the mechanism of how markers are generated and their diagnostic value, a critical factor in drug discovery, remains a challenge. Combining multiple experimental approaches and then integrating the results is a valuable strategy to generalize human cancer's complexity from experimental models [197]. The integration of multiple omics, such as genomics, proteomics, and metabolomics, will help us understand tumours and advance antitumor drug developments [198–200]. In addition, numerous studies have confirmed that developing high-throughput sequencing technologies has revolutionized multi-omics research. It is expected that multi-omics applications will increase in scope with the optimization and maturity of various technologies, making it possible to develop novel biomarkers for CRC due to multi-omics research.

Although multi-omics methods have great potential, there are still limitations and challenges to overcome. The first problem is that omics methods are expensive and require specialized equipment and high-level data analysis skills. There can also be problems in the collection of data and verification of the data because of unreliable data quality, inaccurate data sources, and nonstandard sampling. Currently, there are no standard research platforms or bioinformatics methods for the processing of large-scale omics datasets. Data processing and analysis are the biggest challenges in metabolomics studies because biological organisms contain thousands of metabolites. Additionally, numerous obstacles will need to be overcome in order to translate biomarker discoveries into clinical applications for CRC. It is still difficult to evaluate the specificity and sensitivity of candidate biomarkers due to the absence of strategies and selection panels. Due to the fact that most patient data come from different laboratories, it is also difficult to validate biomarker candidates in large cohorts of patients. To establish potential diagnostic biomarkers, further validation may be obtained through meta-analysis. Another obstacle to overcome is the heterogeneity of the patient population and their sporadic cancers. By performing advanced MS at single-cell resolution, this problem may be tackled by understanding the biological and molecular heterogeneity of disease states.

Author Contributions: Conceptualization, I.U. and X.-J.W.; writing—original draft preparation, I.U.; review and editing X.-J.W., L.Y. and Y.S.; literature search, F.-T.Y., X.-H.L. and J.L.; supervision, X.-J.W., funding acquisition, X.-J.W. All authors have read and agreed to the published version of the manuscript.

Funding: This study was supported by the National Natural Science Foundation of China (No. 81830110, 81903847, 32141005). And UNPYSCT-2020226 Heilongjiang province regular undergraduate higher education institution youth innovation talent training program project.

Conflicts of Interest: The authors declare no conflict of interest.

Abbreviations

CBX8	Chromobox 8
CD96	CD96 Molecule
8MTUS1	Microtubule Associated Scaffold Protein 1
SDC2	Syndecan 2
NDRG4	NDRG Family Member 4
SOX21	SRY-Box Transcription Factor 21
BDNF	Brain-Derived Neurotrophic Factor
PTGS2	Prostaglandin–Endoperoxide Synthase 2
GSK3B	Glycogen Synthase Kinase 3 Beta
CTNNB1	Catenin Beta 1
HPGD	15-Hydroxyprostaglandin Dehydrogenase
YWHAB	Tyrosine 3–Monooxygenase/Tryptophan 5–Monooxygenase Activation Protein Beta
MCM4,	Minichromosome Maintenance Complex Component 4
FBXO46	F-Box Protein 46
DPP7/2	Dipeptidyl Peptidase 7
SDC2	Syndecan 2
TFPI2	Tissue Factor Pathway Inhibitor 2
SNORD15B	Small Nucleolar RNA, C/D Box 15B
SNORA5C	Small Nucleolar RNA, H/ACA Box 5C
GALR1	Galanin Receptor 1
LRRC19	Leucine-rich repeat-containing protein 19
GPR55	G protein-coupled receptor 55
CCAT2	Colon Cancer Associated Transcript 2
CCAT1	Colon Cancer Associated Transcript 1
H19	H19 Imprinted Maternally Expressed Transcript
MALAT1	Metastasis Associated Lung Adenocarcinoma Transcript 1
MEG3	Maternally Expressed 3
HULC	Hepatocellular Carcinoma Up-Regulated Long Non-Coding RNA
HOTAIR	HOX Transcript Antisense RNA
PCAT1	Prostate Cancer Associated Transcript 1
PTENP1	Phosphatase And Tensin Homolog Pseudogene 1
TUSC7	Tumour Suppressor Candidate 7
CHD 9	Chromodomain Helicase DNA Binding Protein 9
ACTBL2	Actin Beta Like 2
CDK3,	Cyclin Dependent Kinase 3
CDK5	Cyclin Dependent Kinase 5
CDK8	Cyclin-dependent kinase 8
STK4 or MST1	serine/threonine kinase 4 or Macrophage Stimulating 1
MRC1	Mannose Receptor C-Type 1
S100A90	S100 Calcium Binding Protein A9
CEACAM-7	CEA Cell Adhesion Molecule 7
CEA	Carcinoembryonic antigen
SPG20	spastic paraplegia 20
STK31	Serine/Threonine Kinase 31
TPM3	Tropomyosin 3
FJX1	Four-Jointed Box Kinase 1
NOP14	Nucleolar protein 14
SPARCL1	Secreted protein acidic and rich in cysteine-like 1

References

1. Siegel, R.L.; Miller, K.D.; Fuchs, H.E.; Jemal, A. Cancer Statistics, 2021. *CA Cancer J. Clin.* **2021**, *71*, 7–33. [CrossRef] [PubMed]
2. Sung, H.; Ferlay, J.; Siegel, R.L.; Laversanne, M.; Soerjomataram, I.; Jemal, A.; Bray, F. Global Cancer Statistics 2020: GLOBOCAN Estimates of Incidence and Mortality Worldwide for 36 Cancers in 185 Countries. *CA Cancer J. Clin.* **2021**, *71*, 209–249. [CrossRef] [PubMed]

3. Yoshino, T.; Arnold, D.; Taniguchi, H.; Pentheroudakis, G.; Yamazaki, K.; Xu, R.H.; Kim, T.W.; Ismail, F.; Tan, I.B.; Yeh, K.H.; et al. Pan-Asian adapted ESMO consensus guidelines for the management of patients with metastatic colorectal cancer: A JSMO-ESMO initiative endorsed by CSCO, KACO, MOS, SSO and TOS. *Ann. Oncol.* **2018**, *29*, 44–70. [CrossRef]
4. Benson, A.B.; Venook, A.P.; Al-Hawary, M.M.; Arain, M.A.; Chen, Y.J.; Ciombor, K.K.; Cohen, S.; Cooper, H.S.; Deming, D.; Farkas, L.; et al. Colon Cancer, Version 2.2021, NCCN Clinical Practice Guidelines in Oncology. *J. Natl. Compr. Cancer Netw.* **2021**, *19*, 329–359. [CrossRef] [PubMed]
5. Iyer, P.; Deng, M.; Handorf, E.A.; Nakhoda, S.; Dotan, E. Assessing Oncologists' Adoption of Biomarker Testing in Metastatic Colorectal Cancer Using Real World Data. *JNCI Cancer Spectr.* **2022**, pkac065. [CrossRef]
6. Amatu, A.; Sartore-Bianchi, A.; Bencardino, K.; Pizzutilo, E.G.; Tosi, F.; Siena, S. Tropomyosin receptor kinase (TRK) biology and the role of NTRK gene fusions in cancer. *Ann. Oncol.* **2019**, *30* (Suppl. S8), viii5–viii15. [CrossRef]
7. Marabelle, A.; Fakih, M.; Lopez, J.; Shah, M.; Shapira-Frommer, R.; Nakagawa, K.; Chung, H.C.; Kindler, H.L.; Lopez-Martin, J.A.; Miller, W.H., Jr.; et al. Association of tumour mutational burden with outcomes in patients with advanced solid tumours treated with pembrolizumab: Prospective biomarker analysis of the multicohort, open-label, phase 2 KEYNOTE-158 study. *Lancet Oncol.* **2020**, *21*, 1353–1365. [CrossRef]
8. Argilés, G.; Tabernero, J.; Labianca, R.; Hochhauser, D.; Salazar, R.; Iveson, T.; Laurent-Puig, P.; Quirke, P.; Yoshino, T.; Taieb, J.; et al. Localised colon cancer: ESMO Clinical Practice Guidelines for diagnosis, treatment and follow-up. *Ann. Oncol.* **2020**, *31*, 1291–1305. [CrossRef]
9. Testing for Colorectal Cancer. How Is Colorectal Cancer Diagnosed? Available online: https://www.cancer.gov/types/colorectal/screening-fact-sheet (accessed on 12 September 2022).
10. Elsafi, S.H.; Alqahtani, N.I.; Zakary, N.Y.; Al Zahrani, E.M. The sensitivity, specificity, predictive values, and likelihood ratios of fecal occult blood test for the detection of colorectal cancer in hospital settings. *Clin. Exp. Gastroenterol.* **2015**, *8*, 279–284. [CrossRef]
11. Chan, S.C.H.; Liang, J.Q. Advances in tests for colorectal cancer screening and diagnosis. *Expert Rev. Mol. Diagn.* **2022**, *22*, 449–460. [CrossRef]
12. Tan, K.C.; Ipcho, S.V.; Trengove, R.D.; Oliver, R.P.; Solomon, P.S. Assessing the impact of transcriptomics, proteomics and metabolomics on fungal phytopathology. *Mol. Plant Pathol.* **2009**, *10*, 703–715. [CrossRef] [PubMed]
13. Dalal, N.; Jalandra, R.; Sharma, M.; Prakash, H.; Makharia, G.K.; Solanki, P.R.; Singh, R.; Kumar, A. Omics technologies for improved diagnosis and treatment of colorectal cancer: Technical advancement and major perspectives. *Biomed. Pharmacother.* **2020**, *131*, 110648. [CrossRef]
14. Jahani-Sherafat, S.; Alebouyeh, M.; Moghim, S.; Ahmadi Amoli, H.; Ghasemian-Safaei, H. Role of gut microbiota in the pathogenesis of colorectal cancer; a review article. *Gastroenterol. Hepatol. Bed Bench* **2018**, *11*, 101–109. [PubMed]
15. Peng, Y.; Nie, Y.; Yu, J.; Wong, C.C. Microbial Metabolites in Colorectal Cancer: Basic and Clinical Implications. *Metabolites* **2021**, *11*, 159. [CrossRef] [PubMed]
16. Kontush, A.; Chapman, M.J. Lipidomics as a tool for the study of lipoprotein metabolism. *Curr. Atheroscler. Rep.* **2010**, *12*, 194–201. [CrossRef] [PubMed]
17. Lam, S.M.; Shui, G. Lipidomics as a principal tool for advancing biomedical research. *J. Genet. Genom.* **2013**, *40*, 375–390. [CrossRef]
18. Mato-Abad, V.; Pazos, A.; Munteanu, C.R.; Liñares-Blanco, J.; Alvarez-Gonzalez, S.; Vázquez-Naya, J.M.; Pedreira, N.; Amigo, J.; Fernandez-Lozano, C. Bioinformatic tools for research in CRC. In *Foundations of Colorectal Cancer*; Elsevier: Amsterdam, The Netherlands, 2022; pp. 231–247.
19. Lopez, G.; Boggio, F.; Ferrero, S.; Fusco, N.; Del Gobbo, A. Molecular and Immunohistochemical Markers with Prognostic and Predictive Significance in Liver Metastases from Colorectal Carcinoma. *Int. J. Mol. Sci.* **2018**, *19*, 3014. [CrossRef]
20. Loomans-Kropp, H.A.; Umar, A. Increasing Incidence of Colorectal Cancer in Young Adults. *J. Cancer Epidemiol.* **2019**, *2019*, 9841295. [CrossRef]
21. Casamassimi, A.; Federico, A.; Rienzo, M.; Esposito, S.; Ciccodicola, A. Transcriptome Profiling in Human Diseases: New Advances and Perspectives. *Int. J. Mol. Sci.* **2017**, *18*, 1652. [CrossRef]
22. Wheeler, D.A.; Wang, L. From human genome to cancer genome: The first decade. *Genome Res.* **2013**, *23*, 1054–1062. [CrossRef]
23. López-Lázaro, M. A new view of carcinogenesis and an alternative approach to cancer therapy. *Mol. Med.* **2010**, *16*, 144–153. [CrossRef] [PubMed]
24. Supplitt, S.; Karpinski, P.; Sasiadek, M.; Laczmanska, I. Current Achievements and Applications of Transcriptomics in Personalized Cancer Medicine. *Int. J. Mol. Sci.* **2021**, *22*, 1422. [CrossRef] [PubMed]
25. Sardo, E.; Napolitano, S.; Della Corte, C.M.; Ciardiello, D.; Raucci, A.; Arrichiello, G.; Troiani, T.; Ciardiello, F.; Martinelli, E.; Martini, G. Multi-Omic Approaches in Colorectal Cancer beyond Genomic Data. *J. Pers. Med.* **2022**, *12*, 128. [CrossRef] [PubMed]
26. Paleri, P. *Revisiting National Security: Prospecting Governance for Human Well-Being*; Springer Nature: Berlin/Heidelberg, Germany, 2022.
27. Bunnik, E.M.; Le Roch, K.G. An Introduction to Functional Genomics and Systems Biology. *Adv. Wound Care* **2013**, *2*, 490–498. [CrossRef]
28. Przybyla, L.; Gilbert, L.A. A new era in functional genomics screens. *Nat. Rev. Genet.* **2022**, *23*, 89–103. [CrossRef]
29. Definition of genomics. NCI Dictionary of Cancer Terms—National Cancer Institute. Available online: https://www.cancer.gov/publications/dictionaries/cancer-terms/def/genomics (accessed on 12 September 2022).

30. Bosch, L.J.; Carvalho, B.; Fijneman, R.J.; Jimenez, C.R.; Pinedo, H.M.; van Engeland, M.; Meijer, G.A. Molecular tests for colorectal cancer screening. *Clin. Colorectal. Cancer* **2011**, *10*, 8–23. [CrossRef]
31. Escobar-Zepeda, A.; Vera-Ponce de León, A.; Sanchez-Flores, A. The Road to Metagenomics: From Microbiology to DNA Sequencing Technologies and Bioinformatics. *Front. Genet.* **2015**, *6*, 348. [CrossRef]
32. Gan, X.; Wang, T.; Chen, Z.Y.; Zhang, K.H. Blood-derived molecular signatures as biomarker panels for the early detection of colorectal cancer. *Mol. Biol. Rep.* **2020**, *47*, 8159–8168. [CrossRef]
33. Ghatak, S.; Mehrabi, S.F.; Mehdawi, L.M.; Satapathy, S.R.; Sjölander, A. Identification of a Novel Five-Gene Signature as a Prognostic and Diagnostic Biomarker in Colorectal Cancers. *Int. J. Mol. Sci.* **2022**, *23*, 793. [CrossRef]
34. Essa, H.Y.S.; Kusaf, G.; Yuruker, O.; Kalkan, R. Epigenetic Alteration in Colorectal Cancer: A Biomarker for Diagnostic and Therapeutic Application. *Glob. Med. Genet.* **2022**, *9*, 258–262. [CrossRef]
35. Esteller, M. Cancer epigenomics: DNA methylomes and histone-modification maps. *Nat. Rev. Genet.* **2007**, *8*, 286–298. [CrossRef] [PubMed]
36. Singh, M.P.; Rai, S.; Suyal, S.; Singh, S.K.; Singh, N.K.; Agarwal, A.; Srivastava, S. Genetic and epigenetic markers in colorectal cancer screening: Recent advances. *Expert Rev. Mol. Diagn.* **2017**, *17*, 665–685. [CrossRef]
37. Zamani, M.; Hosseini, S.V.; Mokarram, P. Epigenetic biomarkers in colorectal cancer: Premises and prospects. *Biomarkers* **2018**, *23*, 105–114. [CrossRef] [PubMed]
38. Berdasco, M.; Esteller, M. Clinical epigenetics: Seizing opportunities for translation. *Nat. Rev. Genet.* **2019**, *20*, 109–127. [CrossRef] [PubMed]
39. Galanopoulos, M.; Tsoukalas, N.; Papanikolaou, I.S.; Tolia, M.; Gazouli, M.; Mantzaris, G.J. Abnormal DNA methylation as a cell-free circulating DNA biomarker for colorectal cancer detection: A review of literature. *World J. Gastrointest. Oncol.* **2017**, *9*, 142–152. [CrossRef] [PubMed]
40. Krishnamurthy, N.; Spencer, E.; Torkamani, A.; Nicholson, L. Liquid Biopsies for Cancer: Coming to a Patient near You. *J. Clin. Med.* **2017**, *6*, 3. [CrossRef]
41. Belkhiri, A.; El-Rifai, W. 5-Methylcytosine hydroxylation-mediated LINE-1 hypomethylation: A novel mechanism of proto-oncogenes activation in colorectal cancer? *Gut* **2014**, *63*, 538–539. [CrossRef]
42. Ichimura, N.; Shinjo, K.; An, B.; Shimizu, Y.; Yamao, K.; Ohka, F.; Katsushima, K.; Hatanaka, A.; Tojo, M.; Yamamoto, E.; et al. Aberrant TET1 Methylation Closely Associated with CpG Island Methylator Phenotype in Colorectal Cancer. *Cancer Prev. Res.* **2015**, *8*, 702–711. [CrossRef]
43. Hur, K.; Cejas, P.; Feliu, J.; Moreno-Rubio, J.; Burgos, E.; Boland, C.R.; Goel, A. Hypomethylation of long interspersed nuclear element-1 (LINE-1) leads to activation of proto-oncogenes in human colorectal cancer metastasis. *Gut* **2014**, *63*, 635–646. [CrossRef]
44. Zheng, R.; Gao, D.; He, T.; Zhang, M.; Zhang, X.; Linghu, E.; Wei, L.; Guo, M. Methylation of DIRAS1 promotes colorectal cancer progression and may serve as a marker for poor prognosis. *Clin. Epigenetics* **2017**, *9*, 50. [CrossRef]
45. Wang, Y.; He, T.; Herman, J.G.; Linghu, E.; Yang, Y.; Fuks, F.; Zhou, F.; Song, L.; Guo, M. Methylation of ZNF331 is an independent prognostic marker of colorectal cancer and promotes colorectal cancer growth. *Clin. Epigenetics* **2017**, *9*, 115. [CrossRef]
46. He, T.; Zhang, M.; Zheng, R.; Zheng, S.; Linghu, E.; Herman, J.G.; Guo, M. Methylation of SLFN11 is a marker of poor prognosis and cisplatin resistance in colorectal cancer. *Epigenomics* **2017**, *9*, 849–862. [CrossRef]
47. Zhu, Y.; Lu, H.; Zhang, D.; Li, M.; Sun, X.; Wan, L.; Yu, D.; Tian, Y.; Jin, H.; Lin, A.; et al. Integrated analyses of multi-omics reveal global patterns of methylation and hydroxymethylation and screen the tumor suppressive roles of HADHB in colorectal cancer. *Clin. Epigenetics* **2018**, *10*, 30. [CrossRef] [PubMed]
48. Vasanthakumar, A.; Godley, L.A. 5-hydroxymethylcytosine in cancer: Significance in diagnosis and therapy. *Cancer Genet.* **2015**, *208*, 167–177. [CrossRef]
49. Wang, J.; Su, Y.; Tian, Y.; Ding, Y.; Wang, X. Characterization of DNA hydroxymethylation profile in cervical cancer. *Artif. Cells Nanomed. Biotechnol.* **2019**, *47*, 2706–2714. [CrossRef]
50. Hlady, R.A.; Sathyanarayan, A.; Thompson, J.J.; Zhou, D.; Wu, Q.; Pham, K.; Lee, J.H.; Liu, C.; Robertson, K.D. Integrating the Epigenome to Identify Drivers of Hepatocellular Carcinoma. *Hepatology* **2019**, *69*, 639–652. [CrossRef] [PubMed]
51. Bhattacharyya, S.; Pradhan, K.; Campbell, N.; Mazdo, J.; Vasantkumar, A.; Maqbool, S.; Bhagat, T.D.; Gupta, S.; Suzuki, M.; Yu, Y.; et al. Altered hydroxymethylation is seen at regulatory regions in pancreatic cancer and regulates oncogenic pathways. *Genome Res.* **2017**, *27*, 1830–1842. [CrossRef] [PubMed]
52. Thomson, J.P.; Ottaviano, R.; Unterberger, E.B.; Lempiäinen, H.; Muller, A.; Terranova, R.; Illingworth, R.S.; Webb, S.; Kerr, A.R.; Lyall, M.J.; et al. Loss of Tet1-Associated 5-Hydroxymethylcytosine Is Concomitant with Aberrant Promoter Hypermethylation in Liver Cancer. *Cancer Res.* **2016**, *76*, 3097–3108. [CrossRef] [PubMed]
53. Jiang, A.C.; Buckingham, L.; Barbanera, W.; Korang, A.Y.; Bishesari, F.; Melson, J. LINE-1 is preferentially hypomethylated within adenomatous polyps in the presence of synchronous colorectal cancer. *Clin. Epigenetics* **2017**, *9*, 25. [CrossRef] [PubMed]
54. Song, X.; Tang, T.; Li, C.; Liu, X.; Zhou, L. CBX8 and CD96 Are Important Prognostic Biomarkers of Colorectal Cancer. *Med. Sci. Monit.* **2018**, *24*, 7820–7827. [CrossRef]
55. Cheng, L.Y.; Huang, M.S.; Zhong, H.G.; Ru, H.M.; Mo, S.S.; Wei, C.Y.; Su, Z.J.; Mo, X.W.; Yan, L.H.; Tang, W.Z. MTUS1 is a promising diagnostic and prognostic biomarker for colorectal cancer. *World J. Surg. Oncol.* **2022**, *20*, 257. [CrossRef]

56. Jin, S.; Ye, Q.; Hong, Y.; Dai, W.; Zhang, C.; Liu, W.; Guo, Y.; Zhu, D.; Zhang, Z.; Chen, S.; et al. A systematic evaluation of stool DNA preparation protocols for colorectal cancer screening via analysis of DNA methylation biomarkers. *Clin. Chem. Lab. Med.* **2020**, *59*, 91–99. [CrossRef]
57. Moradi, K.; Babaei, E.; Hosseinpour Feizi, M.A.; Safaralizadeh, R.; Rezvani, N. Quantitative detection of SRY-Box 21 (SOX21) gene promoter methylation as a stool-based noninvasive biomarker for early diagnosis of colorectal cancer by MethyLight method. *Indian J. Cancer* **2021**, *58*, 217–224. [CrossRef]
58. Ahluwalia, P.; Mondal, A.K.; Bloomer, C.; Fulzele, S.; Jones, K.; Ananth, S.; Gahlay, G.K.; Heneidi, S.; Rojiani, A.M.; Kota, V.; et al. Identification and Clinical Validation of a Novel 4 Gene-Signature with Prognostic Utility in Colorectal Cancer. *Int. J. Mol. Sci.* **2019**, *20*, 3818. [CrossRef]
59. Zhang, W.; Yang, C.; Wang, S.; Xiang, Z.; Dou, R.; Lin, Z.; Zheng, J.; Xiong, B. SDC2 and TFPI2 Methylation in Stool Samples as an Integrated Biomarker for Early Detection of Colorectal Cancer. *Cancer Manag. Res.* **2021**, *13*, 3601–3617. [CrossRef]
60. Shen, L.; Lu, W.; Huang, Y.; He, J.; Wang, Q.; Zheng, X.; Wang, Z. SNORD15B and SNORA5C: Novel Diagnostic and Prognostic Biomarkers for Colorectal Cancer. *Biomed Res. Int.* **2022**, *2022*, 8260800. [CrossRef]
61. Gu, S.; Qian, S.; Lin, S.; Ye, D.; Li, Q.; Yang, J.; Ying, X.; Li, Z.; Tang, M.; Wang, J.; et al. Promoter hypermethylation of GALR1 acts as an early epigenetic susceptibility event in colorectal carcinogenesis. *J. Hum. Genet.* **2022**, *67*, 519–525. [CrossRef] [PubMed]
62. Wang, Y.J.; Liu, M.; Jiang, H.Y.; Yu, Y.W. Downregulation of LRRC19 Is Associated with Poor Prognosis in Colorectal Cancer. *J. Oncol.* **2022**, *2022*, 5848823. [CrossRef] [PubMed]
63. Alizadeh-Sedigh, M.; Mahmoodzadeh, H.; Fazeli, M.S.; Haddadi-Aghdam, M.; Teimoori-Toolabi, L. The potential of PIK3CA, KRAS, BRAF, and APC hotspot mutations as a non-invasive detection method for colorectal cancer. *Mol. Cell. Probes* **2022**, *63*, 101807. [CrossRef] [PubMed]
64. Mantione, K.J.; Kream, R.M.; Kuzelova, H.; Ptacek, R.; Raboch, J.; Samuel, J.M.; Stefano, G.B. Comparing bioinformatic gene expression profiling methods: Microarray and RNA-Seq. *Med. Sci. Monit. Basic Res.* **2014**, *20*, 138–142. [CrossRef]
65. Lowe, R.; Shirley, N.; Bleackley, M.; Dolan, S.; Shafee, T. Transcriptomics technologies. *PLoS Comput. Biol.* **2017**, *13*, e1005457. [CrossRef] [PubMed]
66. Wang, Z.; Gerstein, M.; Snyder, M. RNA-Seq: A revolutionary tool for transcriptomics. *Nat. Rev. Genet.* **2009**, *10*, 57–63. [CrossRef] [PubMed]
67. Jacquier, A. The complex eukaryotic transcriptome: Unexpected pervasive transcription and novel small RNAs. *Nat. Rev. Genet.* **2009**, *10*, 833–844. [CrossRef]
68. Popov, D.V.; Makhnovskii, P.A.; Shagimardanova, E.I.; Gazizova, G.R.; Lysenko, E.A.; Gusev, O.A.; Vinogradova, O.L. Contractile activity-specific transcriptome response to acute endurance exercise and training in human skeletal muscle. *Am. J. Physiol. Endocrinol. Metab.* **2019**, *316*, E605–E614. [CrossRef] [PubMed]
69. Iwata, M.; Yuan, L.; Zhao, Q.; Tabei, Y.; Berenger, F.; Sawada, R.; Akiyoshi, S.; Hamano, M.; Yamanishi, Y. Predicting drug-induced transcriptome responses of a wide range of human cell lines by a novel tensor-train decomposition algorithm. *Bioinformatics* **2019**, *35*, i191–i199. [CrossRef]
70. Jaeger, P.A.; Doherty, C.; Ideker, T. Modeling transcriptome dynamics in a complex world. *Cell* **2012**, *151*, 1161–1162. [CrossRef]
71. Mallardo, M.; Poltronieri, P.; D'Urso, O.F. Non-protein coding RNA biomarkers and differential expression in cancers: A review. *J. Exp. Clin. Cancer Res.* **2008**, *27*, 19. [CrossRef] [PubMed]
72. Bashiardes, S.; Zilberman-Schapira, G.; Elinav, E. Use of Metatranscriptomics in Microbiome Research. *Bioinform. Biol. Insights* **2016**, *10*, 19–25. [CrossRef] [PubMed]
73. Slaby, O.; Svoboda, M.; Michalek, J.; Vyzula, R. MicroRNAs in colorectal cancer: Translation of molecular biology into clinical application. *Mol. Cancer* **2009**, *8*, 102. [CrossRef]
74. Schee, K.; Lorenz, S.; Worren, M.M.; Günther, C.C.; Holden, M.; Hovig, E.; Fodstad, O.; Meza-Zepeda, L.A.; Flatmark, K. Deep Sequencing the MicroRNA Transcriptome in Colorectal Cancer. *PLoS ONE* **2013**, *8*, e66165. [CrossRef]
75. Sun, Y.; Shen, S.; Tang, H.; Xiang, J.; Peng, Y.; Tang, A.; Li, N.; Zhou, W.; Wang, Z.; Zhang, D.; et al. miR-429 identified by dynamic transcriptome analysis is a new candidate biomarker for colorectal cancer prognosis. *Omics* **2014**, *18*, 54–64. [CrossRef] [PubMed]
76. Chen, E.; Li, Q.; Wang, H.; Yang, F.; Min, L.; Yang, J. MiR-92a promotes tumorigenesis of colorectal cancer, a transcriptomic and functional based study. *Biomed Pharmacother.* **2018**, *106*, 1370–1377. [CrossRef] [PubMed]
77. Durán-Vinet, B.; Araya-Castro, K.; Calderón, J.; Vergara, L.; Weber, H.; Retamales, J.; Araya-Castro, P.; Leal-Rojas, P. CRISPR/Cas13-Based Platforms for a Potential Next-Generation Diagnosis of Colorectal Cancer through Exosomes Micro-RNA Detection: A Review. *Cancers* **2021**, *13*, 4640. [CrossRef]
78. Gallardo-Gómez, M.; De Chiara, L.; Álvarez-Chaver, P.; Cubiella, J. Colorectal cancer screening and diagnosis: Omics-based technologies for development of a non-invasive blood-based method. *Expert Rev. Anticancer. Ther.* **2021**, *21*, 723–738. [CrossRef]
79. Chen, E.; Li, Q.; Wang, H.; Zhang, P.; Zhao, X.; Yang, F.; Yang, J. MiR-32 promotes tumorigenesis of colorectal cancer by targeting BMP5. *Biomed. Pharmacother.* **2018**, *106*, 1046–1051. [CrossRef]
80. Yamada, A.; Yu, P.; Lin, W.; Okugawa, Y.; Boland, C.R.; Goel, A. A RNA-Sequencing approach for the identification of novel long non-coding RNA biomarkers in colorectal cancer. *Sci. Rep.* **2018**, *8*, 575. [CrossRef]
81. Zhang, Z.; Jia, H.; Gu, T.; Hu, Q.; Yu, J.; Zang, D.; Song, N.; Wang, H. RNA sequencing and bioinformatics analysis of the long noncoding RNA-mRNA network in colorectal cancer. *J. Cell. Biochem.* **2018**, *119*, 9957–9966. [CrossRef]

82. Jang, J.E.; Kim, H.P.; Han, S.W.; Jang, H.; Lee, S.H.; Song, S.H.; Bang, D.; Kim, T.Y. NFATC3-PLA2G15 Fusion Transcript Identified by RNA Sequencing Promotes Tumor Invasion and Proliferation in Colorectal Cancer Cell Lines. *Cancer Res. Treat.* **2019**, *51*, 391–401. [CrossRef] [PubMed]
83. Wu, Y.; Wang, X.; Wu, F.; Huang, R.; Xue, F.; Liang, G.; Tao, M.; Cai, P.; Huang, Y. Transcriptome profiling of the cancer, adjacent non-tumor and distant normal tissues from a colorectal cancer patient by deep sequencing. *PLoS ONE* **2012**, *7*, e41001. [CrossRef]
84. Kashima, Y.; Sakamoto, Y.; Kaneko, K.; Seki, M.; Suzuki, Y.; Suzuki, A. Single-cell sequencing techniques from individual to multiomics analyses. *Exp. Mol. Med.* **2020**, *52*, 1419–1427. [CrossRef]
85. Haque, A.; Engel, J.; Teichmann, S.A.; Lönnberg, T. A practical guide to single-cell RNA-sequencing for biomedical research and clinical applications. *Genome Med.* **2017**, *9*, 75. [CrossRef] [PubMed]
86. Lee, J.; Hyeon, D.Y.; Hwang, D. Single-cell multiomics: Technologies and data analysis methods. *Exp. Mol. Med.* **2020**, *52*, 1428–1442. [CrossRef]
87. Picelli, S. Single-cell RNA-sequencing: The future of genome biology is now. *RNA Biol.* **2017**, *14*, 637–650. [CrossRef]
88. Pelka, K.; Hofree, M.; Chen, J.H.; Sarkizova, S.; Pirl, J.D.; Jorgji, V.; Bejnood, A.; Dionne, D.; Ge, W.H.; Xu, K.H.; et al. Spatially organized multicellular immune hubs in human colorectal cancer. *Cell* **2021**, *184*, 4734–4752.e4720. [CrossRef]
89. Davis, R.T.; Blake, K.; Ma, D.; Gabra, M.B.I.; Hernandez, G.A.; Phung, A.T.; Yang, Y.; Maurer, D.; Lefebvre, A.; Alshetaiwi, H.; et al. Transcriptional diversity and bioenergetic shift in human breast cancer metastasis revealed by single-cell RNA sequencing. *Nat. Cell Biol.* **2020**, *22*, 310–320. [CrossRef]
90. Shalek, A.K.; Benson, M. Single-cell analyses to tailor treatments. *Sci. Transl. Med.* **2017**, *9*, eaan4730. [CrossRef] [PubMed]
91. Liu, G.H.; Zhou, Z.G.; Chen, R.; Wang, M.J.; Zhou, B.; Li, Y.; Sun, X.F. Serum miR-21 and miR-92a as biomarkers in the diagnosis and prognosis of colorectal cancer. *Tumour. Biol.* **2013**, *34*, 2175–2181. [CrossRef]
92. Wang, J.; Li, X.; Lu, L.; He, L.; Hu, H.; Xu, Z. Circular RNA hsa_circ_0000567 can be used as a promising diagnostic biomarker for human colorectal cancer. *J. Clin. Lab. Anal.* **2018**, *32*, e22379. [CrossRef] [PubMed]
93. Mohammadi, D.; Zafari, Y.; Estaki, Z.; Mehrabi, M.; Moghbelinejad, S. Evaluation of plasma circ_0006282 as a novel diagnostic biomarker in colorectal cancer. *J. Clin. Lab. Anal.* **2022**, *36*, e24147. [CrossRef] [PubMed]
94. Li, J.; Song, Y.; Wang, J.; Huang, J. Plasma circular RNA panel acts as a novel diagnostic biomarker for colorectal cancer detection. *Am. J. Transl. Res.* **2020**, *12*, 7395–7403.
95. Iwasaki, H.; Shimura, T.; Kitagawa, M.; Yamada, T.; Nishigaki, R.; Fukusada, S.; Okuda, Y.; Katano, T.; Horike, S.I.; Kataoka, H. A Novel Urinary miRNA Biomarker for Early Detection of Colorectal Cancer. *Cancers* **2022**, *14*, 461. [CrossRef] [PubMed]
96. Ismail, H.T.H.; AbdelMageed, M.; Lindmark, G.; Hammarström, M.L.; Hammarström, S.; Sitohy, B. Prognostic Significance of GPR55 mRNA Expression in Colon Cancer. *Int. J. Mol. Sci.* **2022**, *23*, 4556. [CrossRef] [PubMed]
97. Kang, E.; Jung, S.C.; Nam, S.K.; Park, Y.; Seo, S.H.; Park, K.U.; Oh, H.K.; Kim, D.W.; Kang, S.B.; Lee, H.S. Tissue miR-200c-3p and circulating miR-1290 as potential prognostic biomarkers for colorectal cancer. *Sci. Rep.* **2022**, *12*, 2295. [CrossRef]
98. Liu, X.; Xu, X.; Pan, B.; He, B.; Chen, X.; Zeng, K.; Xu, M.; Pan, Y.; Sun, H.; Xu, T.; et al. Circulating miR-1290 and miR-320d as Novel Diagnostic Biomarkers of Human Colorectal Cancer. *J. Cancer* **2019**, *10*, 43–50. [CrossRef]
99. Zhang, H.; Zhu, M.; Shan, X.; Zhou, X.; Wang, T.; Zhang, J.; Tao, J.; Cheng, W.; Chen, G.; Li, J.; et al. A panel of seven-miRNA signature in plasma as potential biomarker for colorectal cancer diagnosis. *Gene* **2019**, *687*, 246–254. [CrossRef]
100. Gharib, E.; Nazemalhosseini-Mojarad, E.; Baghdar, K.; Nayeri, Z.; Sadeghi, H.; Rezasoltani, S.; Jamshidi-Fard, A.; Larki, P.; Sadeghi, A.; Hashemi, M.; et al. Identification of a stool long non-coding RNAs panel as a potential biomarker for early detection of colorectal cancer. *J. Clin. Lab. Anal.* **2021**, *35*, e23601. [CrossRef]
101. Ahmed, F.E.; Ahmed, N.C.; Gouda, M.M.; Vos, P.W.; Bonnerup, C. RT-qPCR for Fecal Mature MicroRNA Quantification and Validation. *Methods Mol. Biol.* **2018**, *1765*, 203–215. [CrossRef]
102. Ghazanfar, S.; Fatima, I.; Aslam, M.; Musharraf, S.G.; Sherman, N.E.; Moskaluk, C.; Fox, J.W.; Akhtar, M.W.; Sadaf, S. Identification of actin beta-like 2 (ACTBL2) as novel, upregulated protein in colorectal cancer. *J. Proteom.* **2017**, *152*, 33–40. [CrossRef] [PubMed]
103. Hao, J.J.; Zhi, X.; Wang, Y.; Zhang, Z.; Hao, Z.; Ye, R.; Tang, Z.; Qian, F.; Wang, Q.; Zhu, J. Comprehensive Proteomic Characterization of the Human Colorectal Carcinoma Reveals Signature Proteins and Perturbed Pathways. *Sci. Rep.* **2017**, *7*, 42436. [CrossRef]
104. Yamamoto, T.; Kudo, M.; Peng, W.X.; Takata, H.; Takakura, H.; Teduka, K.; Fujii, T.; Mitamura, K.; Taga, A.; Uchida, E.; et al. Identification of aldolase A as a potential diagnostic biomarker for colorectal cancer based on proteomic analysis using formalin-fixed paraffin-embedded tissue. *Tumour. Biol.* **2016**, *37*, 13595–13606. [CrossRef] [PubMed]
105. Torres, S.; Bartolomé, R.A.; Mendes, M.; Barderas, R.; Fernandez-Aceñero, M.J.; Peláez-García, A.; Peña, C.; Lopez-Lucendo, M.; Villar-Vázquez, R.; de Herreros, A.G.; et al. Proteome profiling of cancer-associated fibroblasts identifies novel proinflammatory signatures and prognostic markers for colorectal cancer. *Clin. Cancer Res.* **2013**, *19*, 6006–6019. [CrossRef]
106. Alnabulsi, A.; Murray, G.I. Proteomics for early detection of colorectal cancer: Recent updates. *Expert Rev. Proteom.* **2018**, *15*, 55–63. [CrossRef]
107. Ganepola, G.A.; Nizin, J.; Rutledge, J.R.; Chang, D.H. Use of blood-based biomarkers for early diagnosis and surveillance of colorectal cancer. *World J. Gastrointest. Oncol.* **2014**, *6*, 83–97. [CrossRef] [PubMed]
108. Ivancic, M.M.; Megna, B.W.; Sverchkov, Y.; Craven, M.; Reichelderfer, M.; Pickhardt, P.J.; Sussman, M.R.; Kennedy, G.D. Noninvasive Detection of Colorectal Carcinomas Using Serum Protein Biomarkers. *J. Surg. Res.* **2020**, *246*, 160–169. [CrossRef] [PubMed]

109. Bhardwaj, M.; Gies, A.; Weigl, K.; Tikk, K.; Benner, A.; Schrotz-King, P.; Borchers, C.H.; Brenner, H. Evaluation and Validation of Plasma Proteins Using Two Different Protein Detection Methods for Early Detection of Colorectal Cancer. *Cancers* **2019**, *11*, 1426. [CrossRef] [PubMed]
110. Peltier, J.; Roperch, J.P.; Audebert, S.; Borg, J.P.; Camoin, L. Quantitative proteomic analysis exploring progression of colorectal cancer: Modulation of the serpin family. *J. Proteom.* **2016**, *148*, 139–148. [CrossRef] [PubMed]
111. Zajkowska, M.; Dulewicz, M.; Kulczyńska-Przybik, A.; Safiejko, K.; Juchimiuk, M.; Konopko, M.; Kozłowski, L.; Mroczko, B. The Significance of Selected C-C Motif Chemokine Ligands in Colorectal Cancer Patients. *J. Clin. Med.* **2022**, *11*, 1794. [CrossRef] [PubMed]
112. Quesada-Calvo, F.; Massot, C.; Bertrand, V.; Longuespée, R.; Blétard, N.; Somja, J.; Mazzucchelli, G.; Smargiasso, N.; Baiwir, D.; De Pauw-Gillet, M.C.; et al. OLFM4, KNG1 and Sec24C identified by proteomics and immunohistochemistry as potential markers of early colorectal cancer stages. *Clin. Proteom.* **2017**, *14*, 9. [CrossRef]
113. Chantaraamporn, J.; Champattanachai, V.; Khongmanee, A.; Verathamjamras, C.; Prasongsook, N.; Mingkwan, K.; Luevisadpibul, V.; Chutipongtanate, S.; Svasti, J. Glycoproteomic Analysis Reveals Aberrant Expression of Complement C9 and Fibronectin in the Plasma of Patients with Colorectal Cancer. *Proteomes* **2020**, *8*, 26. [CrossRef]
114. Fan, N.J.; Chen, H.M.; Song, W.; Zhang, Z.Y.; Zhang, M.D.; Feng, L.Y.; Gao, C.F. Macrophage mannose receptor 1 and S100A9 were identified as serum diagnostic biomarkers for colorectal cancer through a label-free quantitative proteomic analysis. *Cancer Biomark* **2016**, *16*, 235–243. [CrossRef]
115. Łukaszewicz-Zając, M.; Pączek, S.; Mroczko, P.; Kulczyńska-Przybik, A. The Significance of CXCL1 and CXCL8 as Well as Their Specific Receptors in Colorectal Cancer. *Cancer Manag. Res.* **2020**, *12*, 8435–8443. [CrossRef] [PubMed]
116. Pączek, S.; Łukaszewicz-Zając, M.; Gryko, M.; Mroczko, P.; Kulczyńska-Przybik, A.; Mroczko, B. CXCL-8 in Preoperative Colorectal Cancer Patients: Significance for Diagnosis and Cancer Progression. *Int. J. Mol. Sci.* **2020**, *21*, 2040. [CrossRef] [PubMed]
117. Xu, L.; Peng, H.; Huang, X.X.; Xia, Y.B.; Hu, K.F.; Zhang, Z.M. Decreased expression of chromodomain helicase DNA-binding protein 9 is a novel independent prognostic biomarker for colorectal cancer. *Braz. J. Med. Biol. Res.* **2018**, *51*, e7588. [CrossRef]
118. Wang, D.; Zhou, Y.; Hua, L.; Li, J.; Zhu, N.; Liu, Y. CDK3, CDK5 and CDK8 Proteins as Prognostic and Potential Biomarkers in Colorectal Cancer Patients. *Int. J. Gen. Med.* **2022**, *15*, 2233–2245. [CrossRef] [PubMed]
119. Yu, J.; Zhai, X.; Li, X.; Zhong, C.; Guo, C.; Yang, F.; Yuan, Y.; Zheng, S. Identification of MST1 as a potential early detection biomarker for colorectal cancer through a proteomic approach. *Sci. Rep.* **2017**, *7*, 14265. [CrossRef]
120. Tu, C.; Mojica, W.; Straubinger, R.M.; Li, J.; Shen, S.; Qu, M.; Nie, L.; Roberts, R.; An, B.; Qu, J. Quantitative proteomic profiling of paired cancerous and normal colon epithelial cells isolated freshly from colorectal cancer patients. *Proteom. Clin. Appl.* **2017**, *11*, 1600155. [CrossRef]
121. Li, H.; Zhao, S.; Jing, Z.; Li, J.; Shuanying, Y.; Zhang, N. Combination of D-dimer and carcinoembryonic antigen levels as a predictive and prognostic biomarker in advanced colorectal cancer patients. *J. Cell. Biochem.* **2018**, *120*, 8086–8092. [CrossRef] [PubMed]
122. Hassan, N.A.; Idriss, N.K.; Gaber, N.; Ibrahim, A.; Tawfeek, M.A.; Mossad, E.; Mosa, A.A.; Ahmed, E.H.; Sayed, S.A.; Ahmed, H.A.; et al. Spastic Paraplegia 20 and Serine/Threonine Protein Kinase 31 Expression for the Detection of Colorectal Cancer. *Cell. Physiol. Biochem.* **2022**, *56*, 138–149. [CrossRef]
123. Watany, M.M.; Elmashad, N.M.; Badawi, R.; Hawash, N. Serum FBLN1 and STK31 as biomarkers of colorectal cancer and their ability to noninvasively differentiate colorectal cancer from benign polyps. *Clin. Chim. Acta* **2018**, *483*, 151–155. [CrossRef] [PubMed]
124. Liu, L.; Huang, Y.; Li, Y.; Wang, Q.; Hao, Y.; Liu, L.; Yao, X.; Yao, X.; Wei, Y.; Sun, X.; et al. FJX1 as a candidate diagnostic and prognostic serum biomarker for colorectal cancer. *Clin. Transl. Oncol.* **2022**, *24*, 1964–1974. [CrossRef] [PubMed]
125. Lu, C.; Liao, W.; Huang, Y.; Huang, Y.; Luo, Y. Increased expression of NOP14 is associated with improved prognosis due to immune regulation in colorectal cancer. *BMC Gastroenterol.* **2022**, *22*, 207. [CrossRef] [PubMed]
126. Zhang, H.P.; Wu, J.; Liu, Z.F.; Gao, J.W.; Li, S.Y. SPARCL1 Is a Novel Prognostic Biomarker and Correlates with Tumor Microenvironment in Colorectal Cancer. *Biomed Res. Int.* **2022**, *2022*, 1398268. [CrossRef] [PubMed]
127. Liu, Y.; Cheuk-Hay Lau, H.; Cheng, W.Y.; Yu, J. Gut microbiome in colorectal cancer: Clinical diagnosis and treatment. *Genom. Proteom. Bioinform.* **2022**, *in press*. [CrossRef] [PubMed]
128. Wong, S.H.; Yu, J. Gut microbiota in colorectal cancer: Mechanisms of action and clinical applications. *Nat. Rev. Gastroenterol. Hepatol.* **2019**, *16*, 690–704. [CrossRef]
129. Ternes, D.; Karta, J.; Tsenkova, M.; Wilmes, P.; Haan, S.; Letellier, E. Microbiome in Colorectal Cancer: How to Get from Meta-omics to Mechanism? *Trends Microbiol.* **2020**, *28*, 401–423. [CrossRef]
130. Inamura, K. Colorectal Cancers: An Update on Their Molecular Pathology. *Cancers* **2018**, *10*, 26. [CrossRef]
131. Gopalakrishnan, V.; Helmink, B.A.; Spencer, C.N.; Reuben, A.; Wargo, J.A. The Influence of the Gut Microbiome on Cancer, Immunity, and Cancer Immunotherapy. *Cancer Cell* **2018**, *33*, 570–580. [CrossRef]
132. Grenham, S.; Clarke, G.; Cryan, J.F.; Dinan, T.G. Brain-gut-microbe communication in health and disease. *Front. Physiol.* **2011**, *2*, 94. [CrossRef]
133. Holmes, E.; Li, J.V.; Marchesi, J.R.; Nicholson, J.K. Gut microbiota composition and activity in relation to host metabolic phenotype and disease risk. *Cell Metab.* **2012**, *16*, 559–564. [CrossRef]

134. Clos-Garcia, M.; Garcia, K.; Alonso, C.; Iruarrizaga-Lejarreta, M.; D'Amato, M.; Crespo, A.; Iglesias, A.; Cubiella, J.; Bujanda, L.; Falcón-Pérez, J.M. Integrative Analysis of Fecal Metagenomics and Metabolomics in Colorectal Cancer. *Cancers* **2020**, *12*, 1142. [CrossRef]
135. Zeller, G.; Tap, J.; Voigt, A.Y.; Sunagawa, S.; Kultima, J.R.; Costea, P.I.; Amiot, A.; Böhm, J.; Brunetti, F.; Habermann, N.; et al. Potential of fecal microbiota for early-stage detection of colorectal cancer. *Mol. Syst. Biol.* **2014**, *10*, 766. [CrossRef] [PubMed]
136. Zackular, J.P.; Rogers, M.A.; Ruffin, M.T.t.; Schloss, P.D. The human gut microbiome as a screening tool for colorectal cancer. *Cancer Prev. Res.* **2014**, *7*, 1112–1121. [CrossRef] [PubMed]
137. Yang, J.; Li, D.; Yang, Z.; Dai, W.; Feng, X.; Liu, Y.; Jiang, Y.; Li, P.; Li, Y.; Tang, B.; et al. Establishing high-accuracy biomarkers for colorectal cancer by comparing fecal microbiomes in patients with healthy families. *Gut Microbes* **2020**, *11*, 918–929. [CrossRef] [PubMed]
138. Liang, J.Q.; Li, T.; Nakatsu, G.; Chen, Y.X.; Yau, T.O.; Chu, E.; Wong, S.; Szeto, C.H.; Ng, S.C.; Chan, F.K.L.; et al. A novel faecal Lachnoclostridium marker for the non-invasive diagnosis of colorectal adenoma and cancer. *Gut* **2020**, *69*, 1248–1257. [CrossRef] [PubMed]
139. Guo, S.; Li, L.; Xu, B.; Li, M.; Zeng, Q.; Xiao, H.; Xue, Y.; Wu, Y.; Wang, Y.; Liu, W.; et al. A Simple and Novel Fecal Biomarker for Colorectal Cancer: Ratio of Fusobacterium Nucleatum to Probiotics Populations, Based on Their Antagonistic Effect. *Clin. Chem.* **2018**, *64*, 1327–1337. [CrossRef]
140. McQuade, J.L.; Daniel, C.R.; Helmink, B.A.; Wargo, J.A. Modulating the microbiome to improve therapeutic response in cancer. *Lancet Oncol.* **2019**, *20*, e77–e91. [CrossRef]
141. Poore, G.D.; Kopylova, E.; Zhu, Q.; Carpenter, C.; Fraraccio, S.; Wandro, S.; Kosciolek, T.; Janssen, S.; Metcalf, J.; Song, S.J.; et al. Microbiome analyses of blood and tissues suggest cancer diagnostic approach. *Nature* **2020**, *579*, 567–574. [CrossRef] [PubMed]
142. Li, J.; Zhang, A.H.; Wu, F.F.; Wang, X.J. Alterations in the Gut Microbiota and Their Metabolites in Colorectal Cancer: Recent Progress and Future Prospects. *Front. Oncol.* **2022**, *12*, 841552. [CrossRef]
143. Xiao, Q.; Lu, W.; Kong, X.; Shao, Y.W.; Hu, Y.; Wang, A.; Bao, H.; Cao, R.; Liu, K.; Wang, X.; et al. Alterations of circulating bacterial DNA in colorectal cancer and adenoma: A proof-of-concept study. *Cancer Lett.* **2021**, *499*, 201–208. [CrossRef] [PubMed]
144. Wang, X.Q.; Zhang, A.H.; Miao, J.H.; Sun, H.; Yan, G.L.; Wu, F.F.; Wang, X.J. Gut microbiota as important modulator of metabolism in health and disease. *RSC Adv.* **2018**, *8*, 42380–42389. [CrossRef]
145. Chen, F.; Dai, X.; Zhou, C.C.; Li, K.X.; Zhang, Y.J.; Lou, X.Y.; Zhu, Y.M.; Sun, Y.L.; Peng, B.X.; Cui, W. Integrated analysis of the faecal metagenome and serum metabolome reveals the role of gut microbiome-associated metabolites in the detection of colorectal cancer and adenoma. *Gut* **2022**, *71*, 1315–1325. [CrossRef] [PubMed]
146. Jian, H.; Qinglian, Z. Correlation between plasma trimethylamine oxide levels and colorectal neoplastic lesions. *J. Dig. Oncol.* **2020**, *12*, 275–278.
147. Haixia, Z. *Serum Folate and Fecal Short-Chain Fatty Acid Levels and Expression of Tissue Nuclear Stem Factor and Proliferating Cell Nuclear Antigen in Patients with Colorectal Cancer*; Third Military Medical University: Chongqing, China, 2011.
148. Peinan, L. *Study on the Correlation between Left and Right Colon Cancer and Bile Acid Metabolism*; Nanjing Medical University: Nanjing, China, 2018.
149. Yu, H. *The Role of Intestinal Flora in the Pathogenesis of Colorectal Cancer and Its Clinical Diagnostic Value Based on Multi-Omics*; Chinese People's Liberation Army Naval Medical University: Beijing, China, 2019.
150. Weir, T.L.; Manter, D.K.; Sheflin, A.M.; Barnett, B.A.; Heuberger, A.L.; Ryan, E.P. Stool microbiome and metabolome differences between colorectal cancer patients and healthy adults. *PLoS ONE* **2013**, *8*, e70803. [CrossRef] [PubMed]
151. Brown, D.G.; Rao, S.; Weir, T.L.; O'Malia, J.; Bazan, M.; Brown, R.J.; Ryan, E.P. Metabolomics and metabolic pathway networks from human colorectal cancers, adjacent mucosa, and stool. *Cancer Metab.* **2016**, *4*, 11. [CrossRef] [PubMed]
152. Lin, Y.; Ma, C.; Bezabeh, T.; Wang, Z.; Liang, J.; Huang, Y.; Zhao, J.; Liu, X.; Ye, W.; Tang, W.; et al. (1) H NMR-based metabolomics reveal overlapping discriminatory metabolites and metabolic pathway disturbances between colorectal tumor tissues and fecal samples. *Int. J. Cancer* **2019**, *145*, 1679–1689. [CrossRef] [PubMed]
153. Manna, S.K.; Tanaka, N.; Krausz, K.W.; Haznadar, M.; Xue, X.; Matsubara, T.; Bowman, E.D.; Fearon, E.R.; Harris, C.C.; Shah, Y.M.; et al. Biomarkers of coordinate metabolic reprogramming in colorectal tumors in mice and humans. *Gastroenterology* **2014**, *146*, 1313–1324. [CrossRef]
154. Yang, Y.; Misra, B.B.; Liang, L.; Bi, D.; Weng, W.; Wu, W.; Cai, S.; Qin, H.; Goel, A.; Li, X.; et al. Integrated microbiome and metabolome analysis reveals a novel interplay between commensal bacteria and metabolites in colorectal cancer. *Theranostics* **2019**, *9*, 4101–4114. [CrossRef]
155. Wong, S.H.; Kwong, T.N.Y.; Chow, T.C.; Luk, A.K.C.; Dai, R.Z.W.; Nakatsu, G.; Lam, T.Y.T.; Zhang, L.; Wu, J.C.Y.; Chan, F.K.L.; et al. Quantitation of faecal Fusobacterium improves faecal immunochemical test in detecting advanced colorectal neoplasia. *Gut* **2017**, *66*, 1441–1448. [CrossRef]
156. Shah, M.S.; DeSantis, T.Z.; Weinmaier, T.; McMurdie, P.J.; Cope, J.L.; Altrichter, A.; Yamal, J.M.; Hollister, E.B. Leveraging sequence-based faecal microbial community survey data to identify a composite biomarker for colorectal cancer. *Gut* **2018**, *67*, 882–891. [CrossRef] [PubMed]
157. Xu, J.; Yang, M.; Wang, D.; Zhang, S.; Yan, S.; Zhu, Y.; Chen, W. Alteration of the abundance of Parvimonas micra in the gut along the adenoma-carcinoma sequence. *Oncol. Lett.* **2020**, *20*, 106. [CrossRef]

158. Coker, O.O.; Liu, C.; Wu, W.K.K.; Wong, S.H.; Jia, W.; Sung, J.J.Y.; Yu, J. Altered gut metabolites and microbiota interactions are implicated in colorectal carcinogenesis and can be non-invasive diagnostic biomarkers. *Microbiome* **2022**, *10*, 35. [CrossRef]
159. Wu, Y.; Jiao, N.; Zhu, R.; Zhang, Y.; Wu, D.; Wang, A.J.; Fang, S.; Tao, L.; Li, Y.; Cheng, S.; et al. Identification of microbial markers across populations in early detection of colorectal cancer. *Nat. Commun.* **2021**, *12*, 3063. [CrossRef] [PubMed]
160. Eklöf, V.; Löfgren-Burström, A.; Zingmark, C.; Edin, S.; Larsson, P.; Karling, P.; Alexeyev, O.; Rutegård, J.; Wikberg, M.L.; Palmqvist, R. Cancer-associated fecal microbial markers in colorectal cancer detection. *Int. J. Cancer* **2017**, *141*, 2528–2536. [CrossRef] [PubMed]
161. Song, E.M.; Byeon, J.S.; Lee, S.M.; Yoo, H.J.; Kim, S.J.; Lee, S.H.; Chang, K.; Hwang, S.W.; Yang, D.H.; Jeong, J.Y. Fecal Fatty Acid Profiling as a Potential New Screening Biomarker in Patients with Colorectal Cancer. *Dig. Dis. Sci.* **2018**, *63*, 1229–1236. [CrossRef]
162. Nishiumi, S.; Kobayashi, T.; Kawana, S.; Unno, Y.; Sakai, T.; Okamoto, K.; Yamada, Y.; Sudo, K.; Yamaji, T.; Saito, Y.; et al. Investigations in the possibility of early detection of colorectal cancer by gas chromatography/triple-quadrupole mass spectrometry. *Oncotarget* **2017**, *8*, 17115–17126. [CrossRef]
163. Ning, W.; Qiao, N.; Zhang, X.; Pei, D.; Wang, W. Metabolic profiling analysis for clinical urine of colorectal cancer. *Asia Pac. J. Clin. Oncol.* **2021**, *17*, 403–413. [CrossRef] [PubMed]
164. Troisi, J.; Cavallo, P.; Colucci, A.; Pierri, L.; Scala, G.; Symes, S.; Jones, C.; Richards, S. Metabolomics in genetic testing. *Adv. Clin. Chem.* **2020**, *94*, 85–153. [CrossRef] [PubMed]
165. Zhang, A.; Sun, H.; Yan, G.; Wang, P.; Wang, X. Metabolomics for Biomarker Discovery: Moving to the Clinic. *Biomed Res. Int.* **2015**, *2015*, 354671. [CrossRef] [PubMed]
166. Dunn, W.B.; Broadhurst, D.I.; Atherton, H.J.; Goodacre, R.; Griffin, J.L. Systems level studies of mammalian metabolomes: The roles of mass spectrometry and nuclear magnetic resonance spectroscopy. *Chem. Soc. Rev.* **2011**, *40*, 387–426. [CrossRef]
167. Ivanisevic, J.; Want, E.J. From Samples to Insights into Metabolism: Uncovering Biologically Relevant Information in LC-HRMS Metabolomics Data. *Metabolites* **2019**, *9*, 308. [CrossRef]
168. Vignoli, A.; Ghini, V.; Meoni, G.; Licari, C.; Takis, P.G.; Tenori, L.; Turano, P.; Luchinat, C. High-Throughput Metabolomics by 1D NMR. *Angew Chem. Int. Ed. Engl.* **2019**, *58*, 968–994. [CrossRef] [PubMed]
169. Commisso, M.; Strazzer, P.; Toffali, K.; Stocchero, M.; Guzzo, F. Untargeted metabolomics: An emerging approach to determine the composition of herbal products. *Comput. Struct. Biotechnol. J.* **2013**, *4*, e201301007. [CrossRef]
170. Johnson, C.H.; Ivanisevic, J.; Siuzdak, G. Metabolomics: Beyond biomarkers and towards mechanisms. *Nat. Rev. Mol. Cell Biol.* **2016**, *17*, 451–459. [CrossRef]
171. Kim, E.R.; Kwon, H.N.; Nam, H.; Kim, J.J.; Park, S.; Kim, Y.H. Urine-NMR metabolomics for screening of advanced colorectal adenoma and early stage colorectal cancer. *Sci. Rep.* **2019**, *9*, 4786. [CrossRef] [PubMed]
172. Daniluk, U.; Daniluk, J.; Kucharski, R.; Kowalczyk, T.; Pietrowska, K.; Samczuk, P.; Filimoniuk, A.; Kretowski, A.; Lebensztejn, D.; Ciborowski, M. Untargeted Metabolomics and Inflammatory Markers Profiling in Children With Crohn's Disease and Ulcerative Colitis-A Preliminary Study. *Inflamm. Bowel Dis.* **2019**, *25*, 1120–1128. [CrossRef]
173. Wang, H.; Tso, V.K.; Slupsky, C.M.; Fedorak, R.N. Metabolomics and detection of colorectal cancer in humans: A systematic review. *Future Oncol.* **2010**, *6*, 1395–1406. [CrossRef]
174. Nannini, G.; Meoni, G.; Amedei, A.; Tenori, L. Metabolomics profile in gastrointestinal cancers: Update and future perspectives. *World J. Gastroenterol.* **2020**, *26*, 2514–2532. [CrossRef] [PubMed]
175. Gu, J.; Xiao, Y.; Shu, D.; Liang, X.; Hu, X.; Xie, Y.; Lin, D.; Li, H. Metabolomics Analysis in Serum from Patients with Colorectal Polyp and Colorectal Cancer by (1)H-NMR Spectrometry. *Dis. Markers* **2019**, *2019*, 3491852. [CrossRef]
176. Erben, V.; Poschet, G.; Schrotz-King, P.; Brenner, H. Comparing Metabolomics Profiles in Various Types of Liquid Biopsies among Screening Participants with and without Advanced Colorectal Neoplasms. *Diagnostics* **2021**, *11*, 561. [CrossRef]
177. Udo, R.; Katsumata, K.; Kuwabara, H.; Enomoto, M.; Ishizaki, T.; Sunamura, M.; Nagakawa, Y.; Soya, R.; Sugimoto, M.; Tsuchida, A. Urinary charged metabolite profiling of colorectal cancer using capillary electrophoresis-mass spectrometry. *Sci. Rep.* **2020**, *10*, 21057. [CrossRef] [PubMed]
178. Zhang, A.; Sun, H.; Wang, X. Mass spectrometry-driven drug discovery for development of herbal medicine. *Mass Spectrom. Rev.* **2018**, *37*, 307–320. [CrossRef]
179. Yin, F.T.; Zhou, X.H.; Kang, S.Y.; Li, X.H.; Li, J.; Ullah, I.; Zhang, A.H.; Sun, H.; Wang, X.J. Prediction of the mechanism of Dachengqi Decoction treating colorectal cancer based on the analysis method of " into serum components -action target-key pathway". *J. Ethnopharmacol.* **2022**, *293*, 115286. [CrossRef] [PubMed]
180. Wang, R.; Li, B.; Lam, S.M.; Shui, G. Integration of lipidomics and metabolomics for in-depth understanding of cellular mechanism and disease progression. *J. Genet. Genom.* **2020**, *47*, 69–83. [CrossRef] [PubMed]
181. Kyle, J.E. Extracting Biological Insight from Untargeted Lipidomics Data. *Methods Mol. Biol.* **2020**, *2104*, 121–137. [CrossRef] [PubMed]
182. Zaytseva, Y. Lipid Metabolism as a Targetable Metabolic Vulnerability in Colorectal Cancer. *Cancers* **2021**, *13*, 301. [CrossRef] [PubMed]
183. Pakiet, A.; Kobiela, J.; Stepnowski, P.; Sledzinski, T.; Mika, A. Changes in lipids composition and metabolism in colorectal cancer: A review. *Lipids Health Dis.* **2019**, *18*, 29. [CrossRef]

184. Kondo, Y.; Nishiumi, S.; Shinohara, M.; Hatano, N.; Ikeda, A.; Yoshie, T.; Kobayashi, T.; Shiomi, Y.; Irino, Y.; Takenawa, T.; et al. Serum fatty acid profiling of colorectal cancer by gas chromatography/mass spectrometry. *Biomark Med.* **2011**, *5*, 451–460. [CrossRef]
185. Liu, T.; Peng, F.; Yu, J.; Tan, Z.; Rao, T.; Chen, Y.; Wang, Y.; Liu, Z.; Zhou, H.; Peng, J. LC-MS-based lipid profile in colorectal cancer patients: TAGs are the main disturbed lipid markers of colorectal cancer progression. *Anal. Bioanal. Chem.* **2019**, *411*, 5079–5088. [CrossRef]
186. Tevini, J.; Eder, S.K.; Huber-Schönauer, U.; Niederseer, D.; Strebinger, G.; Gostner, J.M.; Aigner, E.; Datz, C.; Felder, T.K. Changing Metabolic Patterns along the Colorectal Adenoma-Carcinoma Sequence. *J. Clin. Med.* **2022**, *11*, 721. [CrossRef] [PubMed]
187. Ecker, J.; Benedetti, E.; Kindt, A.S.D.; Höring, M.; Perl, M.; Machmüller, A.C.; Sichler, A.; Plagge, J.; Wang, Y.; Zeissig, S.; et al. The Colorectal Cancer Lipidome: Identification of a Robust Tumor-Specific Lipid Species Signature. *Gastroenterology* **2021**, *161*, 910–923.e919. [CrossRef]
188. Dehairs, J.; Derua, R.; Rueda-Rincon, N.; Swinnen, J.V. Lipidomics in drug development. *Drug Discov. Today Technol.* **2015**, *13*, 33–38. [CrossRef]
189. Moreno, L.O.; Sánchez, P.N.; Abalo, R. Lipidomics as Tools for Finding Biomarkers of Intestinal Pathology: From Irritable Bowel Syndrome to Colorectal Cancer. *Curr. Drug Targets* **2022**, *23*, 636–655. [CrossRef] [PubMed]
190. Lv, J.; Zhang, L.; Yan, F.; Wang, X. Clinical lipidomics: A new way to diagnose human diseases. *Clin. Transl. Med.* **2018**, *7*, 12. [CrossRef] [PubMed]
191. Subramanian, I.; Verma, S.; Kumar, S.; Jere, A.; Anamika, K. Multi-omics Data Integration, Interpretation, and Its Application. *Bioinform. Biol. Insights* **2020**, *14*, 1177932219899051. [CrossRef] [PubMed]
192. Yan, J.; Risacher, S.L.; Shen, L.; Saykin, A.J. Network approaches to systems biology analysis of complex disease: Integrative methods for multi-omics data. *Brief Bioinform.* **2018**, *19*, 1370–1381. [CrossRef]
193. O'Connor, J.P.; Aboagye, E.O.; Adams, J.E.; Aerts, H.J.; Barrington, S.F.; Beer, A.J.; Boellaard, R.; Bohndiek, S.E.; Brady, M.; Brown, G.; et al. Imaging biomarker roadmap for cancer studies. *Nat. Rev. Clin. Oncol.* **2017**, *14*, 169–186. [CrossRef]
194. Koyande, N.; Gangopadhyay, M.; Thatikonda, S.; Rengan, A.K. The role of gut microbiota in the development of colorectal cancer: A review. *Int. J. Colorectal. Dis.* **2022**, *37*, 1509–1523. [CrossRef]
195. Rebersek, M. Gut microbiome and its role in colorectal cancer. *BMC Cancer* **2021**, *21*, 1325. [CrossRef]
196. Pączek, S.; Łukaszewicz-Zając, M.; Mroczko, B. Granzymes-Their Role in Colorectal Cancer. *Int. J. Mol. Sci.* **2022**, *23*, 5277. [CrossRef] [PubMed]
197. Eylem, C.C.; Yilmaz, M.; Derkus, B.; Nemutlu, E.; Camci, C.B.; Yilmaz, E.; Turkoglu, M.A.; Aytac, B.; Ozyurt, N.; Emregul, E. Untargeted multi-omic analysis of colorectal cancer-specific exosomes reveals joint pathways of colorectal cancer in both clinical samples and cell culture. *Cancer Lett.* **2020**, *469*, 186–194. [CrossRef]
198. Nam, A.S.; Chaligne, R.; Landau, D.A. Integrating genetic and non-genetic determinants of cancer evolution by single-cell multi-omics. *Nat. Rev. Genet.* **2021**, *22*, 3–18. [CrossRef] [PubMed]
199. Vasaikar, S.V.; Straub, P.; Wang, J.; Zhang, B. LinkedOmics: Analyzing multi-omics data within and across 32 cancer types. *Nucleic Acids Res.* **2018**, *46*, D956–D963. [CrossRef] [PubMed]
200. Correa-Aguila, R.; Alonso-Pupo, N.; Hernández-Rodríguez, E.W. Multi-omics data integration approaches for precision oncology. *Mol. Omics* **2022**, *18*, 469–479. [CrossRef] [PubMed]

cancers

MDPI

Article

Association of Multi-Phasic MR-Based Radiomic and Dosimetric Features with Treatment Response in Unresectable Hepatocellular Carcinoma Patients following Novel Sequential TACE-SBRT-Immunotherapy

Lok-Man Ho [1,2], Sai-Kit Lam [3,4], Jiang Zhang [5], Chi-Leung Chiang [6], Albert Chi-Yan Chan [7] and Jing Cai [3,5,*]

1 Faculty of Health and Social Sciences, The Hong Kong Polytechnic University, Hong Kong
2 Radiotherapy and Oncology Centre, Gleneagles Hospital Hong Kong, Hong Kong
3 Research Institute for Smart Ageing, The Hong Kong Polytechnic University, Hong Kong
4 Department of Biomedical Engineering, Faculty of Engineering, The Hong Kong Polytechnic University, Hong Kong
5 Department of Health Technology and Informatics, The Hong Kong Polytechnic University, Hong Kong
6 Department of Clinical Oncology, School of Clinical Medicine, The University of Hong Kong, Hong Kong
7 Department of Surgery, School of Clinical Medicine, The University of Hong Kong, Hong Kong
* Correspondence: jing.cai@polyu.edu.hk

Simple Summary: Hepatocellular carcinoma (HCC) is one of the most prevalent and devastating malignancies worldwide. An ongoing phase-II clinical trial assesses the efficacy of a novel sequential trans-arterial chemoembolization (TACE) plus stereotactic body radiotherapy (SBRT) plus immunotherapy strategy as an induction therapy for unresectable HCC patients. This study aims to investigate the potential association between radiomic features extracted from pre-treatment multi-phasic MR images and treatment response following the novel intervention strategy. In this study, Four DeltaP-derived radiomics that characterize the temporal change in intratumoral randomness and uniformity were identified as the contributors to the treatment response for a 3-month timepoint. Additional arterial phase (AP)-derived radiomic features and tumor morphology were also shown to have strong associations with treatment response for a 6-month timepoint. The success of this study would demonstrate the feasibility of pre-treatment identification of responsive HCC patients, paving the way toward effective and personalized oncology for HCC management.

Abstract: This study aims to investigate the association of pre-treatment multi-phasic MR-based radiomics and dosimetric features with treatment response to a novel sequential trans-arterial chemoembolization (TACE) plus stereotactic body radiotherapy (SBRT) plus immunotherapy regimen in unresectable Hepatocellular Carcinoma (HCC) sub-population. Twenty-six patients with unresectable HCC were retrospectively analyzed. Radiomic features were extracted from 42 lesions on arterial phase (AP) and portal-venous phase (PVP) MR images. Delta-phase (DeltaP) radiomic features were calculated as AP-to-PVP ratio. Dosimetric data of the tumor was extracted from dose-volume-histograms. A two-sided independent Mann–Whitney U test was used to assess the clinical association of each feature, and the classification performance of each significant independent feature was assessed using logistic regression. For the 3-month timepoint, four DeltaP-derived radiomics that characterize the temporal change in intratumoral randomness and uniformity were the only contributors to the treatment response association (p-value = 0.038–0.063, AUC = 0.690–0.766). For the 6-month timepoint, DeltaP-derived radiomic features (n = 4) maintained strong clinical associations with the treatment response (p-value = 0.047–0.070, AUC = 0.699–0.788), additional AP-derived radiomic features (n = 4) that reflect baseline tumoral arterial-enhanced signal pattern and tumor morphology (n = 1) that denotes initial tumor burden were shown to have strong associations with treatment response (p-value = 0.028–0.074, AUC = 0.719–0.773). This pilot study successfully demonstrated associations of pre-treatment multi-phasic MR-based radiomics with tumor response to the novel treatment regimen.

Citation: Ho, L.-M.; Lam, S.-K.; Zhang, J.; Chiang, C.-L.; Chan, A.C.-Y.; Cai, J. Association of Multi-Phasic MR-Based Radiomic and Dosimetric Features with Treatment Response in Unresectable Hepatocellular Carcinoma Patients following Novel Sequential TACE-SBRT-Immunotherapy. *Cancers* **2023**, *15*, 1105. https://doi.org/10.3390/cancers15041105

Academic Editors: Trever G. Bivona and Wei Wu

Received: 10 December 2022
Revised: 27 January 2023
Accepted: 7 February 2023
Published: 9 February 2023

Keywords: magnetic resonance imaging; radiomics; trans-arterial chemoembolization; stereotactic body radiotherapy; immunotherapy; hepatocellular carcinoma

1. Introduction

Hepatocellular carcinoma (HCC) is one of the most prevalent and devastating malignancies worldwide, ranking as the 4th leading cause of cancer-related deaths. It accounts for 80–90% of the sufferers of primary liver cancer, and its highest incidences occur in eastern and southeastern Asia and northern Africa [1,2]. Surgical resection and liver transplantation have been the gold standard curative therapies. Unfortunately, most HCC patients present intermediate to advanced disease at diagnosis [3]. As such, more than 70% of liver cancer patients are considered ineligible for such curative interventions [4], partly due to the presentation of large-sized tumors, poor liver function, or organ shortage. The median survival remains at approximately 16 months for intermediate-stage HCC patients, and half a year for advanced-stage HCC patients, respectively [5], reflecting grievous survivorship in this vulnerable HCC sub-population.

Over the past decades, three key additional regimens have been developed in the hope of serving either as a bridging therapy before liver transplantation or as a curative alternative for unresectable HCC patients; they are Trans-Arterial Chemoembolization (TACE), Stereotactic Body Radiotherapy (SBRT) and Immune Checkpoint Blockade (ICB).

TACE has been widely adopted as a first-line treatment for intermediate-stage HCC [6]. It works by interrupting the major source of oxygen and nutrition supply to the cancer cells from the hepatic arteries, meanwhile selectively delivering cytotoxic chemotherapeutic agents for cancer eradication [7]. However, its efficacy is limited in patients with poor baseline liver function and larger tumor burden, hence TACE alone is often not sufficient for thorough cancer cell elimination in advanced-stage HCC patients [8]. On the other hand, SBRT kills cancer cells non-invasively by delivering an ultra-high radiation dose in a few fractions (usually ≤ 5) to the tumor in a highly precise and conformal manner, under real-time liver and tumor motion monitoring [9]. The survival benefits of SBRT in HCC have been well-documented in the literature for early stage tumors [10,11]. Recently, efforts have been made to investigate the efficacy of sequential TACE-SBRT in intermediate and advanced-stage HCC on the grounds of the reported potential synergism between TACE and radiotherapy [12–18]. For instance, Chiang et al. reported a promising efficacy of combined TACE-SBRT treatments in Barcelona Clinic Liver Cancer (BCLC) system stage B-C HCC patients, yielding an objective response rate of 68% and a 1-year local control rate of 93.6% [19]. However, intra-hepatic and distant dissemination remains a key challenge for managing this subgroup of unresectable HCC patients. Sequential SBRT-immunotherapy has demonstrated improved local tumor control and distant abscopal effect [20], potentially compensating for the deficiency of the sequential TACE-SBRT regimen. Our pilot studies have shown the satisfactory efficacy and safety of combined SBRT and immunotherapy for HCC patients [21,22].

Notably, for the first time in history, there is an ongoing phase-II clinical trial conducted by our group that aims to assess the efficacy of a novel sequential TACE-SBRT-Immunotherapy strategy as an induction therapy for unresectable HCC patients [23], the results are greatly anticipated. While exciting, the potential toxicities associated with this novel aggressive treatment are yet to be reported. In the era of personalized medicine, there is a pressing demand to discriminate between responders and non-responders prior to treatment commencement for the sake of avoiding ineffective and toxic therapies in non-responders and enhancing individualized oncologic care delivery.

The field of radiomics has been caught in the spotlight of attention within the medical community [24–31]. It involves high-throughput extraction of quantitative features from medical images for divulging intrinsic biological and genetic characteristics [32]. The capability of radiomics has been extensively reported in various cancer prediction tasks, including cancer prognosis [33], disease differentiation [34], and treatment response [35,36], highlighting the high potential of radiomics in informing decision-making in a wide spectrum of oncologic care. Apart from this, several research groups have reported improved predictive power when combining both radiomics and radiation dosimetric parameters [37,38]. Particularly, for HCC management, multi-phasic contrast-enhanced magnetic resonance (MR) images are routinely used for obtaining dynamic information on disease pathology and physiology. The role of multi-phasic MR-based radiomics in HCC has been widely studied for predicting micro-vascular invasion [39,40], cancer recurrence [41–43], disease diagnosis [44,45], and treatment response [46–49]. Nevertheless, radiomics studies on treatment response prediction in unresectable HCC patients are scarce [49]. Further, there is no study assessing the association between radiomic features extracted from pre-treatment multi-phasic MR images and treatment response following the novel sequential TACE-SBRT-Immunotherapy regimen.

In this pilot study, we aimed to investigate the association of pre-treatment multi-phasic MR-based radiomics and dosimetric features with treatment response to the sequential TACE-SBRT-Immunotherapy regimen in unresectable HCC sub-population, who were prospectively enrolled in the first-of-its-kind phase-II clinical trial [23]. The success of this study would demonstrate the feasibility of pre-treatment identification of responsive HCC patients for this novel regimen, paving the way towards effective and personalized oncology for HCC management worldwide in the long run.

2. Materials and Methods

2.1. Patient Data

The present study was approved by the Human Subjects Ethics Subcommittee of The Hong Kong Polytechnic University and Institutional Review Board of the University of Hong Kong/Hospital Authority Hong Kong West Cluster. Apart from this, patient data that were analyzed in this study were prospectively enrolled in an ongoing phase-II clinical trial conducted by The University of Hong Kong, entitled "Sequential TransArterial hemoembolization and stereotactic Radiotherapy Followed by ImmunoTherapy for downstaging hepatocellular carcinoma for hepatectomy (START-FIT)" [23]. A total number of 26 newly diagnosed HCC patients, who were treated with sequential TACE-SBRT-Immunotherapy at the Department of Clinical Oncology of Queen Mary Hospital (QMH) between May 2019 to October 2021, were retrospectively analyzed. The inclusion criteria included: (1) diagnosis of unresectable HCC confirmed pathologically according to the American Association for the Study of Liver Diseases (AASLD) practice guideline 2010; (2) male or female between 18 and 80 years old; (3) tumor size between 5 and 15 cm, and the number of lesions less than 3; (4) portal vein involvement; (5) Child–Pugh liver function class A-B7; (6) liver volume minus intrahepatic gross-tumor-volume (GTV) > 700 cc; (7) no prior TACE; and (8) no prior systemic therapy nor immunotherapy, TACE or RT. The specific contraindications of SBRT were: any HCC tumor >15 cm; total maximum sum of HCC diameter >20 cm; more than 3 discrete hepatic nodules; direct tumor extension into the stomach, duodenum, small bowel, large bowel, and main branch of biliary tree.

2.2. Treatment Details

SBRT was performed by using 6 MV or 10 MV photon beams delivered from a linear accelerator within 21–35 days after TACE. The prescribed dose ranged from 27.5 Gy to 50 Gy in 5 fractions, depending on normal tissue constraints. The time interval between fractions was limited to 24 to 72 h, with radiation delivered to all targets within 5 to 15 days. Varian External Beam Planning Software (Varian Medical Systems, Palo Alto, CA, USA) was used for treatment planning. The dosing scheme aimed at using the highest allowable prescription dose for the tumor target, while fulfilling the constraints of surrounding organs-at-risk (OARs). TACE and immunotherapy procedures were performed according to the routine treatment protocol in QMH. For TACE, an emulsion of a mixture of cisplatin with lipiodol in a volume ratio of 1 to 1 was prepared and injected into the tumor by femoral artery puncture. The amount of TACE used was based on tumor size, number, and arterial blood flow. For immunotherapy, Avelumab was administered via IV injection two weeks after SBRT; the amount of dose required was based on the patient's body weight and toxicity.

2.3. Clinical Endpoint

The clinical endpoint of this study was defined as the response rate at 3 and 6 months after SBRT, according to the modified Response Evaluation Criteria in Solid Tumors ((mRECIST) version 1.1) criteria. The response rates were categorized into: (1) Complete response (CR) that represents the disappearance of any intratumoral arterial enhancement in all target lesions; (2) partial response (PR) that represents at least a 30% decrease in the sum of diameters of viable target lesions; (3) stable disease (SD) that represents any cases that do not qualify for either partial response or progressive disease; and (4) progressive disease (PD) that represents an increase of at least 20% in the sum of the diameters of viable (enhancing) target lesions. In this study, the treatment response of each lesion was assessed by a radiologist with 15 years of experience. Prior to subsequent analysis, patients with (1) CR and (2) PR were grouped into a respondent group, while those with (3) SD and (4) PD were grouped into a non-respondent group.

2.4. MRI Acquisition and Segmentation

The pre-intervention gadoxetic acid-enhanced MRI was obtained by using either 1.5T GE Signa system (version: HD16. GE Healthcare, Milwaukee, WI) or Philips 3T MRI Achieva scanner (Philips Healthcare, Best, The Netherlands) with a 12 or 16 channel, phased-array body coil. The image sets were acquired according to the START-FIT and LI-RADS ver. 2017 protocol, including axial arterial phase (AP) and portal venous phase (PVP) T1W image sets. A demonstrative example of AP and PVP MRI is shown in Figure 1a,b.

The segmentations (including gross tumor targets and OARs) were manually delineated on the axial planning CT slice-by-slice by an experienced clinical oncologist (with >15 years of experience). The contours of the lesions were subsequently transformed into other image sets by rigid registration for further processing. The transferred tumor lesions on the pre-intervention MR image sets were defined as the volumes of interest (VOIs).

2.5. Dosimetric Features

Dosimetric features of each lesion were obtained from the dose-volume histograms (DVHs) using the treatment planning system, including volume of GTV and planning-target-volume (PTV), prescription dose, minimum and maximum dose, mean dose, relative GTV volumes (in percentage) receiving specific doses (V5 to V50 in 5 Gy increments), minimum doses to relative liver volumes (D10% to D90% in 10% increments). In total, 33 dosimetric features were calculated for each lesion. A demonstrative example of DVH and dose distribution is shown in Figure 1c,d.

2.6. Image Preprocessing and Radiomic Features Extraction

The MR images with VOIs were imported into a python-based pipeline developed by The Hong Kong Polytechnic University, which was employed previously by other studies [50–53]. Before extracting the radiomics features, multiple pre-processing steps were performed. In order to tackle the parameter variations between image series, isotropic resampling was performed by linear interpolation to obtain a $1 \times 1 \times 1$ mm^3 voxel size. Inhomogeneity correction was performed by the N4 bias field correction algorithm to correct the locally varying intensity. Image intensities were then normalized by shifting and rescaling each image into a mean of zero and a standard deviation of 100 to maintain consistent voxel values across patients. They were further discretized by a fixed bin count of 100 to reduce the noise of image textures.

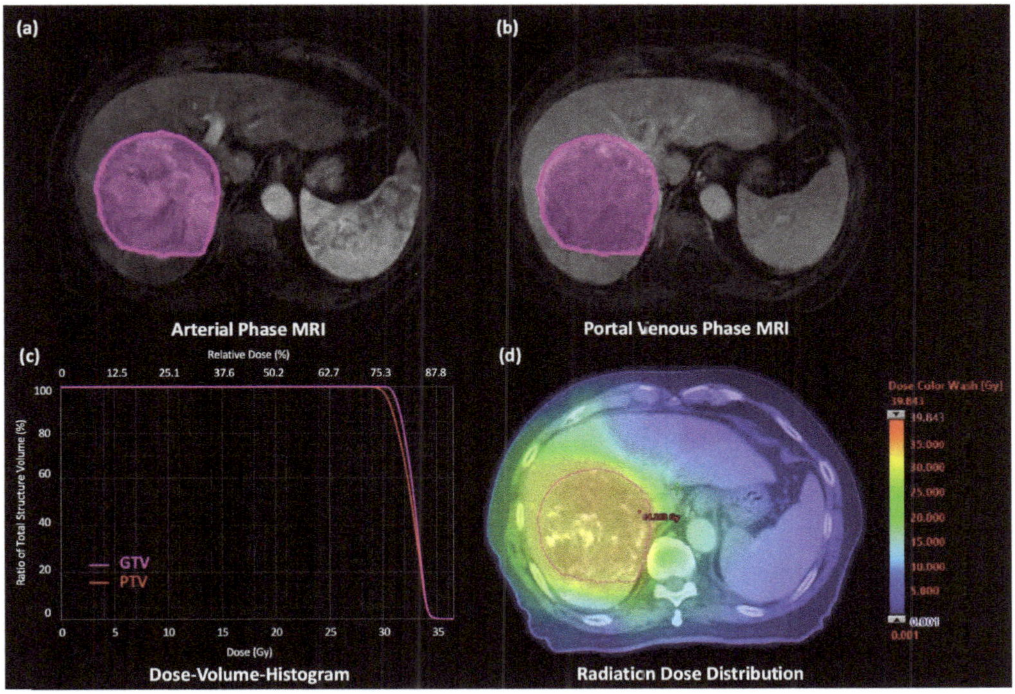

Figure 1. A 76-year-old male patient was diagnosed with advanced-staged HCC. (**a**) Axial AP T1W MR image with the VOI, (**b**) Axial PVP T1W MR image with the VOI. SBRT of 30 Gy was prescribed in 5 fractions to the tumor. (**c**) DVH of gross tumor volume (GTV) and planning target volume (PTV) generated from the treatment planning system and (**d**) Dose distribution of the SBRT treatment plan.

In this study, radiomics features were extracted from each lesion on both arterial phase (AP) T1W and portal venous phase (PVP) T1W MR images using the Pyradiomics (version 2.2.0) package. Delta phase (DeltaP) radiomics features were calculated by dividing feature values extracted from AP images by those extracted from PVP images. Multiple types of Radiomics features were extracted, including shape and size features (n = 14), first-order features (n = 18), and second-order texture features (n = 73). The texture features were calculated from gray-level co-occurrence matrices ((GLCMs): n = 22), gray-level run-length matrices ((GLRLMs): n = 16), gray-level size zone matrix (GLSZM: n = 16), gray-level dependence matrices ((GLDMs) n = 14), and neighboring gray tone difference matrices ((NGTDMs): n = 5). In total, 287 radiomic features were calculated for each lesion. The meaning of each radiomic feature parameter for Pyradiomics can be found in the link: https://pyradiomics.readthedocs.io/en/latest/features.html.

2.7. Statistical Analysis

For each of the studied endpoints, a two-sided independent Mann–Whitney U test was employed to assess the clinical association between treatment response (i.e., respondent or non-respondent) and features in each of the studied feature categories (i.e., AP, PVP, DeltaP radiomic features, and dosimetric features). Features with p-values of <0.05 were considered statistically significant. Moreover, independent endpoint-associated features were identified by using multiple feature selection procedures for each feature category. Ten independent features were first selected by using K-Means clustering from the original feature sets with a cluster number of ten. Cluster centroids were randomly initialized with 100 iterations to reduce potential bias. Independent features were then identified by the most significant features from the Mann–Whitney U test within each cluster. p-values were adjusted with false discovery rate (FDR) by using Benjamini–Hochberg (BH) multiple test correction among the identified ten independent features. Features with an FDR-adjusted p-value of <0.1 were reported. To further assess the predictability of each significant independent feature, logistic regression analysis was applied; the area under the receiver operator characteristic (ROC) curve (AUC), sensitivity, and specificity were then reported. Prior to regression analysis, all features were re-scaled to a mean of 0 and a standard deviation of 1.

Apart from this, a two-sided independent student t-test was used to determine whether there existed a statistically significant difference in patient demographic variables between the respondent and non-respondent groups. All statistical analyses were implemented using R software (version 4.2.1. The R Foundation, Vienna, Austria) and SPSS 26.0 (IBM, Chicago, IL, USA).

3. Results

3.1. Patient Characteristics

Table 1 summarizes patient characteristics. In total, 26 patients (male/female: 25/1; mean age: 67 ± 7.6 years) were included in the present study. Among the 26 patients, 14 had a single lesion, 8 had two lesions, and 4 had three lesions, resulting in a total of 42 lesions. The average diameter of the largest tumor nodule was 9.4 ± 3.7 cm.

Table 1. Patient characteristics.

Total Number of Patients	**26**
Gender	
• Male	25
• Female	1
Age (y), mean ± SD	67.0 ± 7.6
• 50–59	5
• 60–69	10
• 70–79	10
• >79	1
Diameters of largest tumor nodule (cm), mean ± SD	9.4 ± 3.7
Sum of diameter of tumor nodule (cm), mean ± SD	12.6 ± 5.6
Medical History	
• Hepatitis B	17
• Hepatitis C	4
• Alcoholics	2
Vascular invasion	
• Portal vein involvement	6
• Hepatic vein involvement	14
CP Score	
• A5	19
• A6	6
• B7	1
BCLC Stage	
• Stage A	3
• Stage B	7
• Stage C	16
Lesion numbers	
• 1 lesion	14
• 2 lesions	8
• 3 lesions	4

3.2. Lesion Characteristics and Response Rate

Table 2 displays the overall characteristics and response rates of the lesions in respondent and non-respondent groups in both studied timepoints. For the 3-month assessment, respondents were identified in 18 lesions (42.9%, CR: n = 3; PR: n = 15), while non-respondents were identified in 24 lesions (57.1%, SD: n = 16; PD: n = 8). No statistically significant difference in GTV and PTV volumes was observed between groups. For the 6-month assessment, respondents and non-respondents were identified in 28 lesions (66.7%, CR: n = 13; PR: n = 15) and 14 lesions (33.3%, SD: n = 6; PD: n = 8), respectively. There were significant differences between respondent and non-respondent groups in average GTV volumes (295.4 ± 376.2 cm^3 vs. 732.2 ± 728.2 cm^3, $p = 0.014$) and average PTV volumes (418.4 ± 479.6 cm^3 vs. 940.4 ± 857.0 cm^3, $p = 0.015$).

Table 2. Characteristics and response rates of the lesions in respondent and non-respondent groups in both studied timepoints.

	3 Months			6 Months		
	Respondent Group (n = 18)	**Non-Respondent Group (n = 24)**	***p*-Value**	**Respondent Group (n = 28)**	**Non-Respondent Group (n = 14)**	***p*-Value**
GTV size (cc) *, mean ± SD	374.3 ± 426.5	491.0 ± 634.6		295.4 ± 376.2	732.2 ± 728.2	
• <5 cc	3	3		4	2	
• 5–200 cc	5	11	0.504	12	3	0.014
• 200–500 cc	5	1		6	1	
• 500–1000 cc	3	4		4	3	
• >1000 cc	2	5		2	5	
PTV size (cc), mean ± SD	510.1 ± 534.9	654.1 ± 758.3.0	0.496	418.4 ± 479.6	940.4 ± 857.0	0.015
Prescribed Dose for SBRT, mean ± SD	34.3 ± 4.7	32.9 ± 5.2		33.9 ± 4.7	32.7 ± 5.5	
• 27.5 Gy	1	4		2	3	
• 30 Gy	7	11	0.377	12	6	0.449
• 35 Gy	4	3		5	2	
• 40 Gy	6	5		9	2	
• 45 Gy	-	1		-	1	
• 50 Gy	-	-		-	-	
Response Rate						
• CR	3	-		13	-	
• PR	15	-		15	-	
• SD	-	16		-	6	
• PD	-	8		-	8	

* GTV was defined as VOI in this study.

3.3. Clinical Associations between Features and Treatment Response

Table 3 summarizes the statistically significant features identified from the two-sided independent Mann–Whitney U test for 3-month and 6-month response rate, respectively. Based on the results from Table 3, there was an inclination that different modalities of the pre-treatment multi-phasic MR images (AP, PVP, DeltaP) contain specific types of radiomic predictors associated with the response rate in HCC patients treated by the sequential TACE-SBRT-Immunotherapy regimen.

For 3-month response rate assessment, as shown in Table 3a, a total of 17 radiomic features were found to have a significant association with the 3-month response rate ($p < 0.048$), mostly uniformity- and entropy-related features (n = 13/17, 76%). Moreover, DeltaP-derived radiomic features accounted for the largest proportion (n = 15/17, 88%), followed by PVP-derived radiomic features (n = 2/17, 12%). However, it is worth noting that no shape and size features, dosimetric features, and AP-derived radiomic features were found to be significantly different between the respondent and non-respondent groups.

For the 6-month response rate assessment, as indicated in Table 3b, a total of 34 features (Radiomic features: n = 31; Shape and size features: n = 2; Dosimetric features: n = 1) were found to demonstrate significant clinical associations ($p < 0.049$). Among the 31 radiomic features, AP-derived radiomic features accounted for the largest proportion (n = 16/31, 52%), which was absent for the 3-month timepoint (Table 3a); this was followed by DeltaP-derived (n = 14/31, 45%), and PVP-derived (n = 1/31, 3%) radiomic features. Similar to the results from the 3-month response rate assessment, uniformity- and entropy-related features dominated (n = 13/31, 42%), especially in DeltaP-derived features (n = 9/14, 64%). Intriguingly, a considerable proportion of high gray level emphasis-related features (n = 8/31, 25%) demonstrated a statistically significant association with the 6-month response assessment, particularly in AP-derived features (n = 7/16, 43%).

Table 4 showcase a list of independent predictors that were determined to demonstrate the significant clinical association with the 3-month and 6-month response rate assessment, respectively. Based on the results from Table 4, previous findings of the inclination shown in Table 3 remained consistent and valid.

For the 3-month response rate assessment, 4 DeltaP-derived radiomic features were determined as independent predictors (FDR-adjusted p-value < 0.1, ranging from 0.038 to 0.063), as shown in Table 4a. The 4 predictors were mostly uniformity- and entropy-related features (GLCM_Joint Entropy, GLRLM_Run Entropy, GLSZM_Gray-Level Non-Uniformity Normalized and GLSZM_Small Area Emphasis), yielding AUC between 0.690 and 0.766. On the other hand, no shape and size features, dosimetric features, AP-derived, and PVP-derived radiomic features were identified as independent predictors (Table 4a).

Table 3. (a) A list of statistically significant features identified from two-sided independent Mann–Whitney U test for 3-month treatment response assessment. (b) A list of statistically significant features identified from two-sided independent Mann–Whitney U test for 6-month treatment response assessment.

(a)	
Features	***p*-Value**
PVP radiomic features	
First-order feature	
Uniformity	0.048
Second-order feature	
GLCM_Sum Entropy	0.040
DeltaP radiomic features	
First-order feature	
Entropy	0.045
Uniformity	0.029
Second-order feature	
GLCM_Difference Variance	0.011
GLCM_Joint Energy	0.040
GLCM_Joint Entropy	0.037
GLCM_Sum Entropy	0.033
GLRLM_Gray-Level Non-Uniformity	0.031
GLRLM_Gray-Level Non-Uniformity Normalized	0.037
GLRLM_Run Entropy	0.010
GLSZM_Gray-Level Non-Uniformity Normalized	0.010
GLSZM_Size Zone Non-Uniformity Normalized	0.003
GLSZM_Small Area Emphasis	0.003
GLDM_Dependence Entropy	0.029
GLDM_Gray-Level Non-Uniformity	0.019
NGTDM_Contrast	0.029

Table 3. *Cont.*

(b)	
Features	***p*-Value**
Shape and size features	
Major Axis Length	0.018
Maximum 3D Diameter	0.049
AP radiomic features	
First-order feature	
Kurtosis	0.008
Maximum	0.017
Uniformity	0.040
Second-order feature	
GLCM_Auto-correlation	0.004
GLCM_Joint Average	0.004
GLRLM_Gray-Level Non-Uniformity	0.049
GLRLM_Gray-Level Non-Uniformity Normalized	0.046
GLRLM_High Gray-Level Run Emphasis	0.004
GLRLM_Long Run High Gray-Level Emphasis	0.021
GLRLM_Short Run High Gray-Level Emphasis	0.004
GLSZM_Gray-Level Non-Uniformity	0.049
GLSZM_High Gray-Level Zone Emphasis	0.005
GLSZM_Small Area High Gray-Level Emphasis	0.005
GLDM_High Gray-Level Emphasis	0.004
GLDM_Small Dependence High Gray-Level Emphasis	0.021
NGTDM_Coarseness	0.049
PVP radiomic features	
Second-order feature	
GLSZM_High Gray-Level Zone Emphasis	0.046
DeltaP radiomic features	
First-order feature	
Entropy	0.012
Maximum	0.024
Median	0.049
Minimum	0.014
Range	0.038
Uniformity	0.030
Second-order feature	
GLCM_Joint Entropy	0.030
GLCM_Sum Entropy	0.013
GLRLM_Gray-Level Non-Uniformity	0.026
GLRLM_Gray-Level Non-Uniformity Normalized	0.030
GLRLM_Run Entropy	0.009
GLSZM_Gray-Level Non-Uniformity Normalized	0.002
GLDM_Gray-Level Non-Uniformity	0.002
NGTDM_Contrast	0.010
Dosimetric features	
V35Gy Percentage	0.035

For the 6-month response rate assessment, 9 features (Radiomic features: n = 8; Shape and size features: n = 1) were determined as independent predictors (FDR-adjusted *p*-value < 0.1, ranging from 0.028 to 0.074), as shown in Table 4b. Among the radiomic predictors, 4 were DeltaP-derived radiomic features, in which 3 were uniformity- and entropy-related features (GLCM_Joint Entropy, GLRLM_Run Entropy, GLSZM_Gray-Level Non-Uniformity Normalized), yielding AUC between 0.699 and 0.788. On the other hand, 4 radiomic predictors were derived from AP MR images, in which 2 were High Gray Level Emphasis-related radiomic features (GLRLM_Short Run High Gray-Level Emphasis, GLDM_Small Dependence High Gray-Level Emphasis), yielding AUC between 0.719 and 0.773. Moreover, a shape feature of Major Axis Length of the HCC lesions was selected as an independent predictor (FDR-adjusted p-value = 0.074), with an AUC of 0.724. No PVP-derived radiomic features and dosimetric features were identified as independent predictors (Table 4b).

Table 4. A list of independent predictors that demonstrated significant clinical association with the (a) 3-month and (b) 6-month response rate assessment. Statistical significance is indicated by false discovery rate (FDR)-adjusted p-value, after applying the Benjamini–Hochberg (BH) procedure for multiple test corrections. AUC, sensitivity, specificity, and p-value obtained from logistic regression for each of these independent significant predictors are also reported. The superscript [a] denotes FDR-adjusted p-values obtained after applying the BH procedure.

(a)

Features	FDR-Adjusted p-Value [a]	AUC	Sensitivity	Specificity
DeltaP radiomic features				
Second-order feature				
GLCM_Joint Entropy	0.063	0.690 (0.527–0.843)	0.625	0.667
GLRLM_Run Entropy	0.044	0.734 (0.573–0.869)	0.625	0.667
GLSZM_Gray-Level Non-Uniformity Normalized	0.038	0.734 (0.566–0.881)	0.625	0.667
GLSZM_Small Area Emphasis	0.038	0.766 (0.600–0.912)	0.625	0.667

(b)

Features	FDR-Adjusted p-Value [a]	AUC	Sensitivity	Specificity
Shape and size features				
Major Axis Length	0.074	0.724 (0.529–0.891)	0.714	0.607
AP radiomic features				
First-order feature				
Kurtosis	0.028	0.750 (0.589–0.895)	0.786	0.643
Maximum	0.028	0.727 (0.564–0.879)	0.714	0.607
Second-order feature				
GLRLM_Short Run High Gray-Level Emphasis	0.028	0.773 (0.606–0.917)	0.857	0.679
GLDM_Small Dependence High Gray-Level Emphasis	0.055	0.719 (0.533–0.859)	0.786	0.643
DeltaP radiomic features				
First-order feature				
Range	0.047	0.699 (0.518–0.865)	0.786	0.643
Second-order feature				
GLCM_Joint Entropy	0.070	0.707 (0.527–0.877)	0.643	0.571
GLRLM_Run Entropy	0.047	0.747 (0.593–0.883)	0.714	0.607
GLSZM_Gray-Level Non-Uniformity Normalized	0.047	0.788 (0.633–0.909)	0.786	0.643

4. Discussion

Unresectable HCC patients present a vulnerable sub-population of liver cancer patients. Over the years, different forms of combined sequential treatments, such as the TACE-SBRT and SBRT-Immunotherapy have been investigated as either a curative alternative or a bridging therapy for subsequent liver transplantation. For the first time in history, a novel sequential TACE-SBRT-Immunotherapy is being introduced in the hope of integrating the potential synergisms among these three treatment modalities, for this HCC subgroup in an ongoing phase-II clinical trial [23]. While exciting, the clinical benefits may be restricted only to a small portion of patients, especially when it comes to immunotherapy [54,55]. In the era of personalized medicine, there is a tremendous demand for pre-treatment discrimination between responsive and non-responsive candidates in order to avoid ineffective and toxic therapies in non-respondents.

In this pilot study, we successfully identified four independent predictors (All were DeltaP-derived radiomics) for 3-month response rate (FDR-adjusted p-value < 0.1, ranging from 0.038 to 0.063, Table 4a), and nine (four were DeltaP-derived; four were AP-derived; and one HCC shape feature) for 6-month response rate (FDR-adjusted p-value < 0.1, ranging from 0.028 to 0.074, Table 4b) to the sequential TACE-SBRT-Immunotherapy regimen in this prospectively enrolled unresectable HCC sub-population, paving the way towards effective and safe oncologic care delivery in the long run.

Results of the present study underscored that different modality of the pre-treatment multi-phasic MR-based radiomics (AP, PVP, and DeltaP) appears to contain specific types of textural predictors associated with the response rate in unresectable HCC patients treated by this novel aggressive regimen (Tables 3 and 4). Specifically, the DeltaP-derived uniformity-related and entropy-related radiomic features were significantly associated with treatment response rate at both 3-month and 6-month timepoints; and the AP-derived high gray level emphasis-related radiomic features demonstrated a significant clinical association, particularly at the 6-month timepoint (Tables 3 and 4).

The four AP-derived radiomic features that emerged to demonstrate significant association for 6-month response rate ("GLRLM_Short Run High Gray-Level Emphasis", "GLDM_Small Dependence High Gray-Level Emphasis", "Kurtosis", "Maximum") were related to the hyperintense signal intensity (i.e., the arterial-enhanced signal) on the AP image (Table 4b). "GLRLM_Short Run High Gray-Level Emphasis" measures the joint distribution of the short homogeneous runs with high gray-level, with a higher value indicating a greater concentration of high gray-level values within the lesion; "GLDM_Small Dependence High Gray-Level Emphasis" reflects the joint distribution of small dependence with higher gray-level values within the lesion; "Kurtosis" is a first-order statistics that informs the 'peakedness' of the distribution of values within the entire lesion, with a higher value implicating that the mass of the distribution is concentrated towards the tails; and another first-order measure of "Maximum" that tells the maximum gray-level intensity within the lesion. Although the biological meaning of these features in the context of HCC remains to be fully elucidated, they are all related to characteristics and spatial distribution of hyperintense signals within the lesion on AP images. This can be possibly ascribed by the fact that AP hyper-enhancement of tumor is a crucial property of HCC lesion [56], which reflects the tumor's capability in generating new blood vessels (termed as neo-angiogenesis), an ability to drive more nutrition and oxygen supply exclusively from the hepatic arteries. In fact, "Kurtosis" has been frequently correlated to the response rate in various cancer types. For instance, Hou et al. conducted a radiomic study for the prediction of tumor response following systemic treatment of chemoradiotherapy in patients with esophageal carcinoma and reported that "Kurtosis" was identified as one of the most dominant features in their combined radiomic models for PR and CR prediction [57]. Moreover, Wang et al. applied radiomics for predicting tumor response to systemic induction chemotherapy in patients with nasopharyngeal carcinoma and reported "Kurtosis" as one of the predictive biomarkers in their MR-based prediction models [58]. Although this study presents a novel sequential TACE-SBRT-Immunotherapy, we speculated that these radiomic features characterizing baseline arterial-enhanced signal may be indicative of HCC responsiveness to treatment perturbations and deserve further in-depth investigations in the future.

Previous studies on multi-phasic MR-based radiomics for tumor response prediction in HCC patients have mainly focused on TACE treatment [47–49], while those on SBRT and immunotherapy regimens are scarce or absent. For TACE, Kuang et al. conducted a retrospective radiomics study to predict the response of small-sized HCC lesions following TACE and reported 11 AP-derived and 11 T2-weighted radiomic features as final predictors [47]. More recently, Liu et al. examined the predictive power of multi-phasic MRI for predicting HCC response following TACE treatment, and reported that 17 AP-derived and three PVP-derived radiomic features, along with radiomic features from other MR sequences [48]. However, these studies present a varying degree of disparity in study design compared with the present work in terms of target population [47], treatment regimen [47,48], and source images of radiomic features [47,48]. Therefore, a direction comparison of textual predictors between their studies and the present work appears to be infeasible and offers little scientific significance. Nonetheless, it is worth noting that the shape features of HCC seem to be predictive of treatment response despite the mentioned heterogeneity in study design between studies. In this study, the "Maximum 3D Diameter" was also found to demonstrate a significant association with treatment response to the sequential TACE-SBRT-Immunotherapy in the 6-month timepoint (Table 3b). Moreover, the "Major Axis Length", another measure of baseline tumor burden which measures the largest axis length of the HCC-enclosing ellipsoid, was determined as an independent predictor (Table 4b). This finding is in line with the two previous studies where the "Maximum 3D Diameter" [47] and "Maximum 2D Diameter Row" [48] were found to be predictive of TACE treatment. Indeed, this result is also in concordance with the dynamic-CT-based study conducted by Park et al., where smaller tumor size was a significant predictor for complete response in HCC patients following TACE treatment [59].

On the other hand, Delta radiomics is a novel concept that reflects the dynamic variation in radiomic features in longitudinal images, highlighting intratumoral changes in imaging features and hence implicating the underlying tumoral physiological function. Compared with non-Delta radiomic features, Delta radiomics has recently been found to demonstrate higher reproducibility between scanners and institutions in phantom studies, potentially providing a more generalizable predictive capability, which has gained increasing popularity in the research community [60,61]. Nardone et al. provided a comprehensive systematic review of the Delta radiomics studies in the body of literature [60]. In the context of treatment response prediction, it has demonstrated ground-breaking evidence in numerous types of cancer following immunotherapy, including but not limited to renal cell cancer [62], pancreatic cancer [63], and non-small cell lung cancer [64–67]. Notably, to our best understanding, this study is one of the very first to report the potential of MR-based Delta radiomics in associating with treatment response in HCC.

Intriguingly, four DeltaP-derived radiomic features that were selected as independent significant predictors (Table 4b) were all related to randomness and uniformity of the spatial distribution of texture within the lesion. They were "GLCM_Joint Entropy" which measures randomness in signal intensity in neighboring voxels; "GLRLM_Run Entropy" which measures randomness in the distribution of run lengths and gray levels, with a higher value reflecting greater textual heterogeneity; "GLSZM_Gray-Level Non-Uniformity Normalized" which depicts the similarity of normalized gray-level intensity, with a lower value correlating with a greater similarity; and "Range" which tells the range of signal intensities within the lesion. These textual reflect the temporal change in the randomness and uniformity within the lesion between the AP and PVP MR scans, providing a better understanding of the intratumoral heterogeneity in time dimension upon arrival of the imaging contrast agent. Hence, these textual predictors may be indicative of the aggressiveness of HCC lesions and their responsiveness upon treatment perturbation. In fact, the radiomic features of "entropy" and "uniformity" have been well-recognized as of high prognostic value in various cancer types, including but not limited to HCC [68,69], esophageal cancer [70], lung cancer [71], and squamous cell carcinoma of head and neck cancer [72]. For instance, Liu et al. selected "Gray-Level Non-uniformity" as one of the radiomic predictors, which had the largest weight among other predictors, for overall survival prediction in HCC patients [68]. Another study conducted by Ganeshan et al. revealed that a higher hepatic entropy and lower uniformity often reflect a more complex tumor heterogeneity [73]. Along this line of thinking, Mulé et al. analyzed pre-treatment contrast-enhanced CT-based textures for overall survival prediction in advanced HCC patients following treatment with Sorafenib [69]. They reported a significant correlation between tumor heterogeneity and entropy at both AP and PVP phases, and particularly, the PVP-derived entropy was determined to be an independent prognostic factor [69]. These findings may partly explain why the four entropy- and uniformity-related features were determined to be of a significant clinical association with treatment response.

On the other hand, it is worth noting that the studied dosimetric data from the SBRT plan appeared to be not predictive of HCC treatment response prediction in this study. The dose parameter of V35Gy was identified as a significant predictor only for the 6-month timepoint under univariate analysis (Table 3b), which might imply that a dose threshold 35 Gy was required to trigger tumor responses. The biological effective dose for 35 Gy in five fractions was calculated as 59.5 Gy (α/β = 10), which was the minimum effective dose fractionation scheme mentioned in a systematic review and meta-analysis [74]. However, it was not shown predictive after multiple test corrections (Table 4b). In our study, SBRT was prescribed based on the isotoxic principle that the radiation dose was individualized based on tumor size, volume, and proximity of organ-at-risk. Such strategy was commonly adopted in treating large-sized, locally advanced HCC and often resulted in a heterogeneous dose [75,76]. Our study showed that the dose of 27.5–30 Gy was equally effective in terms of tumor response. The potential explanation is that immunotherapy may have sensitized the tumor to radiotherapy and that lower radiation doses can attain similar local control, as demonstrated in the pre-clinical study [77]. With this regard, it is interesting to note that while dosimetric parameters have been predictive mainly in the areas of toxicity prediction [37,38,78–81], there is scarce or none in treatment response prediction. The underlying reasons remain unknown, and it definitely represents an interesting research area for future scrutinization.

From a public health standpoint, the findings of this study demonstrated the feasibility of using cost-effective radiomics techniques in associating with treatment response in a highly vulnerable sub-population of HCC patients following a novel aggressive treatment. Patients in this subgroup often suffer not only from HCC but also other liver-related diseases, such as portal hypertension and ascites due to liver cirrhosis [82]. Taken together with the desperately poor survival rate, tremendous burdens have been placed on this patient subgroup and the healthcare system. Although the pioneering sequential TACE-SBRT-Immunotherapy regimen offers both local-regional and whole-body systemic therapy to this subgroup, the underlying toxicity profile remains unclear. In the long term, the results of this study may provide valuable insights into pre-treatment identification of responding and non-responding candidates for this novel treatment, so as to avoid ineffective, toxic, and costly therapies to refractory patients, while streamlining medical resourcing allocations within the healthcare system. Several limitations of this study should be acknowledged. First, the cohort of patients was small due to the stringent patient inclusion criteria for receiving the novel aggressive treatment in the prospective clinical trial. This limitation diminishes the strength of our results and prevents the possibility of using machine learning, AI algorithms or other sophisticated classification techniques for prediction model development [28]. Despite this, we were able to identify specific types of radiomic predictors from different multi-phasic MR images that can predict tumor response during sequential TACE-SBRT-Immunotherapy regimen, and provide classification performance at the individual feature level. More importantly, the key novelty of this present work lies in that we demonstrated the feasibility of using multi-phasic MR-based radiomics for predicting tumor response to the novel aggressive therapy in a vulnerable subgroup of HCC patients. Notably, patient recruitment in the clinical trial is continuously undertaken, and a larger-cohort study is anticipated and will be part of our future plan. Moreover, the reproducibility of radiomic features against tumor segmentation variability, and the correlation between radiomic features and genetic data were not investigated in this study.

Moving forward with a larger cohort of patients, these should be considered when it comes to building robust predictive models for clinical use in the future.

5. Conclusions

In this pilot study, we successfully demonstrated that four DeltaP-derived radiomic features (characterizing temporal change in intratumoral randomness and uniformity), four AP-derived radiomic features (reflecting baseline tumoral arterial-enhanced signal pattern), and a tumor morphology (denoting initial tumor burden), were determined to be significantly associated with the 6-month response rate in unresectable HCC lesions following aggressive TACE-SBRT-Immunotherapy regimen, while the DeltaP-derived radiomics were the only contributors to the response rate at 3-month timepoint. While results indicated a potential for pre-treatment discrimination between responding and non-responding unresectable HCC candidates for this novel treatment, a larger study cohort is warranted in the future to validate the results of this work.

Author Contributions: Conceptualization, L.-M.H.; methodology, L.-M.H. and S.-K.L.; validation, L.-M.H. and S.-K.L.; formal analysis, L.-M.H., S.-K.L. and J.Z.; writing—original draft preparation, L.-M.H. and S.-K.L.; writing—review and editing, L.-M.H. and S.-K.L.; clinical trial conduction, C.-L.C. and A.C.-Y.C.; supervision, J.C. and C.-L.C. All authors have read and agreed to the published version of the manuscript.

Funding: This research was funded by research grants of Shenzhen-Hong Kong-Macau S and T Program (Category C) (SGDX20201103095002019) and Shenzhen Basic Research Program (JCYJ20210324130209023) from Shenzhen Science and Technology Innovation Committee, and Project of RISA fund (P0043001) from The Hong Kong Polytechnic University.

Institutional Review Board Statement: The current research was approved by the Human Subjects Ethics Subcommittee of The Hong Kong Polytechnic University and Institutional Review Board of the University of Hong Kong/Hospital Authority Hong Kong West Cluster (UW 21-469).

Informed Consent Statement: This study shared the retrospective clinical data from the "Sequential TransArterial chemoembolization and stereotactic Radiotherapy Followed by Immunotherapy for Downstaging Hepatocellular Carcinoma for Hepatectomy (START-FIT)" clinical trial conducted by The University of Hong Kong; the IRB number of the START-FIT trial is UW 18-541. This study was a retrospective study based on analyses of anonymized radiological data, clinical data and dosimetric data. Patient data were accessed retrospectively through hospital electronic medical record systems and RT planning systems. The necessity to obtain individual informed consent was waived.

Data Availability Statement: Not applicable.

Conflicts of Interest: The authors declare no conflict of interest.

References

1. Rumgay, H.; Ferlay, J.; de Martel, C.; Georges, D.; Ibrahim, A.S.; Zheng, R.; Wei, W.; Lemmens, V.E.P.P.; Soerjomataram, I. Global, regional and national burden of primary liver cancer by subtype. *Eur. J. Cancer* **2022**, *161*, 108–118. [CrossRef]
2. Potters, L.; Gaspar, L.E.; Kavanagh, B.; Galvin, J.M.; Hartford, A.C.; Hevezi, J.M.; Kupelian, P.A.; Mohiden, N.; Samuels, M.A.; Timmerman, R.; et al. American Society for Therapeutic Radiology and Oncology (ASTRO) and American College of Radiology (ACR) practice guidelines for image-guided radiation therapy (IGRT). *Int. J. Radiat. Oncol. Biol. Phys.* **2010**, *76*, 319–325. [CrossRef] [PubMed]
3. Pesapane, F.; Nezami, N.; Patella, F.; Geschwind, J.F. New concepts in embolotherapy of HCC. *Med. Oncol.* **2017**, *34*, 58. [PubMed]
4. Bruix, J.; Sherman, M. Practice Guidelines Committee, Management of hepatocellular carcinoma. *Hepatology* **2005**, *42*, 1208–1236. [PubMed]
5. European Association for the Study of the Liver. European Association for the Study of the, EASL Clinical Practice Guidelines: Management of hepatocellular carcinoma. *J. Hepatol.* **2018**, *69*, 182–236.
6. Lo, C.-M.; Ngan, H.; Tso, W.-K.; Liu, C.-L.; Lam, C.-M.; Poon, R.T.-P.; Fan, S.-T.; Wong, J. Randomized controlled trial of transarterial lipiodol chemoembolization for unresectable hepatocellular carcinoma. *Hepatology* **2002**, *35*, 1164–1171.
7. Ghanaati, H.; Mohammadifard, M.; Mohammadifard, M. A review of applying transarterial chemoembolization (TACE) method for management of hepatocellular carcinoma. *J. Fam. Med. Prim. Care* **2021**, *10*, 3553–3560.
8. Shim, S.J.; Seong, J.; Han, K.H.; Chon, C.Y.; Suh, C.O.; Lee, J.T. Local radiotherapy as a complement to incomplete transcatheter arterial chemoembolization in locally advanced hepatocellular carcinoma. *Liver Int.* **2005**, *25*, 1189–1196. [CrossRef]

9. Shampain, K.L.; Hackett, C.E.; Towfighi, S.; Aslam, A.; Masch, W.R.; Harris, A.C.; Chang, S.D.; Khanna, K.; Mendiratta, V.; Gabr, A.M.; et al. SBRT for HCC: Overview of technique and treatment response assessment. *Abdom. Radiol.* **2021**, *46*, 3615–3624.
10. Choi, S.H.; Seong, J. Stereotactic Body Radiotherapy: Does It Have a Role in Management of Hepatocellular Carcinoma? *Yonsei Med. J.* **2018**, *59*, 912–922. [CrossRef]
11. Cárdenes, H.R.; Price, T.R.; Perkins, S.M.; Maluccio, M.; Kwo, P.; Breen, T.E.; Henderson, M.A.; Schefter, T.E.; Tudor, K.; Deluca, J.; et al. Phase I feasibility trial of stereotactic body radiation therapy for primary hepatocellular carcinoma. *Clin. Transl. Oncol.* **2010**, *12*, 218–225. [CrossRef] [PubMed]
12. Huo, Y.R.; Eslick, G.D. Transcatheter Arterial Chemoembolization Plus Radiotherapy Compared With Chemoembolization Alone for Hepatocellular Carcinoma: A Systematic Review and Meta-analysis. *JAMA Oncol.* **2015**, *1*, 756–765. [PubMed]
13. Hawkins, M.A.; Dawson, L.A. Radiation therapy for hepatocellular carcinoma: From palliation to cure. *Cancer* **2006**, *106*, 1653–1663.
14. Aguirre-Ghiso, J.A. Models, mechanisms and clinical evidence for cancer dormancy. *Nat. Rev. Cancer* **2007**, *7*, 834–845. [CrossRef] [PubMed]
15. Steel, P.M. Exploitable mechanisms in combined radiotherapychemotherapy: The concept of additivity. *Int. J. Radiat. Oncol. Biol. Phys.* **1979**, *5*, 85–91. [PubMed]
16. Seong, J.; Kim, S.; Suh, C. Enhancement of tumor radioresponse by combined chemotherapyin murine hepatocarcinoma. *J. Gastroenterol. Hepatol.* **2001**, *16*, 883–889. [PubMed]
17. Chan MK, H.; Lee, V.; Chiang, C.L.; Lee FA, S.; Law, G.; Sin, N.Y.; Siu, K.L.; Wong FC, S.; Tung, S.Y.; Luk, H.; et al. Lipiodol versus Diaphragm in 4D-CBCT-guided Stereotactic Radiotherapy of Hepatocellular Carcinoma. *Strahlenther. Und Onkol.* **2016**, *192*, 92–101.
18. Yue, J.; Sun, X.; Cai, J.; Yin, F.-F.; Yin, Y.; Zhu, J.; Lu, J.; Liu, T.; Yu, J.; Shi, X.; et al. Lipiodol: A potential direct surrogate for cone-beam computed tomography image guidance in radiotherapy of liver tumor. *Int. J. Radiat. Oncol. Biol. Phys.* **2012**, *82*, 834–841. [CrossRef]
19. Chiang, C.L.; Chan, M.K.H.; Yeung, C.S.Y.; Ho, C.H.M.; Lee, F.A.S.; Lee, V.W.Y.; Wong, F.C.S.; Blanck, O. Combined stereotactic body radiotherapy and trans-arterial chemoembolization as initial treatment in BCLC stage B-C hepatocellular carcinoma. *Strahlenther. Onkol.* **2019**, *195*, 254–264.
20. Rhieu, B.-H.; Narang, A.; Meyer, J. Stereotactic Ablative Radiotherapy (SABR/SBRT) for Hepatocellular Carcinoma. *Curr. Hepatol. Rep.* **2018**, *17*, 392–398. [CrossRef]
21. Chiang, C.-L.; Chan, A.C.Y.; Chiu, K.W.H.; Kong, F.-M.S. Combined Stereotactic Body Radiotherapy and Checkpoint Inhibition in Unresectable Hepatocellular Carcinoma: A Potential Synergistic Treatment Strategy. *Front. Oncol.* **2019**, *9*, 1157. [PubMed]
22. Chiang, C.-L.; Chiu, K.W.-H.; Lee, F.A.-S.; Kong, F.-M.S.; Chan, A.C.-Y. Combined Stereotactic Body Radiotherapy and Immunotherapy Versus Transarterial Chemoembolization in Locally Advanced Hepatocellular Carcinoma: A Propensity Score Matching Analysis. *Front. Oncol.* **2021**, *11*, 798832. [PubMed]
23. Chiang, C.L.; Chiu, K.W.H.; Chan, K.S.K.; Lee, F.A.S.; Li, J.C.B.; Wan, C.W.S.; Dai, W.C.; Lam, T.C.; Chen, W.; Wong, N.S.M.; et al. Sequential transarterial chemoembolisation and stereotactic body radiotherapy followed by immunotherapy as conversion therapy for patients with locally advanced, unresectable hepatocellular carcinoma (START-FIT): A single-arm, phase 2 trial. *Lancet Gastroenterol. Hepatol.* **2023**, *8*, 169–178.
24. Lambin, P.; Rios-Velazquez, E.; Leijenaar, R.; Carvalho, S.; van Stiphout, R.G.P.M.; Granton, P.; Zegers, C.M.L.; Gillies, R.; Boellard, R.; Dekker, A.; et al. Radiomics: Extracting more information from medical images using advanced feature analysis *Eur. J. Cancer* **2012**, *48*, 441–446. [CrossRef]
25. Avanzo, M.; Stancanello, J.; El Naqa, I. Beyond imaging: The promise of radiomics. *Phys. Med.* **2017**, *38*, 122–139. [PubMed]
26. Yu, T.-T.; Lam, S.-K.; To, L.-H.; Tse, K.-Y.; Cheng, N.-Y.; Fan, Y.-N.; Lo, C.-L.; Or, K.-W.; Chan, M.-L.; Hui, K.-C.; et al. Pretreatment Prediction of Adaptive Radiation Therapy Eligibility Using MRI-Based Radiomics for Advanced Nasopharyngeal Carcinoma Patients. *Front. Oncol.* **2019**, *9*, 1050. [PubMed]
27. Lam, S.-K.; Zhang, J.; Zhang, Y.-P.; Li, B.; Ni, R.-Y.; Zhou, T.; Peng, T.; Cheung, A.L.-Y.; Chau, T.-C.; Lee, F.K.-H.; et al. A Multi-Center Study of CT-Based Neck Nodal Radiomics for Predicting an Adaptive Radiotherapy Trigger of Ill-Fitted Thermoplastic Masks in Patients with Nasopharyngeal Carcinoma. *Life* **2022**, *12*, 241. [CrossRef]
28. Zhang, Y.; Lam, S.; Yu, T.; Teng, X.; Zhang, J.; Lee, F.K.-H.; Au, K.-H.; Yip, C.W.-Y.; Wang, S.; Cai, J. Integration of an imbalance framework with novel high-generalizable classifiers for radiomics-based distant metastases prediction of advanced nasopharyngeal carcinoma. *Knowl.-Based Syst.* **2022**, *235*, 107649.
29. Rizzo, S.; Botta, F.; Raimondi, S.; Origgi, D.; Fanciullo, C.; Morganti, A.G.; Bellomi, M. Radiomics: The facts and the challenges of image analysis. *Eur. Radiol. Exp.* **2018**, *2*, 36.
30. Lafata, K.; Cai, J.; Wang, C.; Hong, J.; Kelsey, C.R.; Yin, F.-F. Spatial-temporal variability of radiomic features and its effect on the classification of lung cancer histology. *Phys. Med. Biol.* **2018**, *63*, 225003.
31. Teng, X.; Zhang, J.; Ma, Z.; Zhang, Y.; Lam, S.; Li, W.; Xiao, H.; Li, T.; Li, B.; Zhou, T.; et al. Improving radiomic model reliability using robust features from perturbations for head-and-neck carcinoma. *Front. Oncol.* **2022**, *12*, 974467. [PubMed]
32. Pinker-Domenig, K.; Chin, J.; Melsaether, A.N.; Morris, E.A.; Moy, L. Precision Medicine and Radiogenomics in Breast Cancer: New Approaches toward Diagnosis and Treatment. *Radiology* **2018**, *287*, 732–747.

33. Liu, Z.; Meng, X.; Zhang, H.; Li, Z.; Liu, J.; Sun, K.; Meng, Y.; Dai, W.; Xie, P.; Ding, Y.; et al. Predicting distant metastasis and chemotherapy benefit in locally advanced rectal cancer. *Nat. Commun.* **2020**, *11*, 4308. [PubMed]
34. Elshafeey, N.; Kotrotsou, A.; Hassan, A.; Elshafei, N.; Hassan, I.; Ahmed, S.; Abrol, S.; Agarwal, A.; El Salek, K.; Bergamaschi, S.; et al. Multicenter study demonstrates radiomic features derived from magnetic resonance perfusion images identify pseudoprogression in glioblastoma. *Nat. Commun.* **2019**, *10*, 3170. [PubMed]
35. Trebeschi, S.; Drago, S.; Birkbak, N.; Kurilova, I.; Călin, A.; Pizzi, A.D.; Lalezari, F.; Lambregts, D.; Rohaan, M.; Parmar, C.; et al. Predicting response to cancer immunotherapy using noninvasive radiomic biomarkers. *Ann. Oncol.* **2019**, *30*, 998–1004.
36. Colen, R.R.; Rolfo, C.; Ak, M.; Ayoub, M.; Ahmed, S.; Elshafeey, N.; Mamindla, P.; O Zinn, P.; Ng, C.; Vikram, R.; et al. Radiomics analysis for predicting pembrolizumab response in patients with advanced rare cancers. *J. Immunother. Cancer* **2021**, *9*, e001752.
37. Jiang, W.; Song, Y.; Sun, Z.; Qiu, J.; Shi, L. Dosimetric Factors and Radiomics Features Within Different Regions of Interest in Planning CT Images for Improving the Prediction of Radiation Pneumonitis. *Int. J. Radiat. Oncol. Biol. Phys.* **2021**, *110*, 1161–1170. [CrossRef]
38. Bousabarah, K.; Blanck, O.; Temming, S.; Wilhelm, M.-L.; Hoevels, M.; Baus, W.W.; Ruess, D.; Visser-Vandewalle, V.; Ruge, M.I.; Treuer, H.; et al. Radiomics for prediction of radiation-induced lung injury and oncologic outcome after robotic stereotactic body radiotherapy of lung cancer: Results from two independent institutions. *Radiat. Oncol.* **2021**, *16*, 74. [CrossRef]
39. Zhang, Y.; Shu, Z.; Ye, Q.; Chen, J.; Zhong, J.; Jiang, H.; Wu, C.; Yu, T.; Pang, P.; Ma, T.; et al. Preoperative Prediction of Microvascular Invasion in Hepatocellular Carcinoma via Multi-Parametric MRI Radiomics. *Front. Oncol.* **2021**, *11*, 633596. [CrossRef]
40. Yang, L.; Gu, D.; Wei, J.; Yang, C.; Rao, S.; Wang, W.; Chen, C.; Ding, Y.; Tian, J.; Zeng, M. A Radiomics Nomogram for Preoperative Prediction of Microvascular Invasion in Hepatocellular Carcinoma. *Liver Cancer* **2019**, *8*, 373–386. [CrossRef]
41. Luo, J.; Huang, Z.; Wang, M.; Li, T.; Huang, J. Prognostic role of multiparameter MRI and radiomics in progression of advanced unresectable hepatocellular carcinoma following combined transcatheter arterial chemoembolization and lenvatinib therapy. *BMC Gastroenterol.* **2022**, *22*, 108.
42. Zhang, Z.; Jiang, H.; Chen, J.; Wei, Y.; Cao, L.; Ye, Z.; Li, X.; Ma, L.; Song, B. Hepatocellular carcinoma: Radiomics nomogram on gadoxetic acid-enhanced MR imaging for early postoperative recurrence prediction. *Cancer Imaging* **2019**, *19*, 22. [CrossRef] [PubMed]
43. Zhao, Y.; Wu, J.; Zhang, Q.; Hua, Z.; Qi, W.; Wang, N.; Lin, T.; Sheng, L.; Cui, D.; Liu, J.; et al. Radiomics Analysis Based on Multiparametric MRI for Predicting Early Recurrence in Hepatocellular Carcinoma After Partial Hepatectomy. *J. Magn. Reson. Imaging* **2021**, *53*, 1066–1079. [PubMed]
44. Jiang, H.; Liu, X.; Chen, J.; Wei, Y.; Lee, J.M.; Cao, L.; Wu, Y.; Duan, T.; Li, X.; Ma, L.; et al. Man or machine? Prospective comparison of the version 2018 EASL, LI-RADS criteria and a radiomics model to diagnose hepatocellular carcinoma. *Cancer Imaging* **2019**, *19*, 84.
45. Huang, X.; Long, L.; Wei, J.; Li, Y.; Xia, Y.; Zuo, P.; Chai, X. Radiomics for diagnosis of dual-phenotype hepatocellular carcinoma using Gd-EOB-DTPA-enhanced MRI and patient prognosis. *J. Cancer Res. Clin. Oncol.* **2019**, *145*, 2995–3003. [CrossRef]
46. Aujay, G.; Etchegaray, C.; Blanc, J.-F.; Lapuyade, B.; Papadopoulos, P.; Pey, M.-A.; Bordenave, L.; Trillaud, H.; Saut, O.; Pinaquy, J.-B. Comparison of MRI-based response criteria and radiomics for the prediction of early response to transarterial radioembolization in patients with hepatocellular carcinoma. *Diagn. Interv. Imaging* **2022**, *103*, 360–366.
47. Kuang, Y.; Li, R.; Jia, P.; Ye, W.; Zhou, R.; Zhu, R.; Wang, J.; Lin, S.; Pang, P.; Ji, W. MRI-Based Radiomics: Nomograms predicting the short-term response after transcatheter arterial chemoembolization (TACE) in hepatocellular carcinoma patients with diameter less than 5 cm. *Abdom. Radiol.* **2021**, *46*, 3772–3789.
48. Liu, Q.-P.; Yang, K.-L.; Xu, X.; Liu, X.-S.; Qu, J.-R.; Zhang, Y.-D. Radiomics analysis of pretreatment MRI in predicting tumor response and outcome in hepatocellular carcinoma with transarterial chemoembolization: A two-center collaborative study. *Abdom. Radiol.* **2022**, *47*, 651–663. [CrossRef]
49. Sun, Y.; Bai, H.; Xia, W.; Wang, D.; Zhou, B.; Zhao, X.; Yang, G.; Xu, L.; Zhang, W.; Liu, P.; et al. Predicting the Outcome of Transcatheter Arterial Embolization Therapy for Unresectable Hepatocellular Carcinoma Based on Radiomics of Preoperative Multiparameter MRI. *J. Magn. Reson. Imaging* **2020**, *52*, 1083–1090. [CrossRef]
50. Lam, S.-K.; Zhang, Y.; Zhang, J.; Li, B.; Sun, J.-C.; Liu, C.Y.-T.; Chou, P.-H.; Teng, X.; Ma, Z.-R.; Ni, R.-Y.; et al. Multi-organ omics-based prediction for adaptive radiation therapy eligibility in nasopharyngeal carcinoma patients undergoing concurrent chemoradiotherapy. *Front. Oncol.* **2022**, *11*, 5406. [CrossRef]
51. Li, B.; Zheng, X.; Zhang, J.; Lam, S.; Guo, W.; Wang, Y.; Cui, S.; Teng, X.; Zhang, Y.; Ma, Z.; et al. Lung Subregion Partitioning by Incremental Dose Intervals Improves Omics-Based Prediction for Acute Radiation Pneumonitis in Non-Small-Cell Lung Cancer Patients. *Cancers* **2022**, *14*, 4889. [PubMed]
52. Li, B.; Ren, G.; Guo, W.; Zhang, J.; Lam, S.-K.; Zheng, X.; Teng, X.; Wang, Y.; Yang, Y.; Dan, Q.; et al. Function-Wise Dual-Omics analysis for radiation pneumonitis prediction in lung cancer patients. *Comput. Intell. Pers. Med.* **2022**, *110*. [CrossRef] [PubMed]
53. Teng, X.; Zhang, J.; Zwanenburg, A.; Sun, J.; Huang, Y.; Lam, S.; Zhang, Y.; Li, B.; Zhou, T.; Xiao, H.; et al. Building reliable radiomic models using image perturbation. *Sci. Rep.* **2022**, *12*, 1–10.
54. Sangro, B.; Gomez-Martin, C.; de la Mata, M.; Iñarrairaegui, M.; Garralda, E.; Barrera, P.; Riezu-Boj, J.I.; Larrea, E.; Alfaro, C.; Sarobe, P.; et al. A clinical trial of CTLA-4 blockade with tremelimumab in patients with hepatocellular carcinoma and chronic hepatitis C. *J. Hepatol.* **2013**, *59*, 81–88. [CrossRef]

55. El-Khoueiry, A.B.; Sangro, B.; Yau, T.; Crocenzi, T.S.; Kudo, M.; Hsu, C.; Kim, T.-Y.; Choo, S.-P.; Trojan, J.; Welling, T.H., 3rd; et al. Nivolumab in patients with advanced hepatocellular carcinoma (CheckMate 040): An open-label, non-comparative, phase 1/2 dose escalation and expansion trial. *Lancet* **2017**, *389*, 2492–2502. [CrossRef]
56. Cruite, I.; Tang, A.; Sirlin, C.B. Imaging-based diagnostic systems for hepatocellular carcinoma. *AJR Am. J. Roentgenol.* **2013**, *201*, 41–55. [PubMed]
57. Hou, Z.; Ren, W.; Li, S.; Liu, J.; Sun, Y.; Yan, J.; Wan, S. Radiomic analysis in contrast-enhanced CT: Predict treatment response to chemoradiotherapy in esophageal carcinoma. *Oncotarget* **2017**, *8*, 104444–104454.
58. Wang, G.; He, L.; Yuan, C.; Huang, Y.; Liu, Z.; Liang, C. Pretreatment MR imaging radiomics signatures for response prediction to induction chemotherapy in patients with nasopharyngeal carcinoma. *Eur. J. Radiol.* **2018**, *98*, 100–106. [CrossRef]
59. Park, J.H.; Kim, D.H.; Kim, S.H.; Kim, M.Y.; Baik, S.K.; Hong, I.S. The Clinical Implications of Liver Resection Margin Size in Patients with Hepatocellular Carcinoma in Terms of Positron Emission Tomography Positivity. *World J. Surg.* **2018**, *42*, 1514–1522.
60. Nardone, V.; Reginelli, A.; Guida, C.; Belfiore, M.P.; Biondi, M.; Mormile, M.; Buonamici, F.B.; Di Giorgio, E.; Spadafora, M.; Tini, P.; et al. Delta-radiomics increases multicentre reproducibility: A phantom study. *Med. Oncol.* **2020**, *37*, 38. [CrossRef]
61. Plautz, T.E.; Zheng, C.; Noid, G.; Li, X.A. Time stability of delta-radiomics features and the impact on patient analysis in longitudinal CT images. *Med. Phys.* **2019**, *46*, 1663–1676. [PubMed]
62. Goh, V.; Ganeshan, B.; Nathan, P.; Juttla, J.K.; Vinayan, A.; Miles, K.A. Assessment of Response to Tyrosine Kinase Inhibitors in Metastatic Renal Cell Cancer: CT Texture as a Predictive Biomarker. *Radiology* **2011**, *261*, 165–171. [CrossRef] [PubMed]
63. Nasief, H.; Zheng, C.; Schott, D.; Hall, W.; Tsai, S.; Erickson, B.; Li, X.A. A machine learning based delta-radiomics process for early prediction of treatment response of pancreatic cancer. *NPJ Precis. Oncol.* **2019**, *3*, 25. [CrossRef] [PubMed]
64. Dercle, L.; Fronheiser, M.; Lu, L.; Du, S.; Hayes, W.; Leung, D.K.; Roy, A.; Wilkerson, J.; Guo, P.; Fojo, A.T.; et al. Identification of Non-Small Cell Lung Cancer Sensitive to Systemic Cancer Therapies Using Radiomics. *Clin. Cancer Res.* **2020**, *26*, 2151–2162. [PubMed]
65. Fave, X.; Zhang, L.; Yang, J.; Mackin, D.; Balter, P.; Gomez, D.; Followill, D.; Jones, A.K.; Stingo, F.; Liao, Z.; et al. Delta-radiomics features for the prediction of patient outcomes in non-small cell lung cancer. *Sci. Rep.* **2017**, *7*, 588.
66. Khorrami, M.; Prasanna, P.; Gupta, A.; Patil, P.; Velu, P.D.; Thawani, R.; Corredor, G.; Alilou, M.; Bera, K.; Fu, P.; et al. Changes in CT Radiomic Features Associated with Lymphocyte Distribution Predict Overall Survival and Response to Immunotherapy in Non-Small Cell Lung Cancer. *Cancer Immunol. Res.* **2020**, *8*, 108–119. [CrossRef]
67. Liu, Y.; Wu, M.; Zhang, Y.; Luo, Y.; He, S.; Wang, Y.; Chen, F.; Liu, Y.; Yang, Q.; Li, Y.; et al. Imaging Biomarkers to Predict and Evaluate the Effectiveness of Immunotherapy in Advanced Non-Small-Cell Lung Cancer. *Front. Oncol.* **2021**, *11*, 657615. [CrossRef]
68. Liu, Q.; Li, J.; Liu, F.; Yang, W.; Ding, J.; Chen, W.; Wei, Y.; Li, B.; Zheng, L. A radiomics nomogram for the prediction of overall survival in patients with hepatocellular carcinoma after hepatectomy. *Cancer Imaging* **2020**, *20*, 82. [CrossRef]
69. Mulé, S.; Thiefin, G.; Costentin, C.; Durot, C.; Rahmouni, A.; Luciani, A.; Hoeffel, C. Advanced Hepatocellular Carcinoma: Pretreatment Contrast-enhanced CT Texture Parameters as Predictive Biomarkers of Survival in Patients Treated with Sorafenib. *Radiology* **2018**, *288*, 445–455.
70. Yip, C.; Landau, D.; Kozarski, R.; Ganeshan, B.; Thomas, R.; Michaelidou, A.; Goh, V. Heterogeneity as Potential Prognostic Biomarker in Patients Treated with Definitive Chemotherapy and Radiation Therapy. *Radiology* **2014**, *2014*, 141–148.
71. Yoon, S.H.; Park, C.M.; Park, S.J.; Yoon, J.-H.; Hahn, S.; Goo, J.M. Tumor Heterogeneity in Lung Cancer: Assessment with Dynamic Contrast-enhanced MR Imaging. *Radiology* **2016**, *280*, 940–948. [PubMed]
72. Zhang, H.; Graham, C.M.; Elci, O.; Griswold, M.E.; Zhang, X.; Khan, M.A.; Pitman, K.; Caudell, J.J.; Hamilton, R.D.; Ganeshan, B.; et al. Locally Advanced Squamous Cell Carcinoma of the Head and Neck: CT Texture and Histogram Analysis Allow Independent Prediction of Overall Survival in Patients Treated with Induction Chemotherapy. *Radiology* **2013**, *269*, 801–809.
73. Ganeshan, B.; Miles, K.; Young, R.; Chatwin, C. Hepatic entropy and uniformity: Additional parameters that can potentially increase the effectiveness of contrast enhancement during abdominal CT. *Clin. Radiol.* **2007**, *62*, 761–768. [PubMed]
74. Long, Y.; Liang, Y.; Li, S.; Guo, J.; Wang, Y.; Luo, Y.; Wu, Y. Therapeutic outcome and related predictors of stereotactic body radiotherapy for small liver-confined HCC: A systematic review and meta-analysis of observational studies. *Radiat. Oncol.* **2021**, *16*, 68. [PubMed]
75. Bujold, A.; Massey, C.A.; Kim, J.J.; Brierley, J.; Cho, C.; Wong, R.K.; Dinniwell, R.E.; Kassam, Z.; Ringash, J.; Cummings, B.; et al. Sequential phase I and II trials of stereotactic body radiotherapy for locally advanced hepatocellular carcinoma. *J. Clin. Oncol.* **2013**, *31*, 1631–1639.
76. Tse, R.V.; Hawkins, M.; Lockwood, G.; Kim, J.J.; Cummings, B.; Knox, J.; Sherman, M.; Dawson, L.A. Phase I study of individualized stereotactic body radiotherapy for hepatocellular carcinoma and intrahepatic cholangiocarcinoma. *J. Clin. Oncol.* **2008**, *26*, 657–664. [CrossRef]
77. Vanpouille-Box, C.; Alard, A.; Aryankalayil, M.J.; Sarfraz, Y.; Diamond, J.M.; Schneider, R.J.; Inghirami, G.; Coleman, C.N.; Formenti, S.C.; Demaria, S. DNA exonuclease Trex1 regulates radiotherapy-induced tumour immunogenicity. *Nat. Commun.* **2017**, *8*, 15618. [CrossRef]
78. Du, F.; Tang, N.; Cui, Y.; Wang, W.; Zhang, Y.; Li, Z.; Li, J. A Novel Nomogram Model Based on Cone-Beam CT Radiomics Analysis Technology for Predicting Radiation Pneumonitis in Esophageal Cancer Patients Undergoing Radiotherapy. *Front. Oncol.* **2020**, *10*, 596013. [CrossRef]

79. Mostafaei, S.; Abdollahi, H.; Dehkordi, S.K.; Shiri, I.; Razzaghdoust, A.; Moghaddam, S.H.Z.; Saadipoor, A.; Koosha, F.; Cheraghi, S.; Mahdavi, S.R. CT imaging markers to improve radiation toxicity prediction in prostate cancer radiotherapy by stacking regression algorithm. *Radiol. Med.* **2020**, *125*, 87–97. [CrossRef]
80. Bourbonne, V.; Da-Ano, R.; Jaouen, V.; Lucia, F.; Dissaux, G.; Bert, J.; Pradier, O.; Visvikis, D.; Hatt, M.; Schick, U. Radiomics analysis of 3D dose distributions to predict toxicity of radiotherapy for lung cancer. *Radiother. Oncol.* **2021**, *155*, 144–150.
81. Lucia, F.; Bourbonne, V.; Visvikis, D.; Miranda, O.; Gujral, D.; Gouders, D.; Dissaux, G.; Pradier, O.; Tixier, F.; Jaouen, V.; et al. Radiomics Analysis of 3D Dose Distributions to Predict Toxicity of Radiotherapy for Cervical Cancer. *J. Pers. Med.* **2021**, *11*, 398. [PubMed]
82. Puisieux, M.F.; Pellat, A.; Assaf, A.; Ginestet, C.; Brezault, C.; Dhooge, M.; Soyer, P.; Coriat, R. Therapeutic Management of Advanced Hepatocellular Carcinoma: An Updated Review. *Cancers* **2022**, *14*, 2357.

cancers

MDPI

Article

Biomarkers of Tumor Heterogeneity in Glioblastoma Multiforme Cohort of TCGA

Garrett Winkelmaier [1], Brandon Koch [2], Skylar Bogardus [1], Alexander D. Borowsky [3] and Bahram Parvin [1,4,*]

1 Department of Electrical and Biomedical Engineering, College of Engineering, University of Nevada Reno, 1664 N. Virginia St., Reno, NV 89509, USA; gwinkelmaier@unr.edu
2 Department of Biostatics, College of Public Health, Ohio State University, 281 W. Lane Ave., Columbus, OH 43210, USA
3 Department of Pathology, UC Davis Comprehensive Cancer Center, University of California Davis, 1 Shields Ave, Davis, CA 95616, USA
4 Pennington Cancer Institute, Renown Health, Reno, NV 89502, USA
* Correspondence: bparvin@unr.edu

Simple Summary: Identifying biomarkers of survival from a large-scale cohort of Glioblastoma Multiforme (GBM) pathology images is hindered by heterogeneity of tumor signature compounded by age being the single most important confounder in predicting survival in GBM. The main contributions of this manuscript are to define (i) metrics for identifying tumor subtypes of tumor heterogeneity and (ii) relevant statistics for incorporating age for evaluating competing hypotheses. As a result, the GBM cohort are stratified based on interpretable morphometric features with or without preconditioning on published genomic subtypes.

Abstract: Tumor Whole Slide Images (WSI) are often heterogeneous, which hinders the discovery of biomarkers in the presence of confounding clinical factors. In this study, we present a pipeline for identifying biomarkers from the Glioblastoma Multiforme (GBM) cohort of WSIs from TCGA archive. The GBM cohort endures many technical artifacts while the discovery of GBM biomarkers is challenged because "age" is the single most confounding factor for predicting outcomes. The proposed approach relies on interpretable features (e.g., nuclear morphometric indices), effective similarity metrics for heterogeneity analysis, and robust statistics for identifying biomarkers. The pipeline first removes artifacts (e.g., pen marks) and partitions each WSI into patches for nuclear segmentation via an extended U-Net for subsequent quantitative representation. Given the variations in fixation and staining that can artificially modulate hematoxylin optical density (HOD), we extended Navab's Lab method to normalize images and reduce the impact of batch effects. The heterogeneity of each WSI is then represented either as probability density functions (PDF) per patient or as the composition of a dictionary predicted from the entire cohort of WSIs. For PDF- or dictionary-based methods, morphometric subtypes are constructed based on distances computed from optimal transport and linkage analysis or consensus clustering with Euclidean distances, respectively. For each inferred subtype, Kaplan–Meier and/or the Cox regression model are used to regress the survival time. Since age is the single most important confounder for predicting survival in GBM and there is an observed violation of the proportionality assumption in the Cox model, we use both age and age-squared coupled with the Likelihood ratio test and forest plots for evaluating competing statistics. Next, the PDF- and dictionary-based methods are combined to identify biomarkers that are predictive of survival. The combined model has the advantage of integrating global (e.g., cohort scale) and local (e.g., patient scale) attributes of morphometric heterogeneity, coupled with robust statistics, to reveal stable biomarkers. The results indicate that, after normalization of the GBM cohort, mean HOD, eccentricity, and cellularity are predictive of survival. Finally, we also stratified the GBM cohort as a function of EGFR expression and published genomic subtypes to reveal genomic-dependent morphometric biomarkers.

Keywords: Glioblastoma Multiforme; tumor heterogeneity; biomarker; whole slide imaging; TCGA

Citation: Winkelmaier, G.; Koch, B.; Bogardus, S.; Borowsky, A.D.; Parvin, B. Biomarkers of Tumor Heterogeneity in Glioblastoma Multiforme Cohort of TCGA. *Cancers* **2023**, *15*, 2387. https://doi.org/10.3390/cancers15082387

Academic Editor: Trever G. Bivona and Wei Wu

Received: 11 March 2023
Revised: 6 April 2023
Accepted: 14 April 2023
Published: 20 April 2023

1. Introduction

The tumor signature observed in whole slide imaging (WSI) is often heterogeneous, which reflects a complex gene expression program that is unique to each patient. Tumor heterogeneity (TH) can be based on distinct morphological and phenotypic profiles, such as morphology and gene expression. TH is also a strong factor in the tumor burden, with implications for patients' prognosis and treatment. The goal of our study is to investigate whether biomarkers of tumor heterogeneity can be captured based on computed nuclear indices and their organization in WSIs of the Glioblastoma Multiforme (GBM) dataset in The Cancer Genome Atlas (TCGA). GBM is a Grade IV cancer with a five-year survival rate of 9% [1], in which TH should play an important role. The genomic subtypes of GBM have been characterized [2], providing additional constraints for heterogeneity analysis. However, characterizing heterogeneity is not without challenges in a TCGA dataset, as there may be artifacts in WSI (e.g., pen marks), technical variations in sample preparation and staining, and computational strategies for associating heterogeneity to the outcome need to be developed. Furthermore, the confounding factor of age is the single most important variable in predicting survival in GBM. Therefore, any prediction of biomarkers must incorporate rigorous statistical criteria for validation. In fact, learning survival from histology images has been quite challenging, and coupling with CNN has continued to make incremental improvements [3].

Analysis of WSI has benefited from the integration of various technologies, including whole slide scanning, annotations, and filtering to remove artifacts. Recent advances in computational histopathology, which is based on cytological analysis (such as nuclear atypia and cellular density) and automated feature learning, have also contributed significantly to this field. These techniques enable classification, such as tumor grading and detection of micrometastasis, or association via regression to an outcome such as survival. Deep learning is now the preferred method for image-based analysis and representation [4]. For instance, cytological analyses use nuclear segmentation that extends U-Net [5,6] coupled with adversarial training [7], which is highly effective, particularly in identifying vesicular nuclear phenotypes [8] that traditional methods [9] could not detect. Although a thorough review of nuclear segmentation and feature-based representation for computational histopathology is beyond the scope of this manuscript, this article provides a summary of several studies focused on the analysis of low-grade glioma and GBM.

Mobadersany et al. [10] developed a pipeline for training a modified VGG19 model to learn features and associate them with survival using manually selected regions of interest from a WSI. Each image is assigned a risk vector for input to a Cox proportional hazards layer, which computes a loss function for model construction [3]. This approach was applied to diffused gliomas, resulting in a significant prognostic outcome. Chen et al. [11] integrated CNN for feature-based representation, graph-convolutional network following nuclear segmentation, and an attention mechanism to predict tumor grading or survival. Their pipeline was also applied to diffused gliomas with improved statistical analysis. Kong et al. [12] stratified patients based on transcriptomics and morphometric indices computed from tumor biopsy and histology, respectively. Zhang et al. [13] used multi-kernel learning to integrate histopathology and multi-OMIC data (e.g., gene expression, methylation) to perform prognostic tasks. However, the performance of these techniques is hindered as a result of (a) the absence of color normalization across the cohort, (b) not adjusting for age as a strong confounder, (c) failing to incorporate TH in predictive models, and (d) not addressing the sole special needs of GBM. Although some researchers have stratified GBM patients based on aggressive versus non-aggressive therapies and/or integration with molecular data [9,12], the strict utility of histology and TCGA clinical data, such as age, has been lacking. Moreover, from a translational perspective, it would be more valuable to predict the outcome from a low-cost histology section or, at most, coupled with one or two transcripts from an interpretative representation (e.g., nuclear chromatin content, cellularity). These are the issues we aim to investigate.

The overall process and an example of TH are shown in Figure 1. Each WSI underwent artifact and background removal, followed by segmentation of nuclei in each patch, and optical density-based normalization for feature extraction. An important step was to normalize the color images, based on an extension from Navab's Lab, to ensure optical density could be used as a biomarker. Computed indices were then combined to represent TH using two alternative strategies for computing subtypes. These subtypes were then analyzed rigorously using a likelihood ratio and forest plots to predict survival. The premise for using alternative representations is that differences in TH can be quantified, revealing phenotypes that are typically masked by averaging or other higher-order statistics. Section 2 outlines the computational methodologies for representing WSIs, while Section 3 lists statistically significant biomarkers predicted using morphometric analysis, both in the absence and presence of genomic subtypes. Finally, Section 4 provides additional insights into our findings and concludes the manuscript.

2. Methods

2.1. Preprocessing the WSIs

The GBM cohort contains not only technical variations due to sample preparation and staining, but also technical anomalies such as pen marks and out-of-focus images. Initially, we attempted to filter out pen marks using PyHist [14], which utilizes an edge filter and graph cut segmentation. However, this approach did not effectively identify many regions containing pen marks. Therefore, we chose to annotate a dataset of pen marks and create a support vector machine (SVM) classifier that identifies patches of 224-by-224 containing pen marks based on their concatenated RGB pixel histogram (i.e., a vector of 768 $\times$ 1). With 625 annotated images and a 90–10 training and testing split, we achieved a training accuracy of 97%. PyHist was effective in removing other artifacts (e.g., blur) and white regions (e.g., background), and we also used it to partition each WSI into 224-by-224 regions.

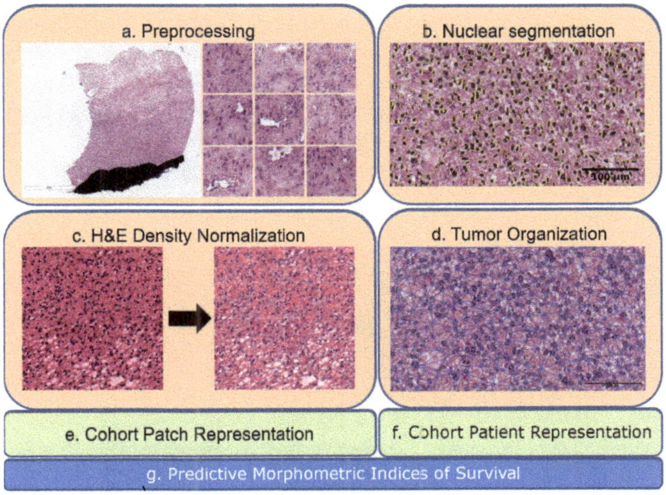

Figure 1. Eash WSI is represented in the context of tumor heterogeneity for biomarker discovery: (**a**) a WSI is partitioned to patches of 224-by-224, where each patch is analyzed for pen marks or other aberrations; (**b**) nuclei are segmented in patches; (**c**) H&E optical density is normalized in each patch; (**d**) nuclei organization is quantified in each patch; (**e**,**f**) computed indices from nuclei and their organizations are used for the dictionary- and PDF-based representations. (**g**) Predictive morphometric indices of survival are identified.

2.2. Nuclear Segmentation

Segmentation of nuclei in H&E stained images can be challenging due to technical variations such as sample preparation and staining, biological heterogeneity such as nuclear atypia and pleomorphism, and variations in Hematoxylin optical density (HOD) and texture such as vesicular phenotype. This important topic has been addressed by numerous researchers [6–8]. In our implementation, we modified an earlier approach for segmenting 3D organoids [15] using the U-Net architecture with a modified loss function that integrates a potential field for delineating touching nuclei. We annotated and extended H&E stained images from previous datasets [8,9] and complemented the data augmentation step with local/global contrast adjustment for nuclei/background optical density. For local contrast adjustment, we randomly selected between 10–20% of nuclei and modulated their color intensities, as well as the background regions. This step was crucial because the GBM cohort is diverse in terms of nuclear chromatin or protein contents, contributing to tumor heterogeneity. We started with 57 annotated training images, each no less than 1000-by-700 pixels. Following data augmentation, the sample size increased to 200,000 patches, each sized 224-by-224. Using the leave-one-out method, we computed an Aggregated Jaccard Index (AJI) [16] of 0.62 for the 57 annotated images.

2.3. Image Normalization

The TCGA histology cohort lacks standardization in terms of staining, which may not be a significant issue for preprocessing or segmentation, but can affect feature extraction. To address this, we proposed that an improved color normalization approach would yield a more reliable HOD and protein readout for biomarker discovery. We applied a state-of-the-art technique in color normalization [17] to normalize WSIs across the entire cohort, which involves incorporating an L1 regularization term in the loss function for non-negative matrix factorization (NMF) to estimate the source image's stain matrix, and then mapping the deconvolved image back into a RGB space using a target image's stain matrix. We also utilized the nuclear mask to aid in rapid convergence and ensure consistent ordering of the two stained channels following NMF. The color normalization method [17] maps each candidate H&E-stained image in the RGB space to a single target image that corresponds to the desired staining, after which we use NMF to estimate HOD. An example of color normalization is shown in Figure 2. In the results section, we compared this approach with a classical method for color decomposition [18] and found that rigorous normalization associates HOD as a statistically validated biomarker predictive of survival.

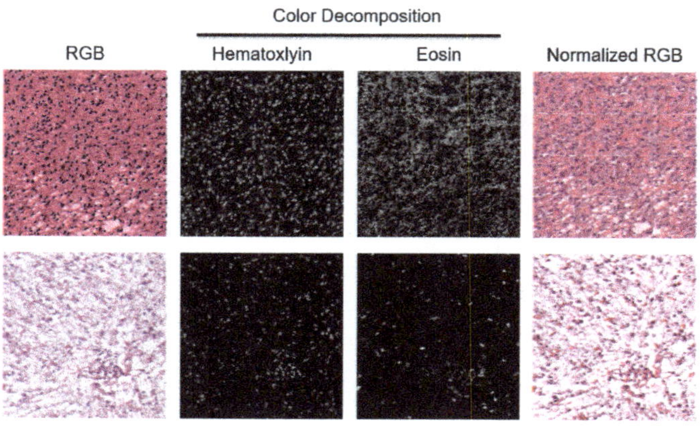

Figure 2. H&E stain is heterogeneous between patients. Two patches from two WSIs indicate a diverse staining signature. They are normalized for quantifying HOD and visualized in the RGB space.

2.4. Computation of the Morphometric Indices

A number of morphometric indices per nucleus were computed following color correction. Indices included were: nuclear size; HOD content; cellularity (a measure of cellular density computed from Delaunay Triangulation); eccentricity (e.g., elongation) (a measure of spindle geometry); and solidity (a measure of nuclear pleomorphism).

2.5. Association of the Morphometric Indices with Survival

Predicting biomarkers, based on heterogeneity, requires a representation and a distance metric for computing stable clusters or tumor subtypes. Each tumor subtype is then examined for whether it is predictive of survival.

2.5.1. Representation

Our motivation is to capture heterogeneity while maintaining reduced dimensionality, for example, using a single morphometric index at a time, in support of interpretation. To achieve this, we chose to represent each index either as a PDF or an ensemble of vocabularies. In the first case, a "cohort PDF" was constructed and binned for each morphometric index. Then, each patient's PDF was projected onto the cohort PDF, ensuring that each patient's PDF was on the same scale for computing distance. In the second case, the median of each computed index per patch and per WSI was first aggregated across the entire cohort to construct stable clusters. Once stable clusters (e.g., vocabularies) across the entire cohort were constructed, each WSI was represented in terms of the frequencies of each cluster.

2.5.2. Distance Metrics and Clustering

Clustering based on the dictionary- and PDF-based methods are summarized below. This is an important step since using the first-order statistics (e.g., mean) of the PDF did not reveal any significant biomarkers.

The dictionary method (e.g., alphabet), which involves using the Euclidean distance to compute distances between computed features, is advantageous due to its simplicity. To cluster data, pairwise distances were computed and consensus clustering was performed by varying the number of clusters from two to four. This particular implementation of consensus clustering injected noise in each iteration, which helped reveal more stable clusters. The clustering results based on optimal transport were visualized using a similarity matrix, cumulative density function (CDF), and Silhouette plots (a visual measure of the quality of clusters), as shown in Figure 3. Pinhole images of each cluster for the eccentricity and cellularity index were shown in Figures 4 and 5, respectively. For clustering, a random subset of 1000 samples was selected and k-means was iteratively performed on them. This process was repeated 10 times to obtain the centroids of the final clusters, which were determined by aggregating the median values corresponding to each sampled dataset for consensus clustering. The stability of each cluster was determined by the change in the CDF between the number of clusters and their silhouette scores. Subsequently, the learned alphabets were projected back into each WSI to create a patient signature based on the frequency of composition of each vocabulary, resulting in a vector (e.g., [0.34, 0.33, 0.33]) for three alphabets. This vector representation was then used as a continuous variable input to a Cox Hazard model to associate an increase in the percentage for each variable of the vector to a hazard ratio. Figure 6 illustrates this entire process.

Using the euclidean distance metric to measure the distance between the PDFs of two WSIs is inaccurate because it ignores the order of probabilities in a vector. In this study, we used the optimal transport method, also known as earth mover distance, to compute the distance between two PDFs. Optimal transport is a linear programming problem that we implemented using the Python Optimal Transport Toolbox. After computing pairwise distances between WSIs (i.e., PDFs), we performed linkage analysis [19,20] to reveal subpopulations. Figure 7 displays similarity matrices based on the optimal transport distance metric from PDF-based representations, and corresponding Kaplan–Meier curves (a probabilistic representation of a patient to survive up to a time) using three computed morphometric indices of nuclear size (e.g., area, left column) solidity (middle column), and total chromatin (right column).

2.5.3. Statistical Analysis of Morphometric Indices for Biomarker Validation

Various statistical techniques, such as Kaplan–Meier curves, can be used to evaluate the predictive strength of each morphometric index on survival. However, the association between a computed index and survival may be biased if there are unaccounted variables (e.g., age) that strongly predict survival but are not balanced among the clusters formed by an index. To avoid such bias, we combined the age confounder with one morphometric index at a time in a regression model. For PDF-based representations of patients, we estimated a Kaplan–Meier curve for each index and then used the Cox regression model to estimate the hazard ratio and its *p*-value. However, when age was included, we observed evidence of a violation of the proportionality assumption in the Cox model (a statistical model for survival outcome with at least one predictor) [21], as hazards were not proportional with a *p*-value of 0.01. By including both age and age-squared, we found no evidence of a violation of the proportionality assumption, and an improved *p*-value. Therefore, we chose to use the likelihood ratio with age and age-squared in all our analyses.

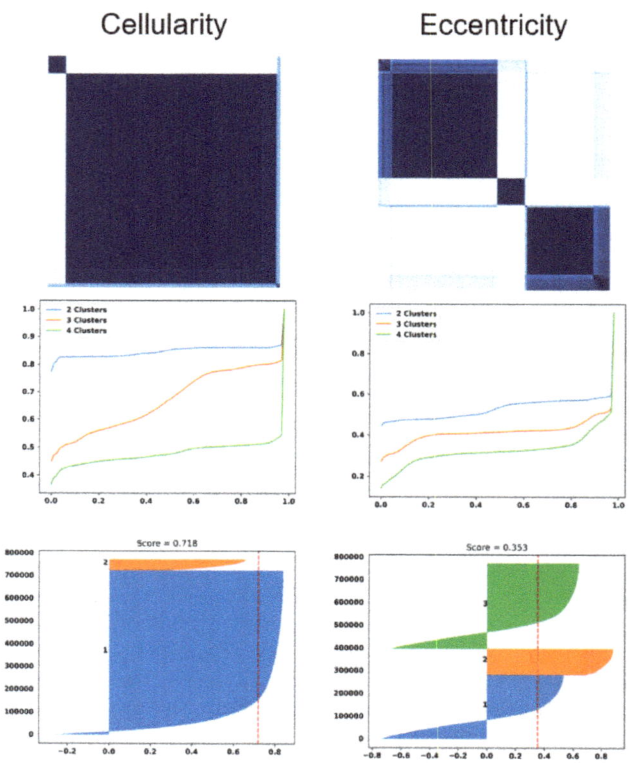

Figure 3. Dictionary-based learning identified two and three subpopulation (e.g., clusters) of patients based on cellularity and eccentricity indices, respectively. (top row): Computed similarity matrices; (middle row) the cumulative Density Function (CDF) of similarity matrices shows the quality of the number of clusters for each index (e.g., a flat horizontal line indicates a low number of misclassified samples between clusters). (bottom row) Silhouette plots of 800,000 randomly sampled nuclei show the similarity of patients within a cluster (e.g., a silhouette score less than 1) and a red dashed indicating the average silhouette score.

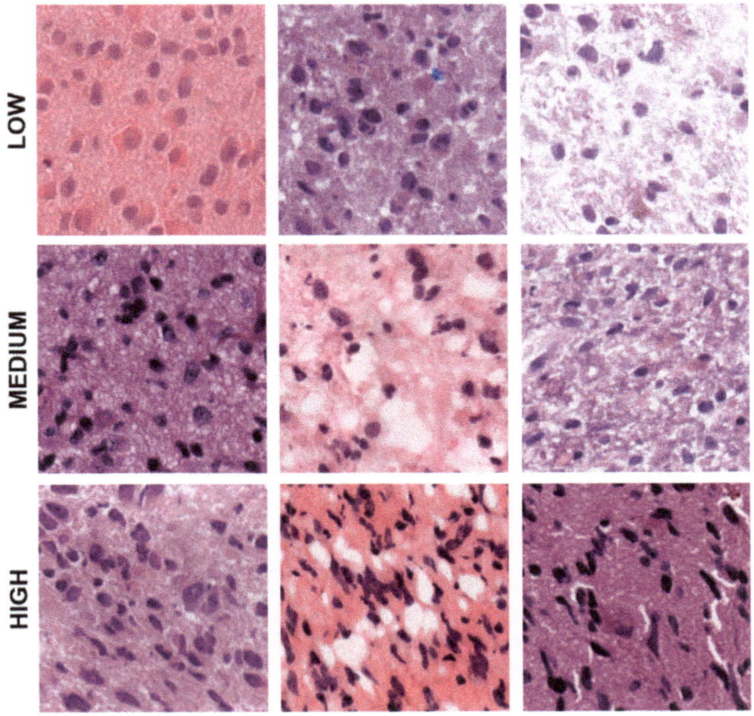

Figure 4. Representative patches showing low, medium, and high eccentricities corresponding to clusters 1, 2, and 3 from the dictionary-based method.

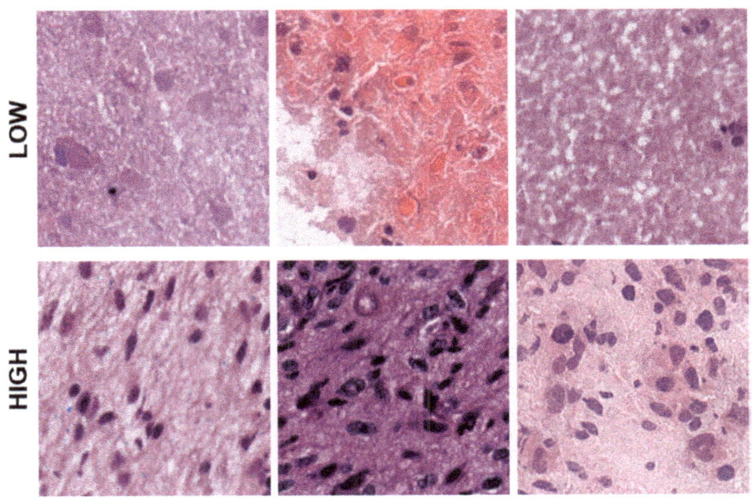

Figure 5. Representative patches showing low, and high cellularities corresponding to clusters 1 and 2 from the dictionary-method.

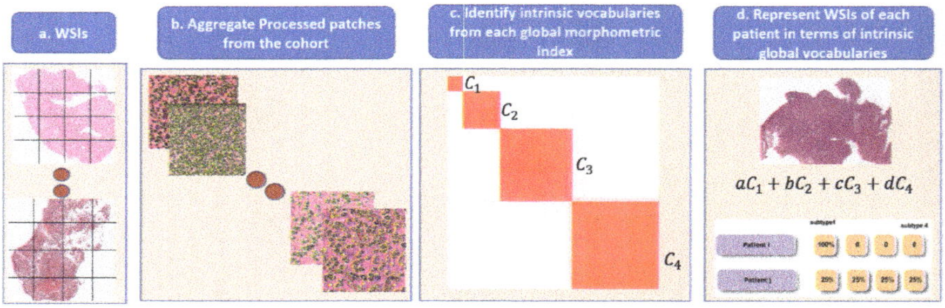

Figure 6. Steps in the dictionary-based method for representing heterogeneity: (**a**) each WSI is partitioned into patches; (**b**) each patch is quantified in terms of nuclear indices and organization; (**c**) each computed index (e.g., HOD content, nuclear size) is aggregated across the entire cohort for dictionary-based learning (e.g., alphabets, which are four in this example); and (**d**) each WSI is then represented as a composition of learned alphabets.

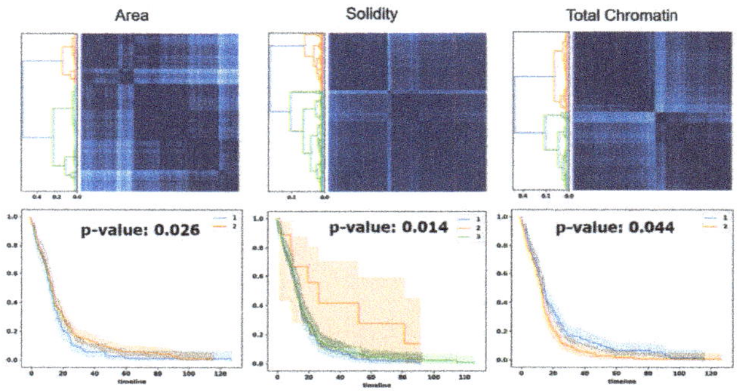

Figure 7. Optimal transport identifies subpopulations of patients, based on PDF representation, for survival analysis. Top row: similarity matrices identified by linkage analysis; Bottom row: Kaplan–Meier plots, hazard ratio, and computed *p*-values for three computed morphometric indices of nuclear size, solidity, and total chromatin.

The Likelihood-ratio test (LRT) is one of the three standard approaches for statistical hypothesis testing that evaluates the goodness of fit of two competing statistical models by comparing their likelihoods. In our study, we used the LRT to compare the Cox regression model, which included only age and age-squared (the "null" model), to the model that included age, age-squared, and the morphometric index (the "alternative" model). If adding an index, such as nuclear size, improved the model's fitness compared to the model without the index, the *p*-value of the likelihood ratio test would be small (<0.05), indicating evidence that the index is significantly associated with survival even after controlling for age. Conversely, if the index did not improve the model's fitness compared to using only age and age-squared, the likelihoods of the null and alternative models would be similar, and the likelihood ratio test would yield a large *p*-value (>0.05). When the LRT yields a small *p*-value, it provides evidence that the index is predictive of survival, but it does not provide information on the size of the effect or whether the range within the index is significantly different. Therefore, for each condition where the LRT *p*-value is less than 0.05, we also computed the 95% confidence intervals and tested pairwise differences between the hazard ratios corresponding to the levels of the variable of interest. In some cases, hazard ratios could not be estimated and were excluded from figures because either the patient's survival time was close to zero or censored.

2.5.4. Computing Resources

The machine learning models were trained on a local server, which was equipped with 8 NVIDIA GeForce RTX 2080 Ti GPUs, each with 12GB of RAM, 256 GB of RAM, and a 64-core CPU. The model development and validation were performed using python 3 and the TensorFlow 2.2 framework. The source code has been made available at https://github.com/gwinkelmaier/GBM-biomarkers (accessed date 31 March 2023).

3. Results

3.1. Biomarker Discovery

3.1.1. Biomarkers of Nuclear Morphometric Indices

Tables 1 display age-adjusted biomarkers based on PDF- and dictionary-based representations, respectively. The PDF representation includes biomarkers such as average chromatin content (e.g., HOD), nuclear size, solidity, and total chromatin (e.g., total HOD) per nucleus. Notably, our results show that the method in [17], from Navab's Lab, yielded a statistically significant biomarker approximation of chromatin content, whereas the classical method based on known densities [18] did not. Additionally, the dictionary method uncovered eccentricity (e.g., elongation) and cellularity. Figure 8 illustrates the forest plots for predicted morphometric indices based on PDF-based methods without preconditioning. Finally, Table 2 displays predicted morphometric indices from both PDF and dictionary-based representations. This was achieved by (a) estimating the parameters of a Cox-Hazard model by integrating morphometric indices from both representations and (b) comparing the learned model with the baseline model of only age and age-squared, leading directly to a p-value.

Table 1. Predicted morphometric biomarkers and their p-values from patients in the TCGA-GBM cohort.

	Morphometric Index	Number of Clusters	p-Value
(**a**) PDF model			
	Area	2	0.026
	Area	3	0.016
	Area	4	0.013
	Mean HOD	3	0.016
	Mean HOD	4	0.006
	Solidity	3	0.014
	Solidity	4	0.007
	Total HOD	2	0.044
	Total HOD	3	0.037
	Total HOD	4	0.025
(**b**) Dictionary model			
	Cellularity	2	0.008
	Cellularity	3	0.040
	Eccentricity	2	0.002
	Eccentricity	3	0.005
	Eccentricity	4	0.011
	Mean HOD	2	0.019

Table 2. Predicted morphometric biomarkers and their *p*-values for the combined model without genomic preconditioning.

Nuclear Morphometric Index	Number of Clusters	*p*-Value
Cellularity	2	0.025
Eccentricity	2	0.007
Eccentricity	3	0.013
Eccentricity	4	0.019
Mean HOD	4	0.028

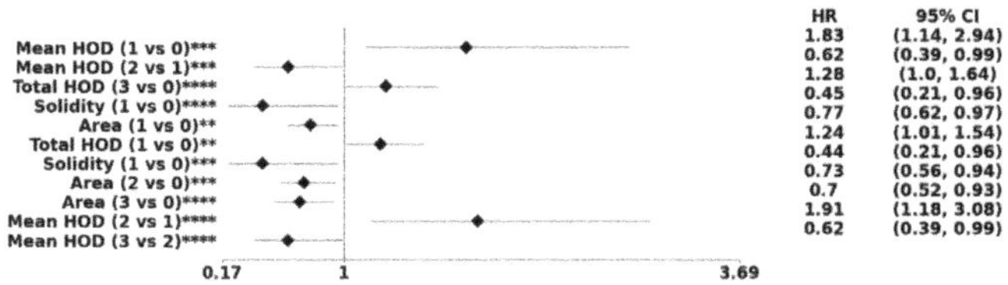

Figure 8. The forest plot indicates biomarkers associated with the subpopulation at risk using the PDF-based representation without any genomic preconditioning. The asterisks **, ***, and **** denote the number of stratifications per morphometric index.

3.1.2. Biomarkers of Morphometric Indices Preconditioned on Genomics Signature

The same set of tests from the previous section was applied following stratification based on published genomics subtypes [2] or EGFR expression (e.g., high versus low expression rendered by linkage analysis).

There are four significant genomic subtypes: mesenchymal, proneural, neural, and classical. Tables 3 display age-adjusted biomarkers based on these subtypes, computed from both PDF- and dictionary-based representations. The corresponding forest plot for the classical subtype in the PDF-based representation is shown in Figure 9, and Table 4 shows predicted biomarkers from the combined representation. For instance, nuclear size is a biomarker for the neural subtype, total HOD is a biomarker for the classical subtype, and solidity is a biomarker for the neural and mesenchymal subtypes. Therefore, there is evidence that each genomic subtype can highlight specific biomarkers, leading to further stratification of the patient population.

Table 3. Predicted morphometric biomarkers for the PDF- and dictionary-based models preconditioned on genomic subtypes.

Nuclear Morphometric Index	Number of Clusters	*p*-Value			
		Neural	Proneural	Mesenchymal	Classical
(**a**) PDF method					
Area	2	0.021	-	-	-
Area	3	0.020	-	-	0.009
Area	4	0.018	-	-	0.006
Mean HOD	4	0.024	-	-	-
Solidity	3	0.006	-	-	-
Solidity	4	<0.001	-	0.009	-
Total HOD	2	-	-	-	0.019
Total HOD	3	-	-	-	0.008
Total HOD	4	-	-	-	0.008
(**b**) Dictionary method					
Area	4	-	-	-	0.040
Total HOD	2	<0.001	-	-	-
Total HOD	3	0.008	-	-	-
Total HOD	4	0.003	-	-	-

Table 4. Predicted morphometric biomarkers and their p-values for the combined model preconditioned on genomic subtypes.

Nuclear Morphometric Index	Number of Clusters	*p*-Value			
		Neural	Proneural	Mesenchymal	Classical
Area	2	0.04	-	-	-
Area	4	-	-	-	0.043
Mean HOD	4	-	0.031	-	-
Solidity	3	0.010	-	-	-
Solidity	4	0.004	-	0.048	-
Total HOD	2	0.001	-	-	-
Total HOD	3	0.012	-	-	-
Total HOD	4	0.004	-	-	0.036

Aberrant overexpression of EGFR is a dominant feature of GBM. As a result, patient data were initially stratified based on low and high EGFR expression, followed by the proposed analysis as described in the Methods section. Tables 5 and 6 present predicted biomarkers for the subpopulation of patients with high or low EGFR expression. Table 7 shows predicted biomarkers based on combined models of the PDF- and dictionary-based method preconditioned on EGFR expression. Note that Table 7 is quite similar to Table 2 as only a subset of patients with matched transcriptome data was used in this analysis. Additionally, Tables 7 do not share a morphometric index, which serves as an internal control. The corresponding forest plot of morphometric biomakers, computed from the PDF representation and preconditioned on low EGFR expression, is shown in Figure 10.

Table 5. Predicted morphometric biomarkers for the PDF- and dictionary-based models preconditioned on patients with high EGFR expression.

	Nuclear Morphometric Index	Number of Clusters	*p*-Value
(**a**) PDF model			
	Total HOD	4	0.048
(**b**) Dictionary model			
	Area	2	0.007
	Area	3	0.009
	Area	4	0.025

Table 6. Predicted morphometric biomarkers for the PDF- and dictionary-based models preconditioned on patients with low EGFR expression.

	Nuclear Morphometric Index	Number of Clusters	*p*-Value
(**a**) PDF model			
	Area	4	0.031
	Cellularity	4	0.018
(**b**) Dictionary model			
	Cellularity	2	0.035
	Total HOD	2	0.001
	Total HOD	3	0.003
	Total HOD	4	0.009

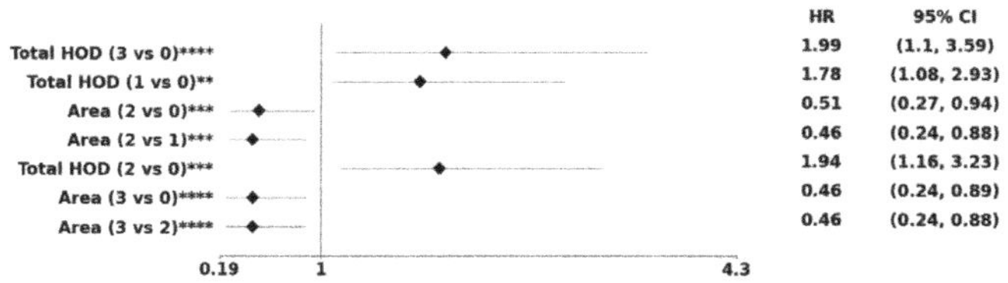

Figure 9. Using the PDF method, pre-conditioned on the classical subtype, the forest plot indicates the subpopulation at risk. The asterisks **, ***, and **** denote the number of stratifications per morphometric index.

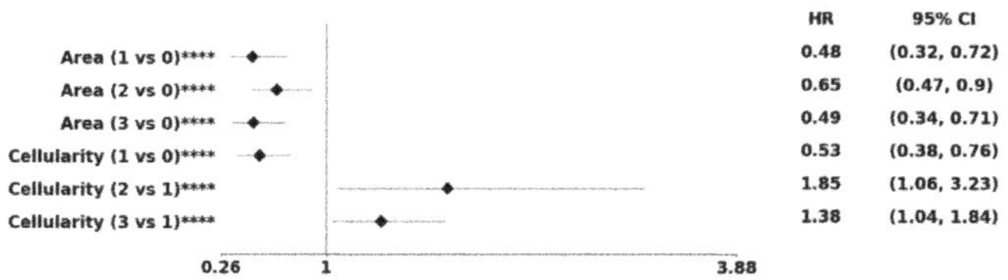

Figure 10. Using the PDF method, pre-conditioned on a high EGFR expression, the forest plot indicates the subpopulation at risk. For example, Area cluster two has an 52% decreased risk of death compared to Area cluster zero. The asterisks **** denote the number of stratifications per morphometric index.

Table 7. Predicted morphometric biomarkers and their *p*-values for the combined model preconditioned on the EGFR transcript.

Nuclear Morphometric Index	Number of Clusters	*p*-Value
(**a**) Biomarkers for patients with matched transcriptome data		
Cellularity	2	0.047
Cellularity	3	0.033
Cellularity	4	0.019
Eccentricity	2	0.040
Mean HOD	3	0.010
Mean HOD	4	0.004
(**b**) Biomarkers of patients stratified with high EGFR expression		
Area	3	0.004
Area	4	0.005
Cellularity	3	0.031
Cellularity	4	0.009
(**c**) Biomarkers of patients with low EGFR expression		
Cellularity	2	0.034
Cellularity	4	0.018
Mean HOD	3	0.015
Mean HOD	4	0.021
Total HOD	2	0.001
Total HOD	3	0.002

4. Discussion

This manuscript presents our extended and applied methodologies for identifying biomarkers in GBM while taking into account tumor heterogeneity. Our approach involves using age-adjusted representations of nuclear morphometric features or their organization in WSIs and utilizing linear associations to improve interpretation while reducing the number of parameters. We suggest that incorporating the Cox Hazard Model in the loss function (as done in [22]) increases the likelihood of finding associations by noise or chance.

Therefore, we advocate for using linear associations instead, as they offer simplicity and using a single computed index at a time improves interpretability and robustness. Lastly, we statistically explored viable patient pathology stratifications by preconditioning on either EGFR expression or genomic subtypes.

The methodological innovations used in this study included image normalization, representation and distance metrics, integration of age as a confounder, and the utility of age-squared to satisfy the proportionality assumption of statistical models. Image normalization is an important step in using HOD as a biomarker because of technical variations in sample preparation and staining. Navab's Lab provided the foundation for normalizing each patch in a WSI to a reference template. NMF enabled the readout of the HOD per nucleus using the nuclear mask computed from the segmentation step. Nuclear masks provided the required initialization for NMF, making convergence rapid without needing a ranking based on the blue channel. Alternatively, image normalization based on classical color deconvolution [18] did not reveal HOD as a biomarker. Because tumors are heterogeneous, the study designed representations based on the PDF or dictionary-based method for each morphometric index in WSIs. The PDF method represents each WSI in terms of its own nuclear morphometric architecture. In contrast, the dictionary-based method represents each WSI in terms of learned alphabets that represent the entire cohort. The PDF method identifies biomarkers that are globally persistent within a WSI because attributes with a low frequency of occurrence can diffuse with the PDF representation. The two methods are complementary. While the distance measures for the dictionary-based method can be Euclidean, a distance measure based on optimal transport is introduced to compute distances between pairwise PDFs followed by linkage analysis. In GBM, age is the single most important predictor of the outcome. In the absence of the utility of age as a confounder, many biomarkers are either erroneously predicted or the proportionality assumption of the Cox Hazard model is violated. However, by using both age and age-squared, a more rigorous statistical analysis can be achieved. For example, since both the dictionary- and PDF-based methods predicted the age-adjusted HOD index as a biomarker, this index is more stable. A larger value of HOD content, a surrogate index for nuclear hyperchromasia, is consistently associated with a higher hazard ratio, which could be due to a higher rate of proliferation. Using the dictionary-based method, increased cellularity (e.g., hypercellularity) is also associated with a higher rate of proliferation and a higher HR. Another biomarker in the PDF method is solidity, which is a surrogate index for pleomorphism. Figure 8 suggests that lower pleomorphism corresponds to a better HR, where higher pleomorphism is associated with a higher tumor grade and worse prognosis. These observations are consistent with key diagnostic points in GBM, including cytological criteria of astrocytoma (e.g., GFAP, spindle-shape nuclei) and anaplasia (e.g., hypercellularity, pleomorphism, nuclear hyperchromasia) [23].

In conclusion, the computational pipeline has stratified tumor heterogeneity within the TCGA GBM cohort and has identified interpretable biomarkers that align with GBM diagnostic criteria [23]. This pipeline provides a valuable platform for evaluating emerging therapies for the treatment GBM patients. The same pipeline can also be applied to other tumor types. For example, it can be applied to WSI collected from Low-Grade Glioma (LGG) because of the similarities of the microanatomy. However, it may require additional extensions if it is to be applied to other organs, such as the breast or pancreas with glandular structures. In these types of organs, normal gland structures and stromal regions need to be delineated prior to the analysis of the tumor regions. This step requires another computational module that can be facilitated by annotation and training of the corresponding regions of the microanatomy.

Author Contributions: B.P. conceived the study and acquired funding. B.P., G.W. and B.K. wrote the paper. G.W. developed the pipeline for the analysis of whole slide images and multiparametric stratification of patient data. B.K. performed detailed statistical analysis for biomarker discovery. S.B. contributed to the quality control and validation. A.D.B. was the project pathologist. All authors have read and agreed to the published version of the manuscript.

Funding: This research was funded by NIH grant number 1R15CA23543.

Institutional Review Board Statement: Not applicable.

Informed Consent Statement: Not applicable.

Data Availability Statement: https://github.com/gwinkelmaier/GBM-biomarkers (accessed on 31 March 2023) contains code and newly annotated images.

Acknowledgments: The authors thank Terrance Speed for insightful exchanges on statistical validations.

Conflicts of Interest: The authors declare no conflict of interest.

References

1. Ostrom, Q.T.; Cioffi, G.; Waite, K.; Kruchko, C.; Barnholtz-Sloan, J.S. CBTRUS statistical report: Primary brain and other central nervous system tumors diagnosed in the United States in 2014–2018. *Neuro-Oncology* **2021**, *23*, iii1–iii105. [CrossRef] [PubMed]
2. Verhaak, R.G.; Hoadley, K.A.; Purdom, E.; Wang, V.; Qi, Y.; Wilkerson, M.D.; Miller, C.R.; Ding, L.; Golub, T.; Mesirov, J.P.; et al. Integrated genomic analysis identifies clinically relevant subtypes of glioblastoma characterized by abnormalities in PDGFRA, IDH1, EGFR, and NF1. *Cancer Cell* **2010**, *17*, 98–110. [CrossRef] [PubMed]
3. Zhu, X.; Yao, J.; Huang, J. Deep convolutional neural network for survival analysis with pathological images. In Proceedings of the 2016 IEEE International Conference on Bioinformatics and Biomedicine (BIBM), Shenzhen, China, 15–18 December 2016; IEEE: Piscataway, NJ, USA, 2016; pp. 544–547.
4. Lin, H.; Chen, H.; Graham, S.; Dou, Q.; Rajpoot, N.; Heng, P.A. Fast scannet: Fast and dense analysis of multi-gigapixel whole-slide images for cancer metastasis detection. *IEEE Trans. Med. Imaging* **2019**, *38*, 1948–1958. [CrossRef] [PubMed]
5. Jung, H.; Lodhi, B.; Kang, J. An automatic nuclei segmentation method based on deep convolutional neural networks for histopathology images. *BMC Biomed. Eng.* **2019**, *1*, 1–12. [CrossRef] [PubMed]
6. Xing, F.; Yang, L. Robust nucleus/cell detection and segmentation in digital pathology and microscopy images: A comprehensive review. *IEEE Rev. Biomed. Eng.* **2016**, *9*, 234–263. [CrossRef] [PubMed]
7. Mahmood, F.; Borders, D.; Chen, R.J.; McKay, G.N.; Salimian, K.J.; Baras, A.; Durr, N.J. Deep adversarial training for multi-organ nuclei segmentation in histopathology images. *IEEE Trans. Med. Imaging* **2019**, *39*, 3257–3267. [CrossRef] [PubMed]
8. Khoshdeli, M.; Winkelmaier, G.; Parvin, B. Fusion of encoder-decoder deep networks improves delineation of multiple nuclear phenotypes. *BMC Bioinform.* **2018**, *19*, 1–11. [CrossRef] [PubMed]
9. Chang, H.; Han, J.; Borowsky, A.; Loss, L.; Gray, J.W.; Spellman, P.T.; Parvin, B. Invariant delineation of nuclear architecture in glioblastoma multiforme for clinical and molecular association. *IEEE Trans. Med Imaging* **2012**, *32*, 670–682. [CrossRef] [PubMed]
10. Mobadersany, P.; Yousefi, S.; Amgad, M.; Gutman, D.A.; Barnholtz-Sloan, J.S.; Velázquez Vega, J.E.; Brat, D.J.; Cooper, L.A. Predicting cancer outcomes from histology and genomics using convolutional networks. *Proc. Natl. Acad. Sci. USA* **2018**, *115*, E2970–E2979. [CrossRef] [PubMed]
11. Chen, R.J.; Lu, M.Y.; Wang, J.; Williamson, D.F.; Rodig, S.J.; Lindeman, N.I.; Mahmood, F. Pathomic fusion: An integrated framework for fusing histopathology and genomic features for cancer diagnosis and prognosis. *IEEE Trans. Med. Imaging* **2020**, *41*, 757–770. [CrossRef] [PubMed]
12. Kong, J.; Cooper, L.A.; Wang, F.; Gutman, D.A.; Gao, J.; Chisolm, C.; Sharma, A.; Pan, T.; Van Meir, E.G.; Kurc, T.M.; et al. Integrative, multimodal analysis of glioblastoma using TCGA molecular data, pathology images, and clinical outcomes. *IEEE Trans. Biomed. Eng.* **2011**, *58*, 3469–3474. [CrossRef] [PubMed]
13. Zhang, Y.; Li, A.; He, J.; Wang, M. A novel MKL method for GBM prognosis prediction by integrating histopathological image and multi-omics data. *IEEE J. Biomed. Health Inform.* **2019**, *24*, 171–179. [CrossRef] [PubMed]
14. Muñoz-Aguirre, M.; Ntasis, V.F.; Rojas, S.; Guigó, R. PyHIST: A histological image segmentation tool. *PLoS Comput. Biol.* **2020**, *16*, e1008349. [CrossRef] [PubMed]
15. Winkelmaier, G.; Parvin, B. An enhanced loss function simplifies the deep learning model for characterizing the 3D organoid models. *Bioinformatics* **2021**, *37*, 3084–3085. [CrossRef] [PubMed]
16. Kumar, N.; Verma, R.; Sharma, S.; Bhargava, S.; Vahadane, A.; Sethi, A. A dataset and a technique for generalized nuclear segmentation for computational pathology. *IEEE Trans. Med. Imaging* **2017**, *36*, 1550–1560. [CrossRef] [PubMed]
17. Vahadane, A.; Peng, T.; Sethi, A.; Albarqouni, S.; Wang, L.; Baust, M.; Steiger, K.; Schlitter, A.M.; Esposito, I.; Navab, N. Structure-preserving color normalization and sparse stain separation for histological images. *IEEE Trans. Med. Imaging* **2016**, *35*, 1962–1971. [CrossRef] [PubMed]
18. Ruifrok, A.C.; Katz, R.L.; Johnston, D.A. Comparison of quantification of histochemical staining by hue-saturation-intensity (HSI) transformation and color-deconvolution. *Appl. Immunohistochem. Mol. Morphol.* **2003**, *11*, 85–91. [CrossRef] [PubMed]
19. Müllner, D. Modern hierarchical, agglomerative clustering algorithms. *arXiv* **2011**, arXiv:1109.2378.
20. Bar-Joseph, Z.; Gifford, D.K.; Jaakkola, T.S. Fast optimal leaf ordering for hierarchical clustering. *Bioinformatics* **2001**, *17*, S22–S29. [CrossRef] [PubMed]
21. Grambsch, P.M.; Therneau, T.M. Proportional hazards tests and diagnostics based on weighted residuals. *Biometrika* **1994**, *81*, 515–526. [CrossRef]

22. Wijethilake, N.; Meedeniya, D.; Chitraranjan, C.; Perera, I. Survival prediction and risk estimation of Glioma patients using mRNA expressions. In Proceedings of the 2020 IEEE 20th International Conference on Bioinformatics and Bioengineering (BIBE), Cincinnati, OH, USA, 26–28 October 2020; IEEE: Piscataway, NJ, USA; 2020, pp. 35–42.
23. Dabbs, D.J. *Diagnostic Immunohistochemistry E-Book: Theranostic and Genomic Applications*; Elsevier Health Sciences: Amsterdam, The Netherlands, 2021.

cancers

MDPI

Article

CDC20 Is Regulated by the Histone Methyltransferase, KMT5A, in Castration-Resistant Prostate Cancer

Zainab A. H. Alebady [1,2], Mahsa Azizyan [1], Sirintra Nakjang [3], Emma Lishman-Walker [1], Dhuha Al-Kharaif [4], Scott Walker [5], Hui Xian Choo [1], Rebecca Garnham [1], Emma Scott [1], Katya L. Johnson [1], Craig N. Robson [6] and Kelly Coffey [1,*]

1 Biosciences Institute, Newcastle Cancer Centre, Newcastle University, Newcastle upon Tyne NE2 4HH, UK
2 Department of Laboratory and Clinical Science, College of Pharmacy, University of AL-Qadisiyah, Al-Diwaniya 58002, Iraq
3 Bioinformatics Support Unit, Newcastle University, Newcastle NE2 4HH, UK
4 Medical Laboratory Technology Department, College of Health Sciences, Public Authority of Applied Education and Training, Safat 13092, Kuwait
5 School of Medicine, Newcastle University, Newcastle upon Tyne NE2 4HH, UK
6 Translational and Clinical Research Institute, Newcastle Cancer Centre, Newcastle University, Newcastle upon Tyne NE2 4HH, UK
* Correspondence: kelly.coffey@newcastle.ac.uk

Simple Summary: The methyltransferase KMT5A is suggested as an oncogene in prostate cancer but the mechanisms underlying its oncogenic properties are poorly understood. This study uncovers genes and cellular pathways which are regulated by KMT5A in prostate cancer to obtain a better understanding of whether or not therapeutic targeting is viable. In particular, we focus on the key cell cycle protein, CDC20, which we reveal to be a KMT5A-regulated gene via two mechanisms; 1. the methylation of histone H4K20 within the *CDC20* promoter to enhance *CDC20* transcription and 2. the inhibition of p53 via direct methylation to release *CDC20* transcriptional repression. Furthermore, we demonstrate that *KMT5A* and *CDC20* are positively correlated in clinical samples of prostate cancer. Due to the roles that KMT5A and CDC20 play in cell cycle regulation and DNA repair processes, we propose that targeting the methylation activity of KMT5A will provide therapeutic benefits where these two oncogenic proteins are overexpressed.

Abstract: The methyltransferase KMT5A has been proposed as an oncogene in prostate cancer and therefore represents a putative therapeutic target. To confirm this hypothesis, we have performed a microarray study on a prostate cancer cell line model of androgen independence following KMT5A knockdown in the presence of the transcriptionally active androgen receptor (AR) to understand which genes and cellular processes are regulated by KMT5A in the presence of an active AR. We observed that 301 genes were down-regulated whilst 408 were up-regulated when KMT5A expression was reduced. KEGG pathway and gene ontology analysis revealed that apoptosis and DNA damage signalling were up-regulated in response to KMT5A knockdown whilst protein folding and RNA splicing were down-regulated. Under these conditions, the top non-AR regulated gene was found to be CDC20, a key regulator of the spindle assembly checkpoint with an oncogenic role in several cancer types. Further investigation revealed that KMT5A regulates CDC20 in a methyltransferase-dependent manner to modulate histone H4K20 methylation within its promoter region and indirectly via the p53 signalling pathway. A positive correlation between KMT5A and CDC20 expression was also observed in clinical prostate cancer samples, further supporting this association. Therefore, we conclude that KMT5A is a valid therapeutic target for the treatment of prostate cancer and CDC20 could potentially be utilised as a biomarker for effective therapeutic targeting.

Keywords: CDC20; biomarker; KMT5A; p53; prostate cancer

Citation: Alebady, Z.A.H.; Azizyan, M.; Nakjang, S.; Lishman-Walker, E.; Al-Kharaif, D.; Walker, S.; Choo, H.X.; Garnham, R.; Scott, E.; Johnson, K.L.; et al. CDC20 Is Regulated by the Histone Methyltransferase, KMT5A, in Castration-Resistant Prostate Cancer. *Cancers* **2023**, *15*, 3597. https://doi.org/10.3390/cancers15143597

Academic Editors: Trever G. Bivona and Wei Wu

Received: 26 May 2023
Revised: 6 July 2023
Accepted: 11 July 2023
Published: 13 July 2023

1. Introduction

Prostate cancer is the most common cancer in men in the UK. Whilst androgen receptor (AR)-targeting therapies have yielded significant patient benefits, relapse to treatment is a significant clinical problem. Hence, there is an urgent need to develop alternative therapeutics to treat advanced disease. The lysine methyltransferase, KMT5A, plays an oncogenic role in a number of cancers [1–3]. Indeed, *KMT5A* siRNA-mediated knockdown inhibits prostate cancer cell proliferation and KMT5A has been identified as an AR-interacting protein that is required for the transcription of the AR-regulated gene, *prostate specific-antigen (PSA)*, via the promotion of mono-methylation on histone H4 at lysine 20 (H4K20Me1) at the *PSA* promoter [4]. Furthermore, KMT5A plays a role in the epithelial–mesenchymal transition (EMT) and enhances the invasiveness of prostate cancer cell line models, independent of the AR through its interplay with ZEB1 [5]. Initially identified as the sole methyltransferase responsible for H4K20Me1, KMT5A was subsequently shown to methylate numerous other non-histone proteins, including p53 [6]. A greater understanding of KMT5A in the context of prostate cancer is required to determine whether or not it is a bona fide therapeutic target.

KMT5A activity is regulated via post-translational mechanisms during specific phases of the cell cycle. During the late S phase and at the G2/M transition, the levels of KMT5A are at their peak and found localised to mitotic chromosomes. As the cell moves through prophase to anaphase, KMT5A is phosphorylated at serine 29 by cdk1/cyclin B. This results in KMT5A dissociation from chromatin and stabilisation via the inhibition of KMT5A association with the APC^{cdh1} E3 ubiquitin ligase [7]. During anaphase, KMT5A is dephosphorylated by cdc14a/b, which in turn permits protein turnover to reduce KMT5A protein levels at G1. During G1, KMT5A levels are sustained, however, during the G1/S transition, SCF^{skp2} ubiquitin ligase targets KMT5A for protein turnover resulting in undetectable KMT5A protein. Interestingly, KMT5A interacts with the proliferating cell nuclear antigen (PCNA) at DNA replication foci and is essential for correct DNA replication [8] suggesting a high turnover rate of chromatin bound KMT5A by $CRL4^{cdt2}$ [7]. The alterations in the levels of KMT5A throughout the cell cycle are mirrored by H4K20Me1 levels suggesting that methyltransferase activity is predominantly regulated by cellular KMT5A levels.

Cell cycle division 20 homologue (CDC20) is a cell cycle regulatory protein implicated in the spindle assembly checkpoint (SAC) and is required for cells to progress through mitosis. Specifically, CDC20 functions as a substrate recognition molecule and activator of APC to result in the ubiquitin-mediated turnover of its substrates. In particular, APC^{CDC20} functions during metaphase to anaphase to result in the destruction of cyclin B and securin, thereby allowing sister chromatids to segregate. CDC20 activity is inhibited by the mitotic checkpoint complex (MCC) and is only released to target its substrates once microtubule binding to the kinetochore and appropriate tension is achieved, thereby preventing genomic instability. Interestingly, there are suggestions that CDC20 may play a role in the DNA damage repair pathway via RAP80 [9] and REV1 [10] down-regulation. Furthermore, DNA damage-induced p53 can directly inhibit the expression of *CDC20* by associating with the *CDC20* promoter region and causing chromatin remodelling [11]. In addition, p21 can inhibit *CDC20* mRNA by associating with CDE-CHR elements in the *CDC20* promoter [12]. The depletion of *PHF8*, an H4K20Me1 demethylase, results in prolonged G2 and defective mitosis and it is itself a substrate of APC^{CDC20} [13] further suggesting that chromatin remodelling can be influenced by CDC20 levels.

CDC20 has been proposed to exhibit an oncogenic role in a number of cancers including prostate cancer [14]. Indeed, biochemical recurrence-free survival is lower in patients with high levels of *CDC20* compared to patients with low *CDC20* expression [15]. CDC20 itself is a target for ubiquitination by the E3 ligase SPOP, which is commonly mutated and non-functional in prostate cancers, providing an explanation for elevated CDC20 levels [16]. Furthermore, *CDC20* expression is associated with resistance to docetaxel [15,17] and is implicated in the wnt/Beta-catenin pathway which is oncogenic in advanced prostate cancer [17,18].

The aim of this study was two-fold; the first aim was to use pathway analysis to provide further evidence that KMT5A regulates oncogenic pathways and is a valid therapeutic target in prostate cancer and the second was to identify individual genes that are regulated by KMT5A in a model of castration-resistant prostate cancer as potential biomarkers for KMT5A activity. Indeed, we show that a number of oncogenic pathways are down-regulated upon *KMT5A* knockdown and we identified and validated *CDC20* as a KMT5A-regulated gene.

2. Materials and Methods

2.1. Antibodies

Antibodies used in this study included KMT5A (cell signalling), CDC20 (Ab190711, and AbCam), PARP1/2 (clone H250, sc-7150, Santa Cruz Biotechnology, Dallas, TX, USA), MDM2 (Clone N-20, sc-813, Santa Cruz Biotechnology), p21 (ab-4, Calbiochem), p53 (pAb-421#OP03, Calbiochem), p53-S15-P (cell signalling), p53-K382-Ac (ab75754, AbCam), H4K20Me1 (Ab9051, AbCam), H4 (07-108, Merck, Darmstadt, Germany), anti-phospho-histone H2AX (Ser139) (clone JBW301, Millipore Corp., Burlington, MA, USA) α-tubulin (clone DM1A, T9026, Sigma, St. Louis, MO, USA), and GAPDH (clone 1E6D9, Proteintech, Rosemont, IL, USA).

2.2. Compounds

Dihydrotestosterone (DHT) (Sigma) was prepared in ethanol at a final concentration of 10 mM and stored at −80 °C. KMT5A inhibitors UNC0379 (S7570, Selleckchem, Houston, TX, USA) and Ryuvidine (2609, R&D Systems, Minneapolis, MN, USA) were purchased in powder form and resuspended in DMSO to a final concentration of 50 mM and 20 mM, respectively. Solutions were stored at −80 °C for no longer than 1 month. Nutlin 3 was provided by Prof. John Lunec (Newcastle Cancer Centre).

2.3. Cell Culture

LNCaP cells, a model of androgen dependence, and AR negative PC3 cells were purchased from American Type Culture Collection (Manassas, VA, USA); LNCaP-AI cells, a model of androgen independence, were generated in-house as described previously [19]. Cells were maintained as previously described [20].

Short tandem repeat profiling was used to authenticate the cell lines used in this study (NewGene, Newcastle upon Tyne, UK). MycoAlert (Lonza, UK) was used to routinely test for the presence of mycoplasma.

2.4. siRNA

The reverse transfection of cell lines with siRNA sequences (25 nM) was carried out using Lipofectamine RNAiMAX (Invitrogen) in accordance with the manufacturer's protocol. Either qPCR or Western blotting confirmed successful knockdown. Non-silencing (N/S): UUCUCCGAACGUGUCACGU[dT][dT]; siKMT5A_1: CCAUGAAGUCCGAGGAACA[dT][dT]; siKMT5A_2: GATGCAACTAGAGAGACA[dT][dT]; siCDC20_1 CGGAAGACCUGCCGUUACA[dT][dT]; siCDC20_2: GGGCCGAACUCCUGGCAAA[dT][dT].

2.5. Western Blotting and Quantitative Polymerase Chain Reaction

Western [21] and qPCR analysis [20] were performed as described previously. Primer sequences are detailed in Supplementary Table S1.

2.6. Microarray

Cellular RNA was extracted using Trizol® (Invitrogen, Waltham, MA, USA) and quality-checked using Agilent Bioanalyzer 2100 prior to analysis using Illumina HT-12 v4.0 Expression BeadChip (Oxford Genomics Centre, The Wellcome Trust Centre for Human Genetics, University of Oxford, Oxford, UK).

R package 'Lumi' was used for array processing, background correction, normalisation and quality control checks. Variance-stabilising transformation was used to convert probe intensity values to VSD (variance-stabilised data). The array normalisation method used was the robust spline normalisation (RSN) method. Outlier samples, poor quality probes (detection threshold < 0.01) and probes that were not detected were removed from downstream analysis. R package 'Limma' was then used to perform a differential expression analysis with p-values adjusted using the Benjamini–Hochberg method [22] to take into account the false discovery rate (FDR). Analysis was performed by the Bioinformatics Support Unit (Newcastle University).

Data can be found at GSE233350.

2.7. RNA-Seq Analysis

Fastq files were downloaded from NCBI GEO (GSE211638, [23]), and RNA-STAR [24] analysis was performed to align raw reads to genome build GRCh37/hg19; QC checks were performed with FastQC. Gene counts were generated using ht-seq count [25] and Gencode v19. Differential expression analysis was carried out using the DESeq2 [26] package (R/Bioconductor) to compare the vehicle versus 10 nM DHT-treated samples.

2.8. Chromatin Immunoprecipitation Assays

LNCaP-AI and LNCaP cells were reverse-transfected with either 25 nM N/S or a pool of 2 *KMT5A*-targeting siRNAs for 72 h in steroid-depleted media followed by chromatin immunoprecipitation as described by Schmidt et al. [27].

For immunoprecipitations, 2 µg of H4K20Me1 (Ab9051, AbCam) or 2 µg of a non-specific isotype control (DAKO) was used. qPCR analysis of immunoprecipitated DNA was performed using primers specific to the *CDC20* promoter (Fwd: 5′-CCGCTAGACTCTCGTG ATAGC-3′; Rev: 5′-TGGCTCCTTCAAAATCCAAC-3′) as previously described [28]. The average fold difference of the % input between experimental arms for at least three independent experiments is presented.

2.9. Sulforhodamine B Growth Analysis

Cellular growth was assessed as described [21].

2.10. Gamma H2AX Assay

Knockdown was carried out over 72 h in LNCaP-AI and LNCaP cells using either N/S or *KMT5A*-targeting siRNA. Cells were harvested and stained for phospho-histone H2AX (Ser139) as previously described [21].

3. Results

3.1. Identification of KMT5A-Regulated Genes in Androgen-Independent Prostate Cancer

KMT5A has been proposed as a therapeutic target in prostate cancer; however, in this context, KMT5A is still largely understudied. Indeed, no study has identified which genes KMT5A can regulate in castration-resistant prostate cancer. To this end, *KMT5A* mRNA was knocked down using two independent siRNA sequences in the LNCaP-AI cell line model of androgen independence. After 72 h of knockdown under steroid-depleted conditions, the androgen, DHT (10 nM), was applied for 24 h prior to RNA isolation and analysis using an Illumina Human HT-12 microarray.

The significant knockdown of *KMT5A*, with both siRNAs, was confirmed within the microarray data set as >80% prior to further analysis (Figure 1A). In the presence of an active AR found after stimulation with DHT for 24 h, we found 408 genes up-regulated and 310 genes down-regulated. (Supplementary Tables S2 and S3). Of these genes, 29% have previously been shown to be AR-regulated in LNCaP cells (Supplementary Tables S2 and S3) [23]. In order to understand which cellular pathways and biological processes are affected under these conditions, the gene lists generated were used in KEGG pathway analysis and gene ontology analysis using DAVID [29,30]. We did observe a level of inconsistency between the siRNA oligos even though the level of *KMT5A* knockdown was consistent (Figure 1A), a common issue when using siRNAs to assess multiple gene expressions, highlighting the importance of further validation studies on any target identified. Hence, all genes which showed a statistically significant change irrespective of the siRNA sequence were included in gene lists for this analysis to enhance confidence in the pathways and genes identified. This analysis revealed the significant up-regulation of PI3K-Akt signalling, apoptosis, p53 signalling and signal transduction whilst the pathways found to be significantly down-regulated included splicing, protein folding, cell division and transcriptional regulation (Supplementary Tables S4–S7). Taken together, the cellular processes and genes altered in this analysis further support our hypothesis that KMT5A is a potential therapeutic target for prostate cancer.

A

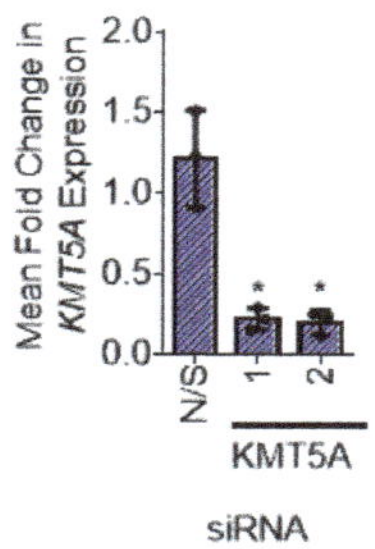

B

KMT5A siRNA	#1		#2		n=310	n=89*
	logfc	adj. p-value	logfc	adj. p-value	mean logfc	AR regulated
KRT18P55	1.92	4.13×10^{-4}	1.22	1.28×10^{-2}	1.57	
SNORD16	1.54	2.02×10^{-6}	1.52	4.01×10^{-6}	1.53	
KRT8	1.78	2.05×10^{-4}	1.27	5.25×10^{-3}	1.53	Yes
FAM189B	1.48	3.04×10^{-2}			1.48	
SPDEF	1.41	1.59×10^{-2}			1.41	Yes
CDC20	1.55	4.98×10^{-3}	1.10	4.71×10^{-2}	1.32	
IMPDH1	1.29	2.62×10^{-2}			1.29	Yes
PAQR4	1.42	7.94×10^{-4}	1.09	7.37×10^{-3}	1.25	Yes
ATP5C1	1.23	1.98×10^{-2}			1.23	
PLEKHF1	1.32	1.56×10^{-3}	1.01	1.16×10^{-2}	1.16	
AK4	1.15	1.52×10^{-2}			1.15	Yes
KRT18	1.28	1.27×10^{-4}	1.02	1.63×10^{-3}	1.15	Yes
ANKRD37	1.14	2.95×10^{-2}			1.14	Yes
SPRYD4	1.12	3.82×10^{-2}			1.12	
RAP1GAP			1.12	2.96×10^{-2}	1.12	Yes
KMT5A	1.17	1.61×10^{-7}	1.05	8.73×10^{-7}	1.11	

C

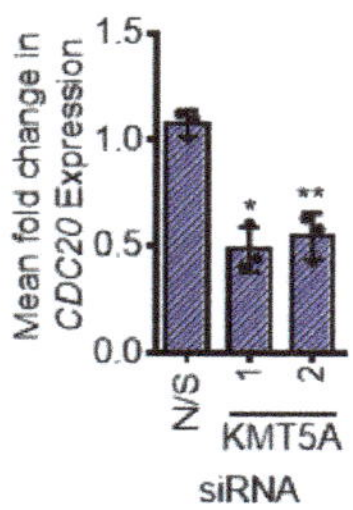

Figure 1. KMT5A-regulated genes in LNCaP-AI cells. (**A**) LNCaP-AI cells reverse-transfected with 25 nM siRNA targeting *KMT5A* or a non-silencing control (N/S) in steroid-depleted media. After 72 h, 10 nM DHT was added to the cells for a further 24 h. RNA was isolated, quality-checked, and its gene expression profiles determined using Illumina HT-12 v4.0 Expression BeadChip Microarray. Three independent experimental repeats were performed. Data analysis confirmed successful *KMT5A* knockdown. (**B**) Table of genes ranked for their down-regulation in response to *KMT5A* knockdown in the presence of DHT stimulation. Genes which were down-regulated more than *KMT5A* are shown (full gene lists can be found in the Supplementary Information). *GSE211638 [23]. (**C**) *CDC20* expression levels in response to *KMT5A* knockdown as determined via microarray analysis. One-way ANOVA with Dunnett's multiple comparisons test. * $p < 0.05$; ** $p < 0.01$.

In terms of individual genes which were down-regulated in response to *KMT5A* knockdown, *CDC20* was identified as the sixth most down-regulated gene after AR-regulated genes such as *KRT8* and *SPDEF* (Figure 1B,C). Due to its role in the cell cycle and previous characterisation as an oncogene [14,17], this gene was chosen for further study as a potential pharmacodynamic biomarker for KMT5A therapeutic targeting and a KMT5A effector protein.

3.2. KMT5A Depletion Reduces CDC20 Expression

In order to validate *CDC20* as a KMT5A-regulated gene, further experiments were conducted in both LNCaP-AI cells and the parental, androgen-sensitive LNCaP cell line. *KMT5A*-targeting siRNAs were transfected into both cell lines in steroid-depleted media for 72 h prior to stimulation with 10 nM DHT or the vehicle for a further 24 h. qPCR confirmed a significant reduction in the expression of *CDC20* in LNCaP-AI cells (Figure 2A) when *KMT5A* was knocked down ($p < 0.01$) (Figure 2B), which is consistent with our microarray data (Figure 1C). In addition, parental LNCaP cells also exhibited a significant reduction in *CDC20* expression (Figure 2C) irrespective of DHT stimulation upon significant *KMT5A* knockdown ($p < 0.001$) (Figure 2D). Furthermore, a robust reduction in CDC20 protein levels was consistently observed in both cell lines (Figure 2E,F) confirming that CDC20 is regulated by KMT5A, at the level of transcription, in both cell lines irrespective of AR activation.

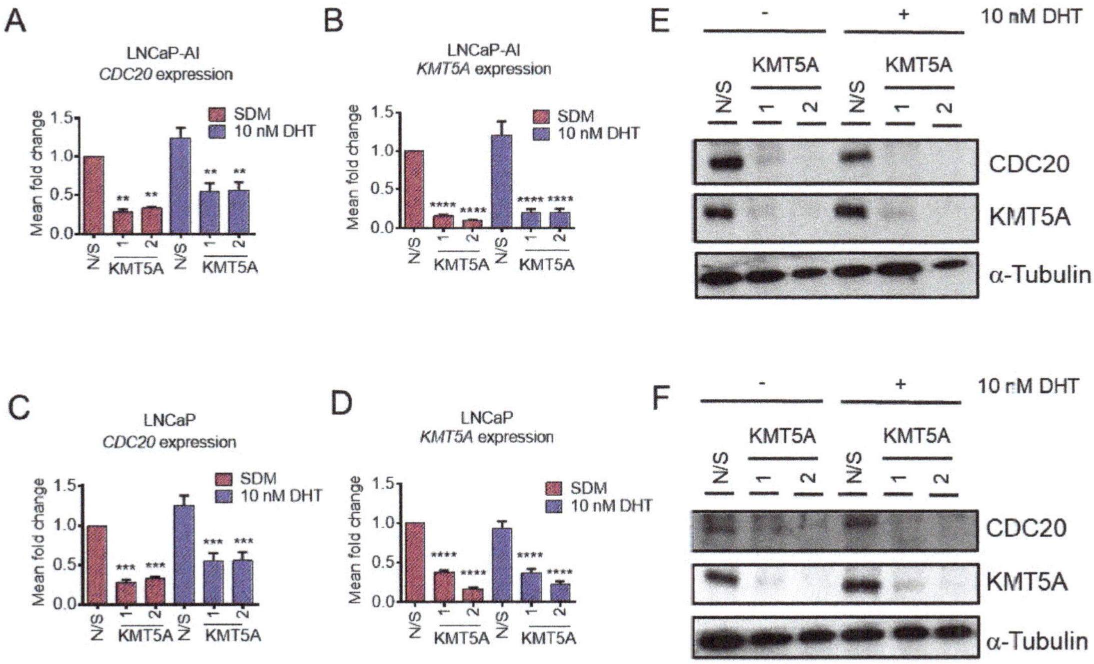

Figure 2. *CDC20* is a KMT5A-regulated gene. (**A**) LNCaP-AI cells were reverse transfected with 25 nM siRNAs targeting *KMT5A* or a non-silencing control (N/S) in steroid-depleted media. After 72 h, 10 nM DHT or a vehicle control was added to the cells for a further 24 h. RNA was isolated and *CDC20* mRNA and (**B**) *KMT5A* mRNA were quantified via qPCR. The same experiment was performed in (**C**) LNCaP cells, and *CDC20* levels and (**D**) *KMT5A* knockdown were confirmed via qPCR. Data are expressed as the mean fold change over 3 independent experiments, +/− SEM. (**E**) Determination of the protein levels of CDC20 and KMT5A via Western blotting in LNCaP-AI and (**F**) LNCaP cells. Alpha-tubulin was used as a loading control. Data shown are representative of 3 independent experiments. Two-way ANOVA with Dunnett's multiple comparisons test; ** $p < 0.01$; *** $p < 0.001$; **** $p < 0.001$. The uncropped blots are shown in File S1.

3.3. CDC20 Depletion Does Not Enhance KMT5A Protein Expression

KMT5A is phosphorylated to protect it from ubiquitin-mediated degradation by APC^{cdh1} during late mitosis [7]. In addition, due to the similarity in recognition mechanisms between CDH1 and CDC20 for targeting proteins to the APC complex, it was suggested that CDC20 may also bind and recognise KMT5A in the absence of phosphorylation [7].

This raised the question of whether or not a feedback mechanism exists between these two proteins to help maintain correct cell cycle progression. To test this theory, *CDC20* was knocked down in our cell line models and KMT5A levels were assessed at both the transcript and protein level. Interestingly, when KMT5A protein levels were examined subsequent to *CDC20* knockdown, no change was observed (Figure 3A,B), suggesting that KMT5A protein turnover does not take place when CDC20 is present in the cell. However, a decrease in *KMT5A* transcripts by ~50% was observed in both cell lines (Figure 3C,D) upon the robust depletion of *CDC20* (Figure 3E,F) although this was not statistically significant. Therefore, it was concluded that CDC20 did not play a significant role in KMT5A protein regulation under our experimental conditions and that KMT5A sits upstream of CDC20.

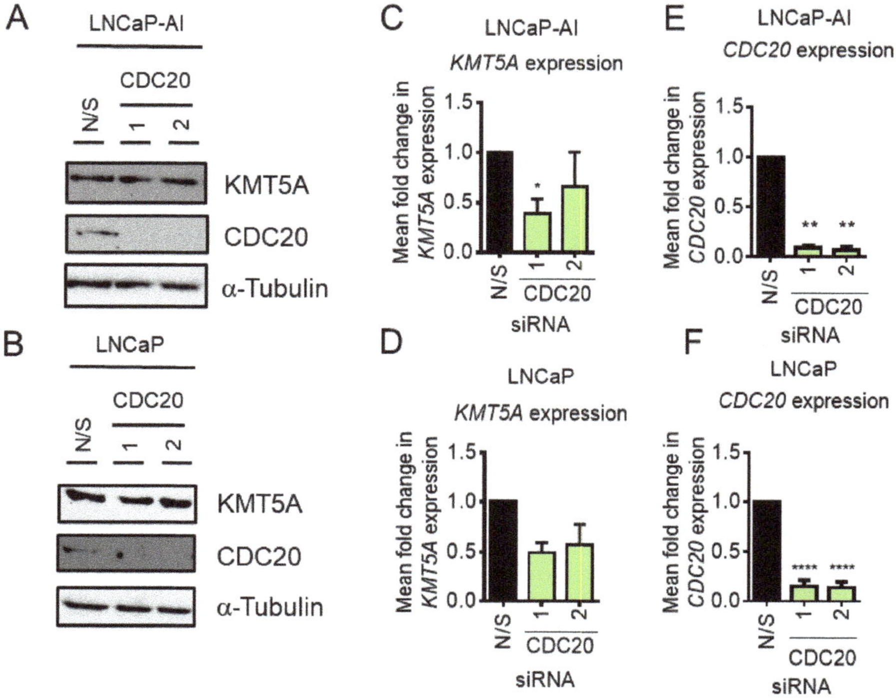

Figure 3. *CDC20* knockdown does not affect KMT5A protein levels. (**A**) LNCaP-AI cells and (**B**) LNCaP cells reverse-transfected with 25 nM siRNA targeting *CDC20* or a non-silencing control (N/S) for 72 h prior to analysis via Western blot analysis. Alpha-tubulin was used as a loading control. (**C**) LNCaP-AI and (**D**) LNCaP cells reverse-transfected with 25 nM siRNA targeting *CDC20* or a non-silencing control (N/S) for 72 h prior to RNA isolation and qPCR analysis for *KMT5A* and (**E**,**F**) *CDC20* mRNA levels. *HPRT1* was used as a housekeeping gene. Data are expressed as mean fold change +/− SEM. One-way ANOVA was used to determine statistical significance. * $p < 0.05$; ** $p < 0.01$; **** $p < 0.0001$. The uncropped blots are shown in File S1.

3.4. KMT5A Expression Correlates with CDC20 Expression in Prostate Cancer Patients

To confirm whether or not our in vitro findings could be translated into clinical specimens of prostate cancer, we interrogated publicly available datasets to confirm a positive correlation between *CDC20* and *KMT5A* transcripts. Upon the interrogation of data sets available in cBioportal [31,32], we found a significant positive correlation between *KMT5A* and *CDC20* transcripts in a number of data sets. In a cohort of 65 treatment-naïve radical prostatectomies [33] a positive correlation between *CDC20* and *KMT5A* transcripts was observed (Spearman = 0.45; $p = 0.0002$) (Figure 4A). Similarly, in the MCTP dataset [34], a positive correlation was also observed (Spearman = 0.23; $p = 0.024$) (Figure 4B). However, this dataset contains samples from primary (blue) and metastatic prostate cancer (red). Upon the correlation analysis of these individual sample types, it was observed that the correlation between *CDC20* and *KMT5A* was strongest in the primary prostate samples (Spearman = 0.28; $p = 0.033$; n = 59) and no statistically significant correlation was observed in the metastatic samples (Spearman = 0.034; $p = 0.85$; n = 35), although the sample numbers were lower. However, in the metastatic cohort reported by Robinson et al. [35] a statistically significant correlation was observed between *CDC20* and *KMT5A* (Spearman = 0.22; $p = 0.016$; n = 118) (Figure 4C). Furthermore, significant correlations were observed in bone metastases (Spearman = 0.41; $p = 0.0003$; n = 72) (Figure 4D) and liver metastases (Spearman = 0.408; $p = 0.011$; n = 38) (Figure 4E) in the samples from Abida et al. [36]. Interestingly, the highest positive correlation between *CDC20* and *KMT5A* expression was seen in prostate neuroendocrine carcinoma samples (Spearman = 0.68; $p < 0.0001$; n = 49) (Figure 4F) [37]. Taken together, this suggests that the positive correlation between *KMT5A* and *CDC20* observed in our cell line models is also observed in advanced prostate cancer.

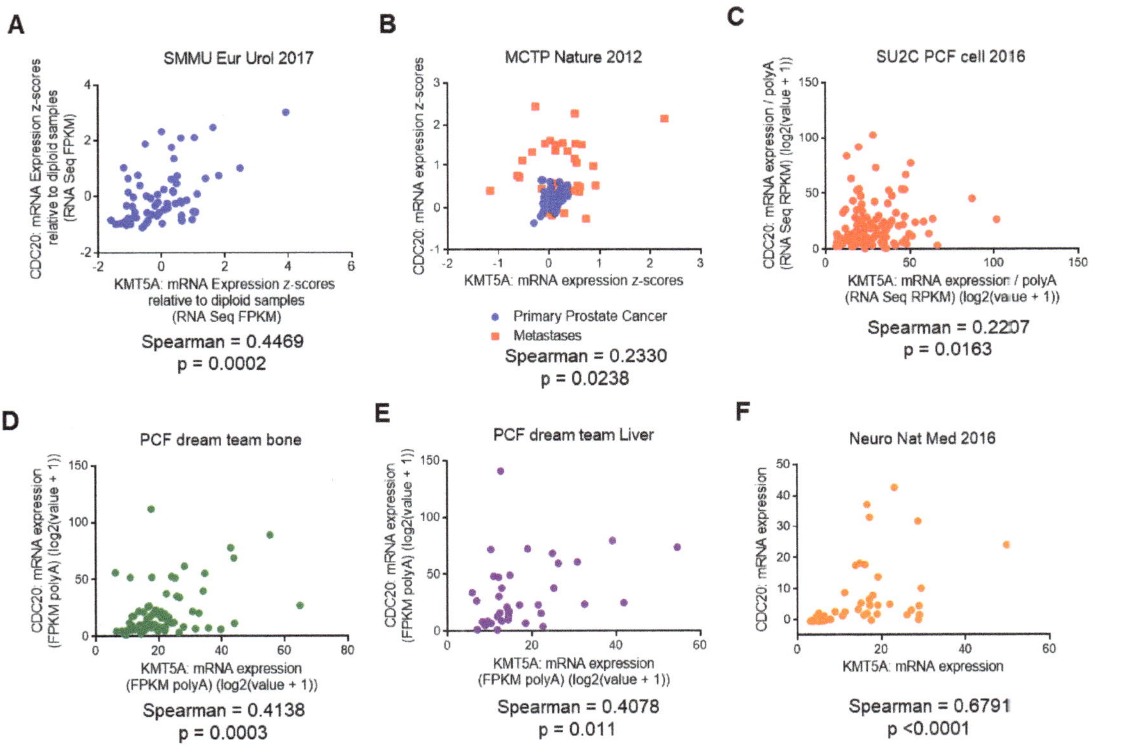

Figure 4. *KMT5A* and *CDC20* are positively correlated in clinical prostate cancer samples. Correlation in expression between *KMT5A* and *CDC20* was carried out in publicly available datasets in cBioportal. (**A**) Treatment naïve radical prostatectomies (n = 65) from [33] (**B**) Both primary and metastatic prostate cancer (n = 94) from [34] (**C**) Metastatic prostate adenocarcinoma samples (n = 150) from [35] (**D**) Bone metastatic prostate adenocarcinoma samples (n = 72/266) and (**E**) Liver metastatic prostate adenocarcinoma samples (n = 38) from [36] (**F**) Prostate neuroendocrine carcinoma samples (n = 49) from [37]. Spearman correlation (two-tailed) was calculated using Graphpad software v6.

3.5. KMT5A Inhibition Reduces CDC20 Expression and Reduces Prostate Cancer Cell Proliferation

KMT5A plays a role in the cell cycle and as such, knockdown of *KMT5A* has been shown to inhibit cellular proliferation [38,39]. Indeed, we observed a reduction in proliferation upon *KMT5A* knockdown in both the cell line models used in this study

(Supplementary Figure S1A–C). In particular, proliferation was most affected under steroid depleted conditions. Furthermore, as expected, knockdown of *CDC20* resulted in a robust and significant inhibition of cellular proliferation (Supplementary Figure S1D,E). Together this provides supporting evidence that both proteins play a role in prostate cancer cell proliferation.

In order to confirm that the methyltransferase activity of KMT5A is important for the regulation of *CDC20* expression in prostate cancer cell lines we used two molecules that have shown inhibitory activity against KMT5A, namely UNC0379 and Ryuvidine [40,41] (Supplementary Figure S2). Firstly, we determined the GI50 values for LNCaP and LNCaP-AI cells (Figure 5A). Interestingly, we found that Ryuvidine was a more potent inhibitor than UNC0379 in LNCaP cells, However, in LNCaP-AI cells there was not such a large difference in efficacy. Secondly, we used a titration of doses of both inhibitors and investigated the dose dependent effects on both KMT5A and its target histone mark, H4K20Me1. We observed that UNC0379 resulted in a robust decrease in H4K20Me1 levels in LNCaP-AI and a modest reduction in LNCaP cells when total H4 levels are taken into account. This coincided with a decrease in KMT5A protein levels in LNCaP cells whilst KMT5A levels showed minimal change in LNCaP-AI cells. Ryuvidine also demonstrated a dose dependent reduction in KMT5A activity in LNCaP cells whilst a decrease in H4K20Me1 was more difficult to achieve in LNCaP-AI cells thereby reflecting the sensitivity differences to Ryuvidine between these two cell lines (Supplementary Figures S2 and S3).

Using the GI50 concentrations for each drug in LNCaP-AI cells we investigated the levels of CDC20 by Western blotting, demonstrating that both drugs result in CDC20 reduction (Figure 5B). Furthermore, when the more potent Ryuvidine was used in LNCaP cells at the GI50 dose a reduction in CDC20 levels was observed (Figure 5C). Together, this led us to conclude that KMT5A enzymatic activity is important in the regulation of CDC20 protein levels.

3.6. KMT5A Knockdown Reduces H4K20Me1 at the CDC20 Promoter

In order to confirm that KMT5A can directly regulate the expression of *CDC20* via the mono-methylation of its only histone target, H4K20, chromatin immunoprecipitation assays were performed in both LNCaP-AI and LNCaP cells subsequent to *KMT5A* knockdown using a siRNA pool of siKMT5A_1 and siKMT5A_2. Upon *KMT5A* knockdown in both cell lines growing in steroid depleted media, a significant reduction in H4K20Me1 was observed at the *CDC20* promoter region (Figure 6A,B). This led us to conclude that KMT5A can directly modulate the expression of *CDC20* via the methylation of H4K20 within the promoter region.

3.7. p53 Mediates KMT5A Regulation of CDC20 Expression

Whilst KMT5A methyltransferase activity is important in regulating the expression of *CDC20* via the regulation of H4K20Me1 within the *CDC20* promoter, KMT5A can also methylate non-histone proteins, including p53, to regulate functional activity. Interestingly, in our pathway analysis we uncovered some pathways within which CDC20 can be modulated. In particular, p53 directly down-regulates *CDC20* expression via association with its promoter in response to DNA damage [11]. Secondly, in the absence of DNA damage, p53 can regulate *CDC20* expression via a CDE-CHR element, independent of p21, when p53 is over-expressed [11,14,42]. To determine whether or not DNA damage in response to KMT5A knockdown was influencing this mechanism, we investigated the levels of γ-H2AX in both LNCaP-AI and LNCaP cells after *KMT5A* knockdown. Consistent with other reports [39], *KMT5A* knockdown resulted in an increased level of DNA damage as denoted by a robust ~3.5 fold and ~2 fold increase in γ-H2AX levels in LNCaP-AI cells and LNCaP cells, respectively (Figure 7A), further suggesting that p53 could be a mediator of KMT5A effects on *CDC20* levels. Indeed, KMT5A is well-known to methylate p53 at K382 to reduce p53 activation [6], and the down-regulation of KMT5A in response to DNA damage has been shown to result in the conversion of this mono-methylation state into a di/tri- methylation state on K382 to increase p53 stability [6,43]. Therefore, we hypothesised that knockdown of *KMT5A* would shift the equilibrium from mono-methylated p53 to acetylated p53, thereby resulting in p53 activation and the subsequent repression of *CDC20* expression. To test this theory, *KMT5A* was knocked down in both LNCaP-AI and LNCaP cells prior to Western blotting for changes in p53 post-translational modifications and MDM2. We observed no alterations in total p53 protein levels in either cell line with siKMT5A_2; however, p53 levels were increased with siKMT5A_1. Nonetheless, a robust increase in acetylation at K382 and an increase in p53-phosphorylation at serine 15 which is associated with enhanced DNA binding was still observed with both siRNA sequences (Figures 7B,C and S6). To confirm that p53 activation results in the down-regulation of CDC20 protein levels, we treated both LNCaP-AI and LNCaP cells with the MDM2 inhibitor, Nutlin 3. In both cell lines, CDC20 protein levels were reduced when p53 was activated (Figure 7D). Taken together, this suggests that p53 activation via *KMT5A* knockdown results in the repression of *CDC20*.

3.8. CDC20 Is Down-Regulated by Protein Turnover in the Absence of p53

As p53 signalling was found to be up-regulated in our KEGG pathway analysis (Supplementary Table S4), we questioned whether or not KMT5A was able to regulate *CDC20* if p53 was not present. As p53 loss is a common phenomenon in cancers, this raised questions regarding the applicability of CDC20 as a KMT5A biomarker for those patients whose tumours lack p53 expression. To investigate further, we performed *KMT5A* knockdowns in p53 null, PC3 cells and performed Western blotting and qPCR analyses of *CDC20* levels. Surprisingly, *KMT5A* knockdown was still able to robustly reduce CDC20 protein levels in this cell line (Figure 8A). However, *CDC20* mRNA levels were unaffected (Figure 8B) suggesting that CDC20 post-translational changes occurred causing alterations in protein turnover in the absence of p53. Therefore, if a protein biomarker read-out could be used then CDC20 may remain as a valid KMT5A activity biomarker.

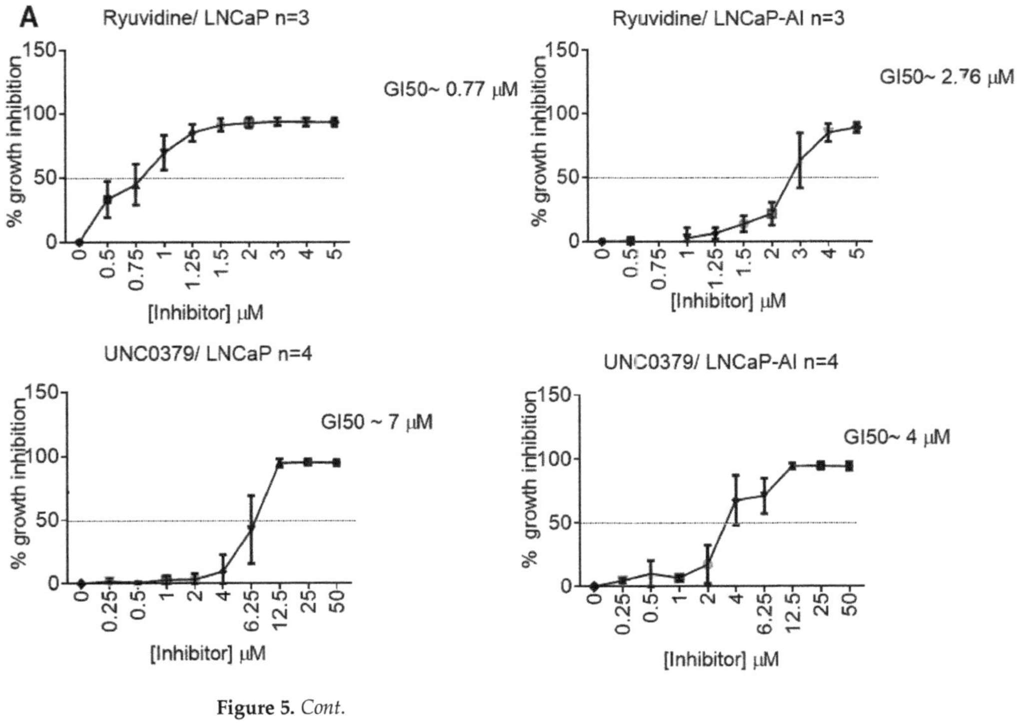

Figure 5. *Cont.*

Cell Line.	UNC0379 (GI50 +/- SD)	Ryuvidine (GI50 +/- SD)
LNCaP	7.06 +/- 1.88	0.77 +/- 0.19
LNCaP-AI	3.87 +/- 1.33	2.77 +/- 0.36

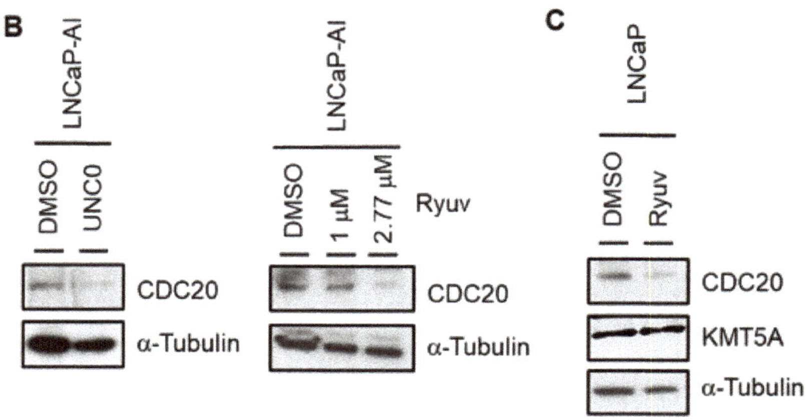

Figure 5. KMT5A inhibitors restrict prostate cancer cell growth and down-regulate CDC20. (**A**) LNCaP-AI and LNCaP cells treated with a dose range of UNC0379 or Ryuvidine and GI50 concentrations determined via SRB assay after 3 doubling times. Data are expressed as mean % growth inhibition +/− SEM from 3 independent experiments. Mean GI50 values are tabulated +/− SD. (**B**) LNCaP-AI cells treated with either UNC0379 (7 μM) or Ryuvidine (1 μM or 2.77 μM) for 48 h prior to protein analysis via Western blotting. (**C**) LNCaP cells were treated with Ryuvidine (0.7 μM) for 48 h prior to protein analysis via Western blotting. Data shown are representative of 3 independent experiments. The uncropped blots are shown in File S1.

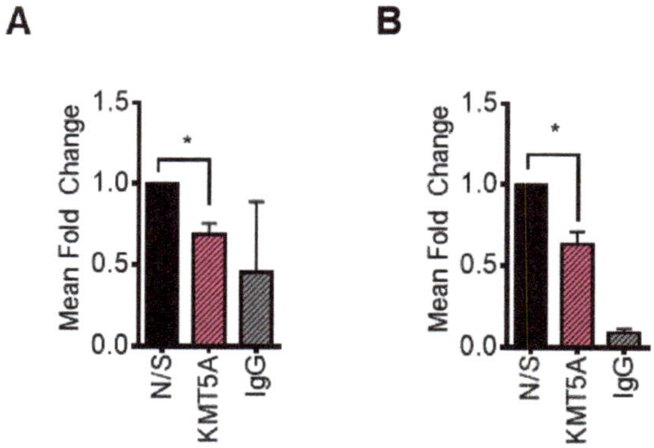

Figure 6. H4K20Me1 is reduced at the CDC20 promoter in response to KMT5A knockdown. *KMT5A* was knocked down in (**A**) LNCaP-AI and (**B**) LNCaP cells growing in steroid depleted media for 72 h. Chromatin was collected, and immunoprecipitation for was H4K20Me1 carried out. Isolated DNA was purified and primers targeting the *CDC20* promoter region were used to determine the levels of H4K20Me1 association with this region. Experiments were performed 3 times and data are expressed as the mean fold change relative to the non-silencing control, +/− SEM. IgG was used as a negative control. Student's *t*-test * $p < 0.05$.

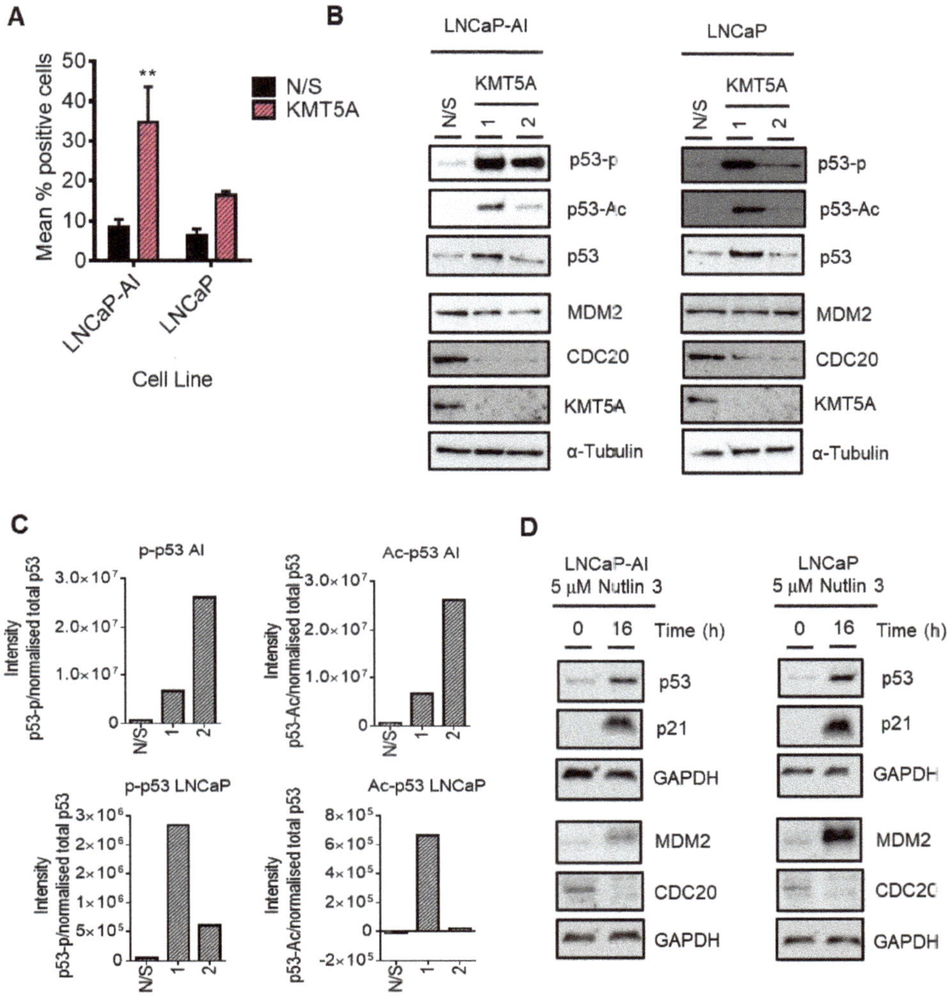

Figure 7. KMT5A regulates *CDC20* via p53 activation. (**A**) LNCaP-AI and LNCaP cells reverse transfected with either a pool of 3 *KMT5A*-targeting siRNAs or a non-silencing (N/S) control for 72 h prior to the assessment of γ-H2AX via flow cytometry. Scatter plots for experimental replicates can be found in Supplementary Figure S4. (**B**) LNCaP-AI and LNCaP cells reverse-transfected with 2 independent siRNA sequences for 72 h prior to Western blotting analysis for CDC20, p53, p-p53, p53-Ac, MDM2, and KMT5A. (**C**) Densitometry of Western blots shown in (**B**). The background was subtracted from the intensity values prior to intensity normalisation against an appropriate loading control. Intensities for post-translationally modified proteins were normalised to total protein intensity. Additional experimental repeats can be found in Supplementary Figure S6. (**D**) LNCaP-AI and LNCaP cells were treated with Nutlin 3 (5 μM) for 0 and 16 h prior to Western blotting analysis. Western blot data shown is representative of 3 independent experiments. Two-way ANOVA test was used to determine statistical significance for data shown in (**A**). ** $p < 0.01$ The uncropped blots are shown in File S1.

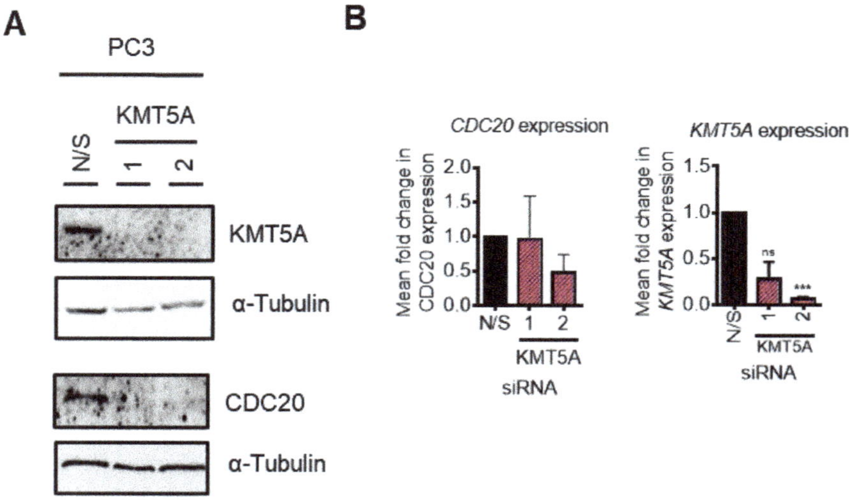

Figure 8. KMT5A regulates protein turnover in the absence of p53. (**A**) *KMT5A* was knocked down with 2 independent siRNAs for 72 h. Protein was collected and analysed via Western blotting for KMT5A and CDC20, and α-tubulin was used as a loading control. (**B**) Expression of *CDC20* and *KMT5A* in a parallel experiment analysed via qPCR. Data are shown as the mean fold change relative to that of N/S controls over 3 experimental repeats, +/− SEM. One-way ANOVA was used to determine statistical significance. *** $p < 0.01$; ns: not significant. The uncropped blots are shown in File S1.

4. Discussion

Alternative therapeutic targets are urgently required for the treatment of advanced prostate cancers which have relapsed after current standard-of-care therapies. As the androgen receptor remains a driver of disease in therapy relapse, proteins which positively modulate the transcriptional activity of the androgen receptor are proposed as putative therapeutic targets. The protein methyltransferase KMT5A has been shown to interact with the androgen receptor [4] and was proposed to offer therapeutic benefits to prostate cancer patients. However, the mechanisms by which KMT5A contributes to prostate cancer progression remains poorly understood.

We uncovered that KMT5A can regulate the levels of the cell cycle regulator protein CDC20 both directly at the chromatin level via the modulation of histone methylation, and indirectly via the methylation of the tumour suppressor protein, p53. This relationship between CDC20 and KMT5A is supported by a significant positive correlation between *KMT5A* and *CDC20* transcripts in prostate cancer patients (Figure 4). Whilst this relationship is independent of the androgen receptor (Figures 2 and S5), both proteins are described as oncogenes in prostate cancer. Critically, there are no reports describing the methylation-specific regulation of CDC20.

KMT5A is the only known methyltransferase to monomethylate histone H4K20. As the H4K20Me1 mark is traditionally associated with a compact chromatin landscape and gene repression [44–47], it is counterintuitive that KMT5A should function to facilitate *CDC20* transcription. However, KMT5A-mediated H4K20Me1 is now well-documented to function as a transcriptional activator for some genes [48,49]. Where this has been observed, there is generally a transcription factor which is implicated, for example TWIST [48]. Furthermore, H4K20Me1 is associated with actively transcribing gene bodies [50] and more recently has been found to result in chromatin accessibility in highly transcribed genes throughout the cell cycle [51]. A role for KMT5A in the pause and release of RNA pol II has also been revealed [52], further supporting the complex role of KMT5A in the positive regulation of gene transcription.

The methylation-dependent regulation of p53 activity by KMT5A is key to ensuring transcriptional activation [6,43], further highlighting the ability of KMT5A to influence gene expression programmes at multiple levels. Consistently, we observed that knockdown of *KMT5A* resulted in enhanced p53 acetylation at K382, which can only occur if this residue is not methylated. Importantly, it is the subsequent phosphorylation event at S15 which facilitates the association of p53 with DNA which is enhanced upon *KMT5A* knockdown. This would permit the recruitment of HDAC1 and mSin3a to the *CDC20* promoter to allow chromatin remodelling to occur and thereby inhibit the transcription of *CDC20* [11]. Importantly, we observed slightly different effects with each KMT5A-targeting siRNA in this experiment with regard to the ability of KMT5A knockdown to stabilise p53, making the interpretation of p53 post-translational modifications more complex. However, densitometry confirmed that both p53 phosphorylation and acetylation do increase with KMT5A knockdown with both siRNA sequences (Figure 7C). Therefore, it appears that there are two complementary mechanisms working together at the *CDC20* promoter modulated by KMT5A to ensure the timely expression of this gene.

Both KMT5A and CDC20 are essential cell cycle regulator proteins. KMT5A is regulated by ubiquitin-mediated protein turnover specifically at the G1/S transition and between metaphase and anaphase [7] whilst CDC20 regulates the SAC to control the progression from metaphase to anaphase and ensure the successful separation of sister chromatids. It is thought that the methylation of H4K20 is key to successful mitosis with the turnover of KMT5A being the major mode of H4K20 methylation regulation. Indeed, H4K20Me-mediated chromosome condensation is important in this process and *KMT5A* knock out studies resulted in chromosome decondensation leading to cell cycle arrest at G2/M [46]. Furthermore, H4K20Me1 is required for kinetochore assembly at centromeres via the recruitment of CENP-T [53]. Hence, it is logical to hypothesise that a lack of KMT5A, resulting in a decrease in H4K20Me1, will result in impaired kinetochore assembly and thereby will invoke the SAC preventing CDC20 from facilitating the onset of the anaphase. Therefore, due to the importance of tightly regulating *CDC20* to ensure effective mitosis, the modulation of *CDC20* levels themselves by KMT5A would provide a failsafe way to prevent a mitotic catastrophe.

CDC20 is required for nuclear movement prior to the anaphase where its activity, as part of the APC^{CDC20} complex, results in the destruction of cyclin B and the inactivation of CDK1. Interestingly, the CDK1-mediated phosphorylation of KMT5A at serine 29 has been reported to occur during metaphase resulting in the removal of KMT5A from chromatin, holding it in a stabilised state without affecting methylase activity. It is not until anaphase that dephosphorylation by cdc14a/b permits KMT5A protein turnover via APC^{cdh1} [7]. Furthermore, APC^{CDC20} targets the H4K20Me1 demethylase, PHF8, for ubiquitin-mediated destruction [13] further highlighting the important relationship between CDC20 and the enzymes which modulate the H4K20 methylation state.

CDC20 has been found to be overexpressed in a number of cancers, including prostate cancer [15,54], and there are a number of studies which demonstrate the relevance of *CDC20* to prostate cancer development and progression. For example, *CDC20* has been identified as a hub gene, alongside *CDK1*, in castration-resistant prostate cancer [55], and contributes to cell migration, disease progression and a poorer prognosis in metastatic prostate cancer [56] with another study showing that CDC20, alongside PLK1 and cyclin A, plays a critical role in prostate cancer metastasis [57]. *CDC20* and *PLK1* are both located at chromosomal region 9p, which is often amplified in cancer. Indeed, high expressions of *CDC20*, *PLK1* and *CDK1* correlate with prostate cancer occurrence [58] and worse biochemical recurrence survival rates [59]. Furthermore, CDC20 is a target protein of the Speckle-type POZ protein (SPOP), which functions to promote ubiquitin-mediated protein turnover. *SPOP* is mutated in up to 15% of prostate cancers [60] and these mutations have been shown to result in an inability of SPOP to associate with CDC20, preventing CDC20 protein turnover and consequently resistance to CDC20 inhibitors [16]. Both SPOP mutation [61] and *CDC20* overexpression are important in docetaxel resistance with the inhibition or knockdown of *CDC20* being able to resensitise cells to docetaxel [62] highlighting the importance of CDC20 as a therapeutic target in prostate cancer at several disease stages. With inhibitors for both CDC20 and KMT5A being developed, it would be important to determine whether or not they are able to synergise with each other in drug-resistant models of prostate cancer.

KMT5A has an important role in the DNA damage repair pathway where it is recruited to double-strand breaks to deposit H4K20Me1 to facilitate Suv4-20-mediated H4K20Me2, which is required for 53BP1 binding and successful repair by NHEJ [63,64]. In addition, the ubiquitination of KMT5A by RNF8 increases KMT5A association with RNF168 which in turn promotes H2A ubiquitination [65]. The ubiquitination of these and other chromatin components results in the recruitment of BRCA1/BARD1/Abraxas and RAP80 to sites of γH2AX to allow the repair process to take place. Interestingly, RAP80 is a target of CDC20 and its overexpression prevents mitotic progression irrespective of DNA damage [9]. This again supports a connection between the functions of KMT5A and CDC20 in cellular processes. Additionally, the role of KMT5A in the suppression of important anti-tumourigenic processes such as the positive regulation of the apoptotic process and the response to gamma and ionising irradiation are also highlighted suggesting the utility of KMT5A inhibition in combination with other DNA damage-inducing therapeutics such as radiotherapy or cytotoxic agents.

The cellular processes regulated by KMT5A identified in this study are consistent with those already described such as genome integrity, cell cycle progression, gene transcription and DNA damage repair. However, some novel processes were identified including RNA splicing and mRNA processing which require further investigation. This is particularly important in prostate cancer where aberrant RNA splicing, particularly of the androgen receptor, is associated with therapy resistance and poor prognosis [66].

5. Conclusions

A number of key oncogenic signalling pathways are regulated by the methyltransferase KMT5A. Here, we provide evidence of the role of KMT5A in both metaphase to anaphase control via the regulation of CDC20 and propose that close links between mitosis and DNA damage repair processes are present via this relationship. As both *CDC20* and *KMT5A* are up-regulated in cancer via a number of mechanisms, the relationship between the two proteins may be dysregulated, thereby promoting genomic instability. This presents an opportunity to identify beneficial therapeutic combinations to treat patients based on a number of criteria such as *SPOP* mutation status, *KMT5A* expression status and therapeutic resistance.

Supplementary Materials: The following supporting information can be downloaded at https://www.mdpi.com/article/10.3390/cancers15143597/s1. Supplementary Figure S1. KMT5A and CDC20 knockdown inhibits proliferation of prostate cancer cells. Supplementary Figure S2. KMT5A inhibition by UNC0379 reduces KMT5A levels and H4K20Me1. Supplementary Figure S3. KMT5A inhibition by Ryuvidine reduces KMT5A levels and H4K20Me1. Supplementary Figure S4. KMT5A knockdown causes DNA damage. Supplementary Figure S5. KMT5A levels remain constant in response to DHT stimulation. Supplementary Figure S6. Acetylation and phosphorylation of p53 occurs upon KMT5A knockdown. Table S1. qPCR primers. Table S2. Genes significantly up-regulated by KMT5A knockdown. Table S3. Genes significantly down-regulated by KMT5A knockdown. Table S4. KEGG pathways down-regulated in response to KMT5A knockdown in the presence of DHT stimulation. Table S5. KEGG pathways up-regulated in response to KMT5A knockdown in the presence of DHT stimulation. Table S6. Biological processes down-regulated in response to KMT5A knockdown in the presence of DHT stimulation. Table S7. Biological processes up-regulated in response to KMT5A knockdown in the presence of DHT stimulation. File S1. Uncut blots.

Author Contributions: Conceptualisation, K.C. and C.N.R.; Methodology, K.C. and C.N.R.; Validation, K.C., Z.A.H.A., E.L.-W. and S.N.; Formal Analysis, S.N. and Z.A.H.A.; Investigation, K.C., Z.A.H.A., M.A., D.A.-K., K.L.J., H.X.C., R.G. and S.W.; Resources, K.C. and C.N.R.; Data Curation, Z.A.H.A.; Writing—Original Draft Preparation, Z.A.H.A., M.A. and K.C.; Writing—Review and Editing, K.C., C.N.R., E.L.-W. and E.S.; Visualisation, K.C.; Supervision, K.C. and C.N.R.; Project Administration, K.C.; Funding Acquisition, K.C., Z.A.H.A. and D.A.-K. All authors have read and agreed to the published version of the manuscript.

Funding: This work was supported by a Movember funded Prostate Cancer UK Career Development Fellowship (CDF12-006) to K.C. which supported S.W., The Higher Committee for Education Development in Iraq, as a PhD studentship to Z.A.H.A, and The Public Authority of Applied Education and Training, Kuwait to D.A.-K. E.S. is supported by a Prostate Cancer UK Travelling Prize fellowship [TLD-PF19-002]. E.L.-W. is supported by a Prostate Cancer UK Research Innovation Award (RIA19-ST2-005). R.G. is supported by a William Edmond Harker Foundation Studentship. C.N.R is supported by Cancer Research UK (C27826/A15994). For the purpose of Open Access, the author has applied a Creative Commons Attribution (CC BY) licence to any Author Accepted Manuscript (AAM) version arising from this submission.

Institutional Review Board Statement: Not applicable.

Informed Consent Statement: Not applicable.

Data Availability Statement: The microarray data presented in this study are openly available in GEO: GSE233350. Publicly available datasets were analysed in this project. This data can be found in cBioportal.

Acknowledgments: Nutlin 3 was provided as a gift from John Lunec (Newcastle Cancer Centre, Newcastle University). We thank the High-Throughput Genomics Group at the Wellcome Trust Centre for Human Genetics (funded by Wellcome Trust grant reference 090532/Z/09/Z and MRC Hub grant G0900747 91070) for the generation of the Gene Expression data. We thank Oswald To for contributions to chromatin generation for this study, Kendal Cagney for assistance with KMT5A inhibitor studies, Victoria Harle for flow cytometry assistance, and Maddie Blackham for technical assistance.

Conflicts of Interest: The authors declare no conflict of interest.

References

1. Veschi, V.; Liu, Z.; Voss, T.C.; Ozbun, L.; Gryder, B.; Yan, C.; Hu, Y.; Ma, A.; Jin, J.; Mazur, S.J.; et al. Epigenetic siRNA and Chemical Screens Identify SETD8 Inhibition as a Therapeutic Strategy for p53 Activation in High-Risk Neuroblastoma. *Cancer Cell* **2017**, *31*, 50–63. [CrossRef] [PubMed]
2. Wu, J.; Qiao, K.; Du, Y.; Zhang, X.; Cheng, H.; Peng, L.; Guo, Z. Downregulation of histone methyltransferase SET8 inhibits progression of hepatocellular carcinoma. *Sci. Rep.* **2020**, *10*, 4490. [CrossRef] [PubMed]
3. Zhang, J.; Hou, W.; Chai, M.; Zhao, H.; Jia, J.; Sun, X.; Zhao, B.; Wang, R. MicroRNA-127-3p inhibits proliferation and invasion by targeting SETD8 in human osteosarcoma cells. *Biochem. Biophys. Res. Commun.* **2016**, *469*, 1006–1011. [CrossRef]
4. Yao, L.; Li, Y.; Du, F.; Han, X.; Li, X.; Niu, Y.; Ren, S.; Sun, Y. Histone H4 Lys 20 methyltransferase SET8 promotes androgen receptor-mediated transcription activation in prostate cancer. *Biochem. Biophys. Res. Commun.* **2014**, *450*, 692–696. [CrossRef] [PubMed]
5. Hou, L.; Li, Q.; Yu, Y.; Li, M.; Zhang, D. SET8 induces epithelialmesenchymal transition and enhances prostate cancer cell metastasis by cooperating with ZEB1. *Mol. Med. Rep.* **2016**, *13*, 1681–1688. [CrossRef]
6. Shi, X.; Kachirskaia, I.; Yamaguchi, H.; West, L.E.; Wen, H.; Wang, E.W.; Dutta, S.; Appella, E.; Gozani, O. Modulation of p53 function by SET8-mediated methylation at lysine 382. *Mol. Cell* **2007**, *27*, 636–646. [CrossRef]
7. Wu, S.; Wang, W.; Kong, X.; Congdon, L.M.; Yokomori, K.; Kirschner, M.W.; Rice, J.C. Dynamic regulation of the PR-Set7 histone methyltransferase is required for normal cell cycle progression. *Genes Dev.* **2010**, *24*, 2531–2542. [CrossRef]
8. Wu, S.; Rice, J.C. A new regulator of the cell cycle: The PR-Set7 histone methyltransferase. *Cell Cycle* **2011**, *10*, 68–72. [CrossRef]
9. Cho, H.J.; Lee, E.H.; Han, S.H.; Chung, H.J.; Jeong, J.H.; Kwon, J.; Kim, H. Degradation of human RAP80 is cell cycle regulated by Cdc20 and Cdh1 ubiquitin ligases. *Mol. Cancer Res.* **2012**, *10*, 615–625. [CrossRef]
10. Chun, A.C.; Kok, K.H.; Jin, D.Y. REV7 is required for anaphase-promoting complex-dependent ubiquitination and degradation of translesion DNA polymerase REV1. *Cell Cycle* **2013**, *12*, 365–378. [CrossRef]
11. Banerjee, T.; Nath, S.; Roychoudhury, S. DNA damage induced p53 downregulates Cdc20 by direct binding to its promoter causing chromatin remodeling. *Nucleic Acids Res.* **2009**, *37*, 2688–2698. [CrossRef] [PubMed]

12. Kidokoro, T.; Tanikawa, C.; Furukawa, Y.; Katagiri, T.; Nakamura, Y.; Matsuda, K. CDC20, a potential cancer therapeutic target, is negatively regulated by p53. *Oncogene* **2008**, *27*, 1562–1571. [CrossRef] [PubMed]
13. Lim, H.J.; Dimova, N.V.; Tan, M.K.; Sigoillot, F.D.; King, R.W.; Shi, Y. The G2/M regulator histone demethylase PHF8 is targeted for degradation by the anaphase-promoting complex containing CDC20. *Mol. Cell Biol.* **2013**, *33*, 4166–4180. [CrossRef]
14. Wang, L.; Zhang, J.; Wan, L.; Zhou, X.; Wang, Z.; Wei, W. Targeting Cdc20 as a novel cancer therapeutic strategy. *Pharmacol. Ther.* **2015**, *151*, 141–151. [CrossRef] [PubMed]
15. Mao, Y.; Li, K.; Lu, L.; Si-Tu, J.; Lu, M.; Gao, X. Overexpression of Cdc20 in clinically localized prostate cancer: Relation to high Gleason score and biochemical recurrence after laparoscopic radical prostatectomy. *Cancer Biomark.* **2016**, *16*, 351–358. [CrossRef]
16. Wu, F.; Dai, X.; Gan, W.; Wan, L.; Li, M.; Mitsiades, N.; Wei, W.; Ding, Q.; Zhang, J. Prostate cancer-associated mutation in SPOP impairs its ability to target Cdc20 for poly-ubiquitination and degradation. *Cancer Lett.* **2017**, *385*, 207–214. [CrossRef]
17. Li, K.; Mao, Y.; Lu, L.; Hu, C.; Wang, D.; Si-Tu, J.; Lu, M.; Peng, S.; Qiu, J.; Gao, X. Silencing of CDC20 suppresses metastatic castration-resistant prostate cancer growth and enhances chemosensitivity to docetaxel. *Int. J. Oncol.* **2016**, *49*, 1679–1685. [CrossRef]
18. Li, J.; Karki, A.; Hodges, K.B.; Ahmad, N.; Zoubeidi, A.; Strebhardt, K.; Ratliff, T.L.; Konieczny, S.F.; Liu, X. Cotargeting Polo-Like Kinase 1 and the Wnt/beta-Catenin Signaling Pathway in Castration-Resistant Prostate Cancer. *Mol. Cell Biol.* **2015**, *35*, 4185–4198. [CrossRef]
19. Coffey, K.; Rogerson, L.; Ryan-Munden, C.; Alkharaif, D.; Stockley, J.; Heer, R.; Sahadevan, K.; O'Neill, D.; Jones, D.; Darby, S.; et al. The lysine demethylase, KDM4B, is a key molecule in androgen receptor signalling and turnover. *Nucleic Acids Res.* **2013**, *41*, 4433–4446. [CrossRef]
20. Bainbridge, A.; Walker, S.; Smith, J.; Patterson, K.; Dutt, A.; Ng, Y.M.; Thomas, H.D.; Wilson, L.; McCullough, B.; Jones, D.; et al. IKBKE activity enhances AR levels in advanced prostate cancer via modulation of the Hippo pathway. *Nucleic Acids Res.* **2020**, *48*, 5366–5382. [CrossRef]
21. Coffey, K.; Blackburn, T.J.; Cook, S.; Golding, B.T.; Griffin, R.J.; Hardcastle, I.R.; Hewitt, L.; Huberman, K.; McNeill, H.V.; Newell, D.R.; et al. Characterisation of a Tip60 specific inhibitor, NU9056, in prostate cancer. *PLoS ONE* **2012**, *7*, e45539. [CrossRef] [PubMed]
22. Benjamini, Y.; Hochberg, Y. Controlling the false discovery rate: A practical and powerful approach to multiple hypothesis testing. *J. R. Stat. Soc. B* **1995**, *57*, 289–300. [CrossRef]
23. Labaf, M.; Li, M.; Ting, L.; Karno, B.; Zhang, S.; Gao, S.; Patalano, S.; Macoska, J.A.; Zarringhalam, K.; Han, D.; et al. Increased AR expression in castration-resistant prostate cancer rapidly induces AR signaling reprogramming with the collaboration of EZH2. *Front. Oncol.* **2022**, *12*, 1021845. [CrossRef]
24. Dobin, A.; Davis, C.A.; Schlesinger, F.; Drenkow, J.; Zaleski, C.; Jha, S.; Batut, P.; Chaisson, M.; Gingeras, T.R. STAR: Ultrafast universal RNA-seq aligner. *Bioinformatics* **2013**, *29*, 15–21. [CrossRef] [PubMed]
25. Anders, S.; Pyl, P.T.; Huber, W. HTSeq—A Python framework to work with high-throughput sequencing data. *Bioinformatics* **2015**, *31*, 166–169. [CrossRef]
26. Love, M.I.; Huber, W.; Anders, S. Moderated estimation of fold change and dispersion for RNA-seq data with DESeq2. *Genome Biol.* **2014**, *15*, 550. [CrossRef]
27. Schmidt, D.; Wilson, M.D.; Spyrou, C.; Brown, G.D.; Hadfield, J.; Odom, D.T. ChIP-seq: Using high-throughput sequencing to discover protein-DNA interactions. *Methods* **2009**, *48*, 240–248. [CrossRef]
28. Xie, Q.; Wu, Q.; Mack, S.C.; Yang, K.; Kim, L.; Hubert, C.G.; Flavahan, W.A.; Chu, C.; Bao, S.; Rich, J.N. CDC20 maintains tumor initiating cells. *Oncotarget* **2015**, *6*, 13241–13254. [CrossRef]
29. Huang, D.W.; Sherman, B.T.; Tan, Q.; Collins, J.R.; Alvord, W.G.; Roayaei, J.; Stephens, R.; Baseler, M.W.; Lane, H.C.; Lempicki, R.A. The DAVID Gene Functional Classification Tool: A novel biological module-centric algorithm to functionally analyze large gene lists. *Genome Biol.* **2007**, *8*, R183. [CrossRef]
30. Huang, D.W.; Sherman, B.T.; Tan, Q.; Kir, J.; Liu, D.; Bryant, D.; Guo, Y.; Stephens, R.; Baseler, M.W.; Lane, H.C.; et al. DAVID Bioinformatics Resources: Expanded annotation database and novel algorithms to better extract biology from large gene lists. *Nucleic Acids Res.* **2007**, *35*, W169–W175. [CrossRef]
31. Cerami, E.; Gao, J.; Dogrusoz, U.; Gross, B.E.; Sumer, S.O.; Aksoy, B.A.; Jacobsen, A.; Byrne, C.J.; Heuer, M.L.; Larsson, E.; et al. The cBio cancer genomics portal: An open platform for exploring multidimensional cancer genomics data. *Cancer Discov.* **2012**, *2*, 401–404. [CrossRef] [PubMed]
32. Gao, J.; Aksoy, B.A.; Dogrusoz, U.; Dresdner, G.; Gross, B.; Sumer, S.O.; Sun, Y.; Jacobsen, A.; Sinha, R.; Larsson, E.; et al. Integrative analysis of complex cancer genomics and clinical profiles using the cBioPortal. *Sci. Signal* **2013**, *6*, pl1. [CrossRef] [PubMed]
33. Ren, S.; Wei, G.H.; Liu, D.; Wang, L.; Hou, Y.; Zhu, S.; Peng, L.; Zhang, Q.; Cheng, Y.; Su, H.; et al. Whole-genome and Transcriptome Sequencing of Prostate Cancer Identify New Genetic Alterations Driving Disease Progression. *Eur. Urol.* **2018**, *73*, 322–339. [CrossRef] [PubMed]
34. Grasso, C.S.; Wu, Y.M.; Robinson, D.R.; Cao, X.; Dhanasekaran, S.M.; Khan, A.P.; Quist, M.J.; Jing, X.; Lonigro, R.J.; Brenner, J.C.; et al. The mutational landscape of lethal castration-resistant prostate cancer. *Nature* **2012**, *487*, 239–243. [CrossRef]
35. Robinson, D.; Van Allen, E.M.; Wu, Y.M.; Schultz, N.; Lonigro, R.J.; Mosquera, J.M.; Montgomery, B.; Taplin, M.E.; Pritchard, C.C.; Attard, G.; et al. Integrative clinical genomics of advanced prostate cancer. *Cell* **2015**, *161*, 1215–1228. [CrossRef]

36. Abida, W.; Cyrta, J.; Heller, G.; Prandi, D.; Armenia, J.; Coleman, I.; Cieslik, M.; Benelli, M.; Robinson, D.; Van Allen, E.M.; et al. Genomic correlates of clinical outcome in advanced prostate cancer. *Proc. Natl. Acad. Sci. USA* **2019**, *116*, 11428–11436. [CrossRef]
37. Beltran, H.; Prandi, D.; Mosquera, J.M.; Benelli, M.; Puca, L.; Cyrta, J.; Marotz, C.; Giannopoulou, E.; Chakravarthi, B.V.; Varambally, S.; et al. Divergent clonal evolution of castration-resistant neuroendocrine prostate cancer. *Nat. Med.* **2016**, *22*, 298–305. [CrossRef]
38. Jorgensen, S.; Elvers, I.; Trelle, M.B.; Menzel, T.; Eskildsen, M.; Jensen, O.N.; Helleday, T.; Helin, K.; Sorensen, C.S. The histone methyltransferase SET8 is required for S-phase progression. *J. Cell Biol.* **2007**, *179*, 1337–1345. [CrossRef] [PubMed]
39. Tardat, M.; Murr, R.; Herceg, Z.; Sardet, C.; Julien, E. PR-Set7-dependent lysine methylation ensures genome replication and stability through S phase. *J. Cell Biol.* **2007**, *179*, 1413–1426. [CrossRef]
40. Blum, G.; Ibanez, G.; Rao, X.; Shum, D.; Radu, C.; Djaballah, H.; Rice, J.C.; Luo, M. Small-molecule inhibitors of SETD8 with cellular activity. *ACS Chem. Biol.* **2014**, *9*, 2471–2478. [CrossRef] [PubMed]
41. Ma, A.; Yu, W.; Xiong, Y.; Butler, K.V.; Brown, P.J.; Jin, J. Structure-activity relationship studies of SETD8 inhibitors. *Medchemcomm* **2014**, *5*, 1892–1898. [CrossRef] [PubMed]
42. Nath, S.; Chowdhury, A.; Dey, S.; Roychoudhury, A.; Ganguly, A.; Bhattacharyya, D.; Roychoudhury, S. Deregulation of Rb-E2F1 axis causes chromosomal instability by engaging the transactivation function of Cdc20-anaphase-promoting complex/cyclosome. *Mol. Cell Biol.* **2015**, *35*, 356–369. [CrossRef] [PubMed]
43. Williams, K.; Christensen, J.; Rappsilber, J.; Nielsen, A.L.; Johansen, J.V.; Helin, K. The histone lysine demethylase JMJD3/KDM6B is recruited to p53 bound promoters and enhancer elements in a p53 dependent manner. *PLoS ONE* **2014**, *9*, e96545. [CrossRef] [PubMed]
44. Trojer, P.; Li, G.; Sims, R.J., 3rd; Vaquero, A.; Kalakonda, N.; Boccuni, P.; Lee, D.; Erdjument-Bromage, H.; Tempst, P.; Nimer, S.D.; et al. L3MBTL1, a histone-methylation-dependent chromatin lock. *Cell* **2007**, *129*, 915–928. [CrossRef]
45. Shoaib, M.; Walter, D.; Gillespie, P.J.; Izard, F.; Fahrenkrog, B.; Lleres, D.; Lerdrup, M.; Johansen, J.V.; Hansen, K.; Julien, E.; et al. Histone H4K20 methylation mediated chromatin compaction threshold ensures genome integrity by limiting DNA replication licensing. *Nat. Commun.* **2018**, *9*, 3704. [CrossRef]
46. Oda, H.; Hubner, M.R.; Beck, D.B.; Vermeulen, M.; Hurwitz, J.; Spector, D.L.; Reinberg, D. Regulation of the histone H4 monomethylase PR-Set7 by CRL4(Cdt2)-mediated PCNA-dependent degradation during DNA damage. *Mol. Cell* **2010**, *40*, 364–376. [CrossRef]
47. Nishioka, K.; Rice, J.C.; Sarma, K.; Erdjument-Bromage, H.; Werner, J.; Wang, Y.; Chuikov, S.; Valenzuela, P.; Tempst, P.; Steward, R.; et al. PR-Set7 is a nucleosome-specific methyltransferase that modifies lysine 20 of histone H4 and is associated with silent chromatin. *Mol. Cell* **2002**, *9*, 1201–1213. [CrossRef]
48. Yang, F.; Sun, L.; Li, Q.; Han, X.; Lei, L.; Zhang, H.; Shang, Y. SET8 promotes epithelial-mesenchymal transition and confers TWIST dual transcriptional activities. *EMBO J.* **2012**, *31*, 110–123. [CrossRef]
49. Li, Z.; Nie, F.; Wang, S.; Li, L. Histone H4 Lys 20 monomethylation by histone methylase SET8 mediates Wnt target gene activation. *Proc. Natl. Acad. Sci. USA* **2011**, *108*, 3116–3123. [CrossRef]
50. Vakoc, C.R.; Sachdeva, M.M.; Wang, H.; Blobel, G.A. Profile of histone lysine methylation across transcribed mammalian chromatin. *Mol. Cell Biol.* **2006**, *26*, 9185–9195. [CrossRef]
51. Shoaib, M.; Chen, Q.; Shi, X.; Nair, N.; Prasanna, C.; Yang, R.; Walter, D.; Frederiksen, K.S.; Einarsson, H.; Svensson, J.P.; et al. Histone H4 lysine 20 mono-methylation directly facilitates chromatin openness and promotes transcription of housekeeping genes. *Nat. Commun.* **2021**, *12*, 4800. [CrossRef] [PubMed]
52. Kapoor-Vazirani, P.; Vertino, P.M. A dual role for the histone methyltransferase PR-SET7/SETD8 and histone H4 lysine 20 monomethylation in the local regulation of RNA polymerase II pausing. *J. Biol. Chem.* **2014**, *289*, 7425–7437. [CrossRef] [PubMed]
53. Hori, T.; Shang, W.H.; Toyoda, A.; Misu, S.; Monma, N.; Ikeo, K.; Molina, O.; Vargiu, G.; Fujiyama, A.; Kimura, H.; et al. Histone H4 Lys 20 monomethylation of the CENP-A nucleosome is essential for kinetochore assembly. *Dev. Cell* **2014**, *29*, 740–749. [CrossRef] [PubMed]
54. Wu, F.; Sun, Y.; Chen, J.; Li, H.; Yao, K.; Liu, Y.; Liu, Q.; Lu, J. The Oncogenic Role of APC/C Activator Protein Cdc20 by an Integrated Pan-Cancer Analysis in Human Tumors. *Front. Oncol.* **2021**, *11*, 721797. [CrossRef]
55. Liang, X.; Hu, K.; Li, D.; Wang, Y.; Liu, M.; Wang, X.; Zhu, W.; Wang, X.; Yang, Z.; Lu, J. Identification of Core Genes and Potential Drugs for Castration-Resistant Prostate Cancer Based on Bioinformatics Analysis. *DNA Cell Biol.* **2020**, *39*, 836–847. [CrossRef]
56. Dai, L.; Song, Z.X.; Wei, D.P.; Zhang, J.D.; Liang, J.Q.; Wang, B.B.; Ma, W.T.; Li, L.Y.; Dang, Y.L.; Zhao, L.; et al. CDC20 and PTTG1 are Important Biomarkers and Potential Therapeutic Targets for Metastatic Prostate Cancer. *Adv. Ther.* **2021**, *38*, 2973–2989. [CrossRef]
57. Gu, P.; Yang, D.; Zhu, J.; Zhang, M.; He, X. Bioinformatics analysis identified hub genes in prostate cancer tumorigenesis and metastasis. *Math. Biosci. Eng.* **2021**, *18*, 3180–3196. [CrossRef]
58. Wei, J.; Yin, Y.; Deng, Q.; Zhou, J.; Wang, Y.; Yin, G.; Yang, J.; Tang, Y. Integrative Analysis of MicroRNA and Gene Interactions for Revealing Candidate Signatures in Prostate Cancer. *Front. Genet.* **2020**, *11*, 176. [CrossRef]
59. Luo, C.; Chen, J.; Chen, L. Exploration of gene expression profiles and immune microenvironment between high and low tumor mutation burden groups in prostate cancer. *Int. Immunopharmacol.* **2020**, *86*, 106709. [CrossRef]

60. Barbieri, C.E.; Baca, S.C.; Lawrence, M.S.; Demichelis, F.; Blattner, M.; Theurillat, J.P.; White, T.A.; Stojanov, P.; Van Allen, E.; Stransky, N.; et al. Exome sequencing identifies recurrent SPOP, FOXA1 and MED12 mutations in prostate cancer. *Nat. Genet.* **2012**, *44*, 685–689. [CrossRef]
61. Shi, Q.; Zhu, Y.; Ma, J.; Chang, K.; Ding, D.; Bai, Y.; Gao, K.; Zhang, P.; Mo, R.; Feng, K.; et al. Prostate Cancer-associated SPOP mutations enhance cancer cell survival and docetaxel resistance by upregulating Caprin1-dependent stress granule assembly. *Mol. Cancer* **2019**, *18*, 170. [CrossRef] [PubMed]
62. Wu, F.; Lin, Y.; Cui, P.; Li, H.; Zhang, L.; Sun, Z.; Huang, S.; Li, S.; Huang, S.; Zhao, Q.; et al. Cdc20/p55 mediates the resistance to docetaxel in castration-resistant prostate cancer in a Bim-dependent manner. *Cancer Chemother. Pharmacol.* **2018**, *81*, 999–1006. [CrossRef] [PubMed]
63. Tuzon, C.T.; Spektor, T.; Kong, X.; Congdon, L.M.; Wu, S.; Schotta, G.; Yokomori, K.; Rice, J.C. Concerted activities of distinct H4K20 methyltransferases at DNA double-strand breaks regulate 53BP1 nucleation and NHEJ-directed repair. *Cell Rep.* **2014**, *8*, 430–438. [CrossRef]
64. Dulev, S.; Tkach, J.; Lin, S.; Batada, N.N. SET8 methyltransferase activity during the DNA double-strand break response is required for recruitment of 53BP1. *EMBO Rep.* **2014**, *15*, 1163–1174. [CrossRef]
65. Lu, X.; Xu, M.; Zhu, Q.; Zhang, J.; Liu, G.; Bao, Y.; Gu, L.; Tian, Y.; Wen, H.; Zhu, W.G. RNF8-ubiquitinated KMT5A is required for RNF168-induced H2A ubiquitination in response to DNA damage. *FASEB J.* **2021**, *35*, e21326. [CrossRef]
66. Jimenez-Vacas, J.M.; Herrero-Aguayo, V.; Montero-Hidalgo, A.J.; Gomez-Gomez, E.; Fuentes-Fayos, A.C.; Leon-Gonzalez, A.J.; Saez-Martinez, P.; Alors-Perez, E.; Pedraza-Arevalo, S.; Gonzalez-Serrano, T.; et al. Dysregulation of the splicing machinery is directly associated to aggressiveness of prostate cancer. *EBioMedicine* **2020**, *51*, 102547. [CrossRef] [PubMed]

cancers

MDPI

Article

Multimodal Data Integration to Predict Severe Acute Oral Mucositis of Nasopharyngeal Carcinoma Patients Following Radiation Therapy

Yanjing Dong [1], Jiang Zhang [1], Saikt Lam [2,3], Xinyu Zhang [1], Anran Liu [1], Xinzhi Teng [1], Xinyang Han [1], Jin Cao [1], Hongxiang Li [4], Francis Karho Lee [5], Celia Waiyi Yip [5], Kwokhung Au [5], Yuanpeng Zhang [6] and Jing Cai [1,2,7,*]

1 Department of Health Technology and Informatics, The Hong Kong Polytechnic University, Hong Kong SAR, China
2 Research Institute for Smart Ageing, The Hong Kong Polytechnic University, Hong Kong SAR, China
3 Department of Biomedical Engineering, Faculty of Engineering, The Hong Kong Polytechnic University, Hong Kong SAR, China
4 Department of Radiology, Fujian Medical University Union Hospital, Fujian Medical University, Fuzhou 350000, China
5 Department of Clinical Oncology, Queen Elizabeth Hospital, Hong Kong SAR, China
6 Department of Medical Informatics, Nantong University, Nantong 226000, China
7 The Hong Kong Polytechnic University Shenzhen Research Institute, Shenzhen 518000, China
* Correspondence: jing.cai@polyu.edu.hk

Simple Summary: The acute oral mucositis (AOM) is a prevalent side effect of radiation therapy for nasopharyngeal carcinoma (NPC). Severe AOM could impair the survival and quality of life for NPC patients. Accurate method to predict the incidence of severe AOM can aid clinicians in adjusting the treatment plan to improve the outcomes for NPC patients. We integrated multi-modalities, multi-omics and multi-regions data with two methods, integrate the original data or combine data after feature selection. The performance of models using each data integration method with different modalities, types of data and VOIs were analyzed. We developed a best-performing model with mean AUC at 0.81 ± 0.10 to predict the incidence of severe AOM for NPC patients following radiation therapy.

Abstract: (1) Background: Acute oral mucositis is the most common side effect for nasopharyngeal carcinoma patients receiving radiotherapy. Improper or delayed intervention to severe AOM could degrade the quality of life or survival for NPC patients. An effective prediction method for severe AOM is needed for the individualized management of NPC patients in the era of personalized medicine. (2) Methods: A total of 242 biopsy-proven NPC patients were retrospectively recruited in this study. Radiomics features were extracted from contrast-enhanced CT (CECT), contrast-enhanced T1-weighted (cT1WI), and T2-weighted (T2WI) images in the primary tumor and tumor-related area. Dosiomics features were extracted from 2D or 3D dose-volume histograms (DVH). Multiple models were established with single and integrated data. The dataset was randomized into training and test sets at a ratio of 7:3 with 10-fold cross-validation. (3) Results: The best-performing model using Gaussian Naive Bayes (GNB) (mean validation AUC = 0.81 ± 0.10) was established with integrated radiomics and dosiomics data. The GNB radiomics and dosiomics models yielded mean validation AUC of 0.6 ± 0.20 and 0.69 ± 0.14, respectively. (4) Conclusions: Integrating radiomics and dosiomics data from the primary tumor area could generate the best-performing model for severe AOM prediction.

Keywords: multimodal data integration; radiomics; dosiomics; nasopharyngeal carcinoma; acute mucositis

Citation: Dong, Y.; Zhang, J.; Lam, S.; Zhang, X.; Liu, A.; Teng, X.; Han, X.; Cao, J.; Li, H.; Lee, F.K.; et al. Multimodal Data Integration to Predict Severe Acute Oral Mucositis of Nasopharyngeal Carcinoma Patients Following Radiation Therapy. *Cancers* **2023**, *15*, 2032. https://doi.org/10.3390/cancers15072032

Academic Editors: Wei Wu and Trever G. Bivona

Received: 23 February 2023
Revised: 21 March 2023
Accepted: 26 March 2023
Published: 29 March 2023

1. Introduction

Nasopharyngeal carcinoma (NPC) is a kind of malignant epithelial head and neck (H&N) cancer. It originates from the nasopharyngeal mucosal lining with high incidence in Southeast Asia [1]. Over 95% of NPC cases are non-keratinizing squamous cell carcinoma that is highly sensitive to radiation [2]. Radiotherapy with/without chemotherapy (CRT or RT, respectively) is the primary treatment for NPC patients to achieve disease control. Although advanced radiation strategies such as intensity-modulated RT (IMRT) are widely applied to reduce the recurrence rate of tumor with decreased radiation-induced side effects [3,4], radiation toxicity is still a trouble for NPC patients undergoing radiation therapy [5,6]. The radiation damages DNA and cellular components, which could cause mitotic cell death, apoptosis, and cytokine cascade in the human body. These cascades of events could eventually and inevitably lead to toxicity effects [7]. Most NPC patients suffer from dermatitis, mucositis, dysphagia, and xerostomia during and after RT/CRT. Among these acute toxicity effects, acute oral mucositis (AOM) is the most common painful symptomatic complication for NPC patients [8].

More than 60% of H&N patients have experienced AOM following RT-based anti-cancer therapies. Approximately 65% of these patients have developed severe AOM ($\geq$grade 3) [8]. The AOM is typically characterized by atrophy, swelling, erythema, and ulceration. It often impairs patients' functional status and quality of life (QoL) [9]. The soreness of AOM initially leads to open-mouth difficulty, which further causes decreased food and water intake, loss of weight, and systematic infection. Patients who have developed severe AOM with painful experiences would receive a reduced dose of chemotherapy. Some patients even tend to break the RT regime. Severe AOM can exacerbate the morbidity of patients, which may finally contribute to worsen QoL and increase mortality [10]. It therefore is necessary to analyze the critical contributors to the oral mucositis.

Previous studies have focused on analyzing one type of data, such as genetic data and clinical variables, for predicting severe AOM. Various factors have been identified to correlate with the incidence and severity of oral mucositis, for instance, genetic background [11,12], dose of RT [13,14], chemo-drugs [15–17], and nutritional status [18]. Among these factors, the dose of RT is a critical factor influencing the severity of AOM. Additional chemotherapy, especially with some AOM-associated agents, for instance, alkylating agents and antimetabolites, could exacerbate the events. Besides, AOM patients are more likely to have weight loss during the treatment scheme [19,20]. Saito et al. [20,21] reported that low BMI is a risk factor for severe AOM. Andy et al. [21] indicated that patients with advanced tumors are prone to experience AOM. Moreover, a recent two-stage genome-wide association study [12] showed that four single nucleotide polymorphisms (SNPs) might be correlated with acute mucositis. However, they failed to validate their results in the validation stage. Clinicogenomic variables alone are inadequate to accurately predict the incidence, correlations, and severity of AOM for NPC patients after RT.

In addition to the genomics information, contrast-enhanced computed tomography (CECT), magnetic resonance imaging (MRI), and dose files routinely acquire clinical data for NPC patients with RT plans [22,23]. Radiomics and dosiomics are two quantitative information-extraction methods to provide minable texture and dose-distribution information for clinical prognosis prediction. Traditional experiences have demonstrated that single or double sources of data had limited prediction power for acute AOM. Integration of complementary data from multiple types of datasets can lead to an intricate outcome than a simple summation of information [24]. Integration of multimodal data from multiple sources, for instance, clinical, radiomics, and dosiomics for NPC patients, has the potential to overcome the boundaries of conventional medical analysis [25–31]. Unfortunately, few studies have reported the possibilities of radiomics or dosiomics for AOM prediction [13,32,33]. To the best of our knowledge, there are no studies in the literature assessing whether the data fusion of multi-regions and multimodalities could enhance their capability of severe AOM prediction.

In this study, we aimed to analyze multimodal data, including clinical, radiomics, and dosiomics information, to predict the incidence of severe AOM in NPC patients following RT/CRT. The radiomics and dosiomics data were extracted from multi-regions related to RT treatment. These data were also extracted from multimodalities of images (CECT, contrast-enhanced T1-weighted, and T2-weighted MRI). In daily clinical practice, clinicians could benefit from adjusting treatment plans before RT for patients with a higher possibility of developing severe AOM to achieve personalized diagnosis and treatment.

2. Materials and Methods

Patient data. All patient data were retrospectively collected from NPC patients who underwent CRT or RT at Hong Kong Queen Elizabeth Hospital from 2012 to 2015. Informed consent of patients was waived due to the nature of the retrospective study. NPC patients were recruited based on the following inclusion and exclusion criteria. The inclusion criteria were: (1) NPC patients with pathological validation and absence of distant metastasis and co-existing tumors of other types at diagnosis, (2) patients treated with a total RT dose of 60–70 Gy, and (3) patients with a completed set of clinical, image, and radiation dosimetry data. The exclusion criteria were: (1) patients aged less than 18, (2) patients without original image or clinical data; and (3) patients for whom exact standard terminology criteria for adverse events (CTCAE) evaluation for AOM had not been recorded. Symptoms in grades 1 and 2 were defined as mild AOM, and grades 3 and 4 as severe AOM [34] All the patients were negative of oral mucositis according to the CTCAE grading system before receiving radiation therapy.

Clinical variables included (1) treatment information: TNM stage, treatment, past health condition, allergy history, vision condition, hearing condition, and CTCAE evaluation for AOM and (2) demographic data: age, gender, body weight, height, body mass index (BMI), and smoking and drinking habits. All clinical variables were acquired one week before RT commencement, except the CTCAE evaluation results, which were recorded 4–5 weeks after RT commencement. The equation for BMI is defined as follows:

$$\text{Body mass index (BMI)} = \text{weight}/\text{height}^2 \quad (1)$$

Patients were maintained in a supine position during the imaging examination. Details of imaging acquisition are summarized in Tables A1 and A2.

Image pre-processing. In this study, the imaging pre-processing steps were based on our previous work [28] and are in accordance with the Image Biomarker Standardisation Initiative (IBSI) guidelines [35]. Specifically, (1) voxel size resampling: all images (CECT and MRI) were resampled to a voxel size of $1 \times 1 \times 1$ mm^3; (2) volume of interest (VOI) re-segmentation: CECT images were re-segmented to confine the Hounsfield unit (HU) to (−150,180) to eliminate the non-soft tissue in the VOI; (3) image filtering: a Laplacian of Gaussian (LoG) filter with three levels of Gaussian radius parameter was used under fine (1 mm), medium (3 mm), and coarse (6 mm) scales; (4) quantization of gray levels: gray-level intensities of the images were fixed to 50 bins; and (5) inhomogeneity correction of image pixel value: N4B bias correction in the "N4 Bias Field Correction Image Filter" in SimpleITK (v1.2.4) was implemented, in particular, to MRI images.

Radiomics and dosiomics feature extraction. Feature extraction was performed using our in-house platform based on publicly available SimpleITK (v1.2.4) and PyRadiomics (v2.2.0) [36,37]. All VOIs were delineated by an experienced senior clinician [33]. The gross tumor volumes (GTVs) were contoured based on CECT with the assistance of MRI images.

Radiomics. The gross tumor volume of the NPC primary tumor (GTVnp) and the gross tumor volume of nodal lesions (GTVn) were selected as the main VOIs for radiomics feature extraction. Features with or without LoG filters were both involved. All these features were extracted from CECT, contrast-enhanced T1 weighted (cT1WI), and T2 weighted (T2WI) images (for details, please refer to Figure 1). Meaning of each VOI for different image modalities were listed in Table 1.

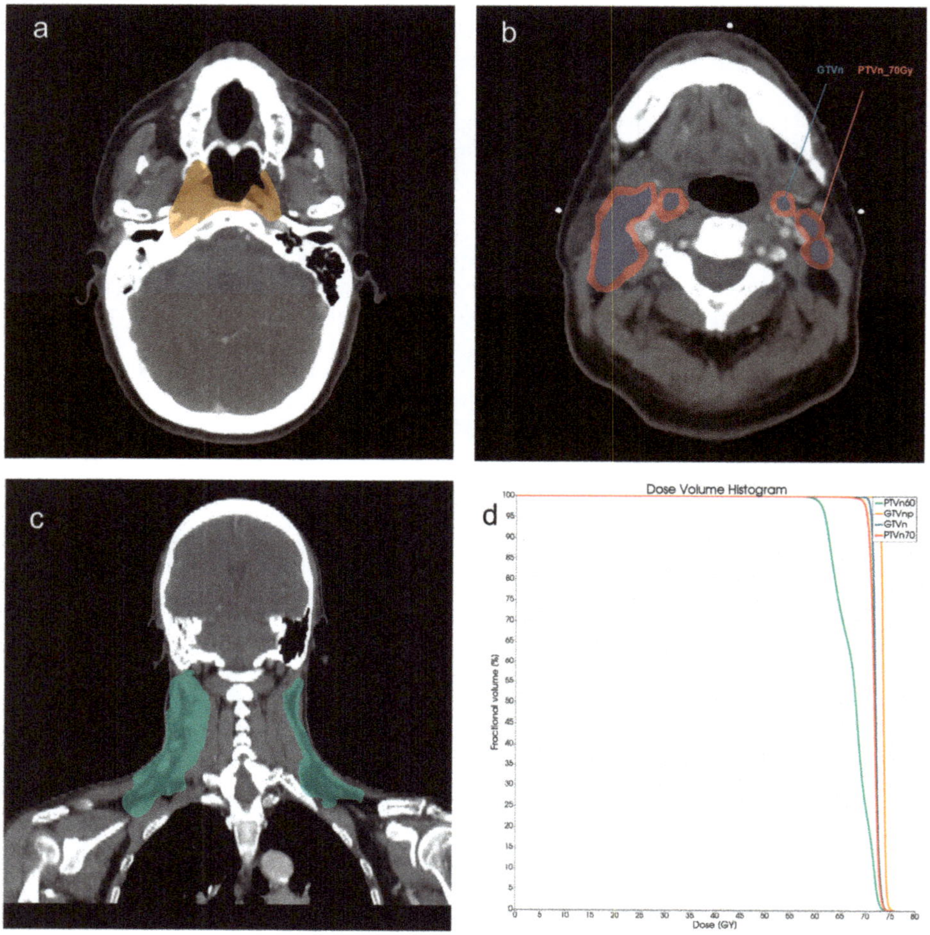

Figure 1. VOI examples for NPC patients with CECT examination. (**a**) Region of GTVnp (orange), axial view. (**b**) Region of GTVn (blue) and PTVn_70 Gy (red), axial view. (**c**) Region of PTVn_60 Gy (green), coronal view. (**d**) DVH curve of four VOIs.

Table 1. VOIs and image modalities.

VOIs	Descriptions of VOI	Imaging Modalities/Images
GTVnp	Gross tumor volume of primary NPC tumor	CECT, cT1WI, T2WI, DVH
GTVn	Metastatic lymph nodes area	CECT, DVH
PTVn_70Gy	Regions of nodal planning target volume with the prescribed dose level of 70Gy	DVH
PTVn_60Gy	PTVn with the prescribed dose level of 60Gy	DVH

Three categories of radiomics features were extracted: shape, first-order statistics, and texture features. The texture features can be further categorized into gray-level difference matrix (GLDM), gray-level cooccurrence matrix (GLCM), gray-level run-length matrix (GLRLM), gray-level size-zone matrix (GLSZM), and neighboring gray-tone difference matrix (NGTDM) classes.

Dosiomics. Except for GTVnp and GTVn, the region of the high-dose nodal planning target volume (PTVn_70Gy) and region of the low-dose nodal planning target volume (PTVn_60Gy) were also added to the dosiomics analysis (please refer to Figures 1 and A2 and Table 1 for more details).

Two-dimensional (2D) and three-dimensional (3D) dose–volume histograms (DVHs) of each studied VOI were extracted from dose files for dosiomics feature extraction. All dosiomics features were extracted based on Gabry et al.'s previous study [39]. Features that reflect dose distribution, for instance, mean dose, spatial dose gradient, and spatial dose spread were extracted accordingly. All the calculation algorithms have been listed in a previous publication by Buettner et al. [40].

Model development and evaluation. The statistical analysis, model training, and evaluation were conducted in Jupyter 6.4.12 and SPSS 25. The receiver operating characteristic (ROC) curve and area under the ROC curve (AUC) with 10-fold validation was performed to evaluate model performance. The CTCAE grade scale of patients in mucositis was dichotomized between severe AOM (grade $\geq$ 3) and mild AOM (grade $\leq$ 2) as the prediction endpoint. Patients were stratified based on CTCAE grade to training and testing groups at a 7:3 ratio (details in Figure 2).

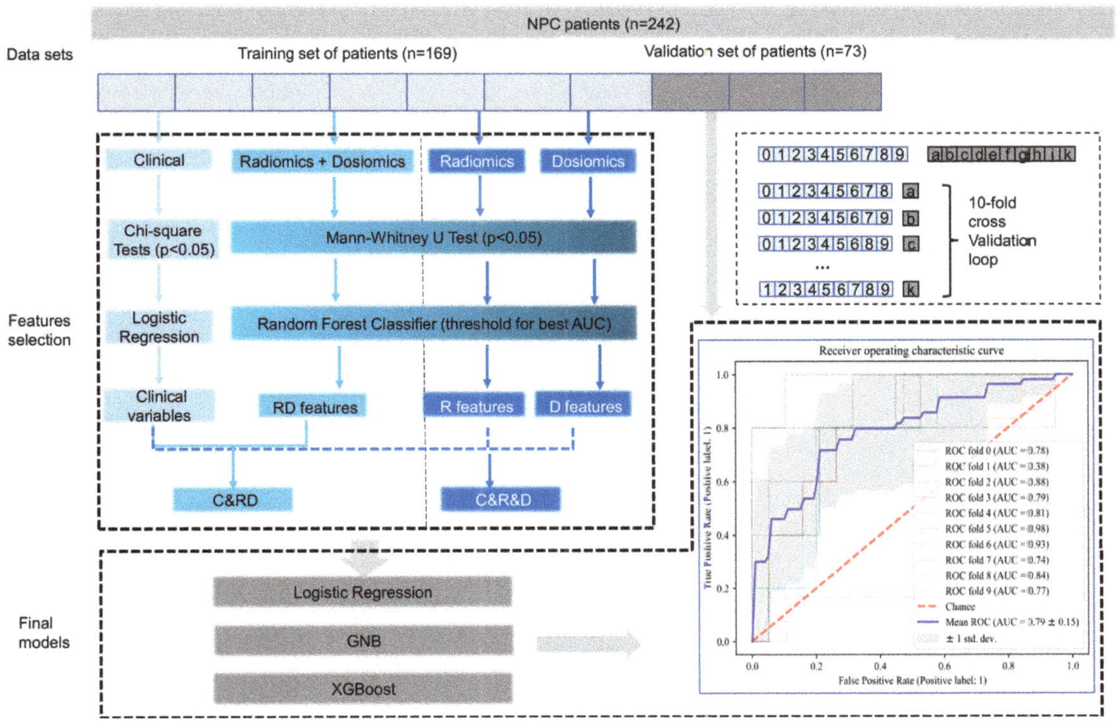

Figure 2. Scheme of feature selection and modeling. Training and validation sets were separated before data analysis. The training set of data was used for feature selection. The validation set of data was used for model evaluation. To further manipulate the numerical and categorical data, reduce the interactions, and solve the collinearity problems, random forest (RF) selection was applicated for radiomics, dosiomics, and integrated data. Three linear or non-linear models were developed with independent validation data sets with selected features. The area under the curve (AUC) was set as the main evaluation method for the model performance.

Single-modal models. The data sources for single-modal models were restricted to single modality of data (radiomics, dosiomics, or clinics), single modality of imaging (CECT, cT1WI, T2WI, and DVH), and single region of patients (GTVnp, GTVn, PTVn_60Gy, and PTVn_70Gy). Each single data set had two steps in this stage: (1) feature selection and model training in the training group set and (2) AUC evaluation in validation groups.

For clinical data, chi-square and Mann–Whitney U tests were employed for binary and non-binary variables for univariate analysis. p values < 0.05 was considered to be statistically significant. All radiomics and dosiomics data were standardized with the MinMax scaler before selection. For radiomics and dosiomics data, we first identified significant features between severe and mild AOM patients in the training set with Mann–Whitney U tests. After that, random forest (RF) was used to rank the importance of the significant features considering both feature interactions and nonlinearities. The optimal feature number was set according to the best RF training model score. Three models, including logistic regression (LR), Gaussian Naïve Bayes (GNB), and extreme gradient boosting (XGBoost), were applied to evaluate the combined predictive value of these selected features in the independent validation set. All VOIs data were analyzed separately at the single model stage.

Multimodal data integration. Clinical data after multivariant analysis (LR) with p value < 0.05 were selected for data integration. Dosiomics and radiomics data from different VOIs and image modalities were integrated with two methods: (1) dosiomics and radiomics data were combined together before feature selection and (2) the features selected from the RF model were merged and directly combined without a further feature-selection step (please refer to Figure 2 for more details). All the data-integration methods are listed in Table 2.

Table 2. Data resources and integration/combination methods.

Name of Model	Methods
GTVnp_RD	Integration of radiomics and dosiomics GTVnp data before feature selection
GTVnp_R_CECTcT1T2	Integration of radiomics GTVnp data from CECT, cT1WI, T2WI before feature selection
GTVnp_R_CECTcT1	Integration of radiomics GTVnp data from CECT and cT1WI before feature selection
GTVnp_R_cT1T2	Integration of radiomics GTVnp data from cT1WI and T2WI before feature selection
GTVnp_R_cT1	Single radiomics data from CT1WI
GTVnp_R_CECT	Single radiomics data from CECT
GTVnp_R_T2	Single radiomics data from T2WI
GTVnp_D	Single dosiomics data from GTVnp
GTVn_RD	Integration of radiomics and dosiomics data from GTVn before feature selection
GTVn_R	Single radiomics data from GTVn
GTVn_D	Single dosiomics data from GTVn
PTVn_D	Integration of 60 and 70 Gy dosiomics data before feature selection
PTVn_60Gy_D	Single dosiomics data from PTVn_60Gy
PTVn_70Gy_D	Single dosiomics data from PTVn_70Gy
R	Integration of all radiomics data before feature selection
D	Integration of all dosiomics data before feature selection
C	Single clinical data
C&D	Combine selected clinical and dosiomics data for modeling
C&R	Combine selected clinical and radiomics data for modeling
RD	Integration of radiomics and dosiomics data before feature selection
C&RD	Combine selected clinical and RD data for modeling
C&GTVnp RD	Combine selected clinical and GTVnp RD data for modeling
R&D	Combine selected radiomics and dosiomics data for modeling
C&R&D	Combine selected clinical, radiomics and dosiomics data for modeling

Shapley Additive Explanations (SHAP), an explainable artificial intelligence (AI)-based tool, was applied for further explanation of feature importance for the model with the best AUC result and specific features [41].

3. Results

3.1. Patient Characteristics

A total of 397 continuous patients were collected based on their final diagnosis with pathological validation. Of these patients, with a median age of 54 (range 26–86 years), 242 were enrolled for further analysis following the inclusion and exclusion criteria (details in Figure A1). All patients were negative for oral mucositis with CTCAE graded 0 before radiation therapy. Univariate analysis results of demographic and clinical characteristics for those patients are listed in Table 3.

Table 3. Demographic and clinical characteristics for all patients.

Characteristics	AOM < Grade 3 (Mild AOM)	AOM ≥ Grade 3 (Severe AOM)	*p* Value
Total Number	191 (78.9%)	51 (21.1%)	
Age, mean ± SD, years	54.89 ± 12.25	50.9 ± 10.60	0.036 *
18–65	149 (61.6%)	44 (18.1%)	
≥65	42 (17.4%)	7 (2.9%)	0.192
Gender			
Male	135 (55.8%)	41 (16.9%)	
Female	56 (23.1%)	10 (4.1%)	0.167
Treatment			0.004 *
RT alone	27 (11.2%)	0	
CRT	164 (67.8%)	51 (21.1%)	
T stage			0.031 *
T1	15 (6.2%)	3 (0.1%)	
T2	8 (3.3%)	5 (2.1%)	
T3	137 (56.6%)	28 (11.6%)	
T4	31 (12.8%)	15 (6.2%)	
N stage			0.091
N1	28 (11.2%)	1 (0.4%)	
N2	142 (58.7%)	45 (18.6%)	
N3	20 (8.2%)	5 (2.1%)	
Pathology			
Non-keratinizing squamous cell	175 (72.3%)	48 (19.8%)	0.556
Keratinizing squamous-cell carcinoma	16 (6.6%)	3 (1.3%)	0.487
Past health condition			
Past health good	92 (38.0%)	27 (11.2%)	
Basic diseases/cancer	99 (40.9%)	24 (9.9%)	0.545
Allegory of History			
No known drug allergies	176 (72.7%)	46 (19.0%)	
Allergy history	15 (6.2%)	5 (2.1%)	0.653
Vision			
Normal	189 (78.1%)	51 (21.1%)	
With eye impairment	2 (0.8%)	0	0.463
Hearing			
Normal	186 (76.9%)	48 (19.8%)	
With hearing impairment	5 (2.1%)	3 (1.2%)	0.247
Habits			
Smoking	9 (3.7%)	6 (2.5%)	0.044 *
Non-smoker	182 (75.2%)	45 (18.6%)	
Drinking	4 (1.7%)	1 (0.4%)	
No alcohol consumption	187 (77.3%)	50 (20.7%)	0.953

Table 3. *Cont.*

Characteristics	AOM < Grade 3 (Mild AOM)	AOM ≥ Grade 3 (Severe AOM)	*p* Value
Height, mean ± SD, cm	163.4 ± 8.5	165.0 ± 8.0	0.561
Body weight, mean ± SD, kg			
1st week of RT	63.1 ± 11.9	66.2 ± 14.6	1.599
2nd week of RT	62.0 ± 11.8	64.9 ± 14.5	1.5
3rd week of RT	61.2 ± 11.4	63.9 ± 14.1	0.116
4th week of RT	60.2 ± 11.3	62.8 ± 14.0	1.418
BMI			
1st week of RT			
<25	131 (54.1%)	32 (13.2%)	
≥25	60 (24.8%)	19 (7.9%)	0.429
2nd week of RT			
<25	131 (54.1%)	51 (21.1%)	
≥25	60 (24.8%)	22 (9.1%)	0.116
3rd week of RT			
<25	131 (54.1%)	31 (12.8%)	
≥25	55 (22.7%)	20 (8.3%)	0.153
4th week of RT			
<25	142 (58.7%)	34 (14.0%)	
≥25	49 (20.2%)	17 (7.0%)	0.274

* $p < 0.05$. All the above data are derived from biopsy-proven primary NPC patients without the existence of distant metastasis or co-existing tumors of other type at diagnosis.

3.2. Feature Extraction and Model Development

3.2.1. Feature Extraction

In this study, a total of 1544 radiomics features, 386 features each for four modalities of imaging, were extracted from raw and LoG-filtered images. A total of 836 dosiomics features (210 for GTVn, 211 for GTVnp, 204 for PTVn_60Gy, and 211 for PTVn_70Gy) were extracted from dose images.

3.2.2. Models

For the clinical data, four variables, including age, RT treatment alone, T stage, and smoking habits, were selected after univariate analysis. The logistic regression (LR) model was established with these variables. T stage and smoking habits had statistical significance in the LR model with a *p*-value < 0.05 (details in Table 4).

Table 4. Logistic regression results for single clinical data model.

Variables	*p*-Value	95% Confidence Interval	
		Lower 95% Bound	Upper 95% Bound
Age (18, 65)	0.802	0.345	2.274
T	0.007 *		
T 1	0.591	0.149	2.96
T 2	0.069	0.881	29.854
T 3	0.024 *	0.195	0.891
RT alone	0.998	0	.
Smoker	0.043 *	1.037	10.683

* $p < 0.05$.

Radiomics and dosiomics features extracted from various VOIs were put into Mann–Whitney U tests and RF classifier step by step. RF selection results of the threshold and feature numbers are listed in Table A3.

Nine categories of single-modal models (C, PTVn_70Gy_D, PTVn_60Gy_D, GTVn_D, GTVnp_D, GTVn_R, GTVnp_R_T2, GTVnp_R_CECT, and GTVnp_R_cT1) were established with single modal, single modality, and single VOI data. The best validation AUC was at 0.75 ± 0.12 (training AUC = 0.73 ± 0.01) of a GNB model (GTVnp_R_cT1) with radiomics data from GTVnp of cT1WI. Seven groups of models with data integrated before feature selection (raw-data integration) were generated with the best AUC of a GNB model (GTVnp_RD) at 0.81 ± 0.01 (training AUC = 0.79 ± 0.01). This best-performing model was constructed with features selected from radiomics and dosiomics data in the region of GTVnp. In addition, six sets of combined data after feature selection were also used for modeling. A best LR model (C&R&D) with AUC at 0.79 ± 0.14 (training AUC = 0.81 ± 0.02) was set with the simply combined data of selected clinical, dosiomics, and radiomics features (details of mean 10-fold validation AUC results are listed in Figure 3)

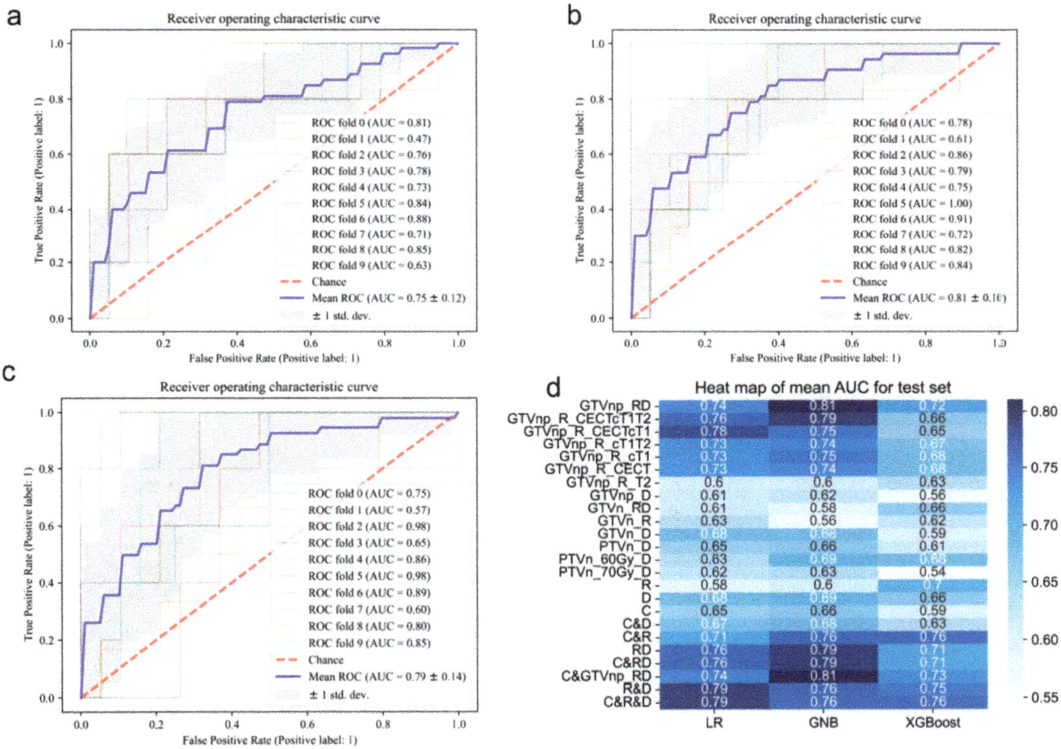

Figure 3. 10-fold validation AUC results for the test set. (**a**) The AUC plot of GNB model for the GTVnp_R_cT1 data set. (**b**) The AUC plot of GNB model for the GTVnp_RD data set. (**c**) The AUC plot of LR model for the C&R&D data set. (**d**) The heatmap of mean AUC results for all models.

The SHAP analysis showed the importance of the five features in the GTVnp_RD model for prediction of severe AOM. Four of the five features were derived from cT1WI. All five features are texture features. No dosiomics features were selected after the feature selections (details in Figure 4).

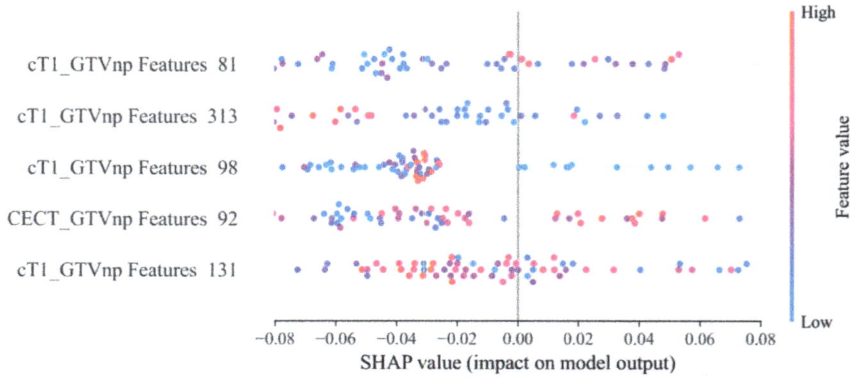

Figure 4. Feature importance of SHAP for XGBoost model of GTVnp_RD. From the highest to the lowest level, the features are categorized in GLSZM, log sigma 60 mm 3D GLCM, original GLDM, GLDM, and log sigma 10 mm 3D GLCM.

4. Discussion

In our study, we used simply combined and data-fusion methods to manage multimodalities of data (clinical, radiomics, and dosiomics), multimodalities of imaging (CECT, cT1WI, and T2WI), and multi-regional information (GTVn, GTVnp, and PTVn) to predict the incidence of severe AOM. Multiple models were established to evaluate and determine which method was effective for clinical decision-making. Comparison of the AUC between models showed that the simple combination of single-modal data of selected features had the most stable performance (C&R&D), with an average AUC of 0.77 ± 0.17. In addition, data-fusion methods, integrating radiomics and dosiomics data before selection procedures, resulted in the best-performing model (GTVnp_RD), with the best test AUC of 0.81 ± 0.01. This is also the best AUC among the existing AOM prediction models from previous studies.

The feature numbers in the C&R&D model and GTVnp RD were 29 and 5, respectively. Obviously, data fusion was more efficient for training a model with one-sixth the number of features to achieve stronger model predictability. To better explain the correlations of the selected features and severe AOM for NPC patients, a SHAP plot was applied for the GTVnp_RD XGBoost model. In this model, radiomics features extracted from GTVnp in cT1WI images yielded the highest and majority prediction value for severe AOM.

Poolakkad and his colleagues established a machine learning (ML) model of 253 H&N patients' clinical data with the best AUC of 0.79 for AOM prediction [42]. Most clinical data selected in their study were late after the CRT scheme, for example, the anti-neoplastic chemotherapy-induced pancytopenia, co-morbidity score, and agranulocytosis. It is worth noting that the features and variables selected in our study were all from the data collected before implementation of the RT regimen. Clinicians could predict the severe AOM before the commencement of RT planning. Personalized treatment strategies adjustment could be achieved using the developed prediction model.

Strictly speaking, the concept of dosiomics is originated from radiomics. The data for dosiomics and radiomics are similar in terms of the feature calculation algorithm [43]. The clinical data are different from the "-omics" data in nature. Therefore, instead of integrating raw clinical data with other data, the combination of selected clinical data was only applied in this study. Compared with less increase of AUC for the combination of clinical data with integrated RD data, the clinical data could enhance the prediction capability of single-modal models. The single radiomics models (R) and single clinical models (C) have limited prediction performance with average AUC of three models (LR, GNB, and XGBoost) at 0.63 ± 0.06 and 0.63 ± 0.64, respectively. When combining the clinical data with the selected radiomics features, the model (C&R) outperformed both R and C models with average AUC at 0.74 ± 0.03.

The single dosiomics models yielded poor performance, with most AUCs under 0.7 in the validation data set. In a previous study, dose distribution correlated with the incidence and severity of AOM [44]. Dean et al. [45] developed an RF model with a testing AUC of 0.71 ± 0.09, using a dose–volume histogram, spatial dose metrics from the oral cavity, and clinical data. In the current study, the best dosiomics model had the mean testing AUC of 0.69 ± 0.14. Different tumor-related VOIs may present different prediction value for severe AOM. The difference in VOI selection between the two studies might shed some light on the discrepancy in the findings. The oral cavity directly represents the dose distribution in the oral mucosa, which might be more accurate than the GTVn, GTVnp, PTVn 60 Gy, or PTVn 70 Gy. The VOI of the oral cavity requires specific contouring. It is worth noting that contouring of the oral cavity is not a common practice in the participating hospital of this study. Extra contouring is labor-intensive work in daily clinical practice. Our study only selected the routine VOI broadly used for RT planning, which could support our model to be applied from bench to bedside for clinical decision-making. Besides, the DVH is prone to over-simplifying the dose distribution [46]. It is recommended to combine or integrate dosiomics data with other modalities of data. When incorporating dosiomics data with other data types, the best mean validation AUC could surge to 0.81 ± 0.01.

At present, there exists no effective preventive measures for the occurrence of severe AOM in NPC patients undergoing RT. Nevertheless, it is feasible to mitigate the severity of this affliction: (1) Use of alternative radiation techniques, such as proton therapy, may be considered to reduce the risk of oral mucositis while maintaining treatment efficacy [34,47]. (2) Shortening the duration of chemotherapy. For advanced NPC patients who need to accept both radiotherapy and chemotherapy, shortening the exposition time to chemotherapy agents has shown lower mucosal toxicity [48]. (3) Photobiomodulation is a supportive treatment for the protection of high-risk mucositis patients [49]. (4) Supportive care interventions: preemptive or proactive use of supportive care interventions, such as oral hygiene measures, pain management, or nutritional support, may be considered to prevent or reduce the severity of AOM [50].

The limitations of our study were: (1) The mucositis grade levels of our patients had an imbalanced distribution. This might have had a negative influence on the data analysis work. The imbalanced results were the nature of the clinical situation. Patients were stratified into the training and validation groups according to the severity of OM, which could offset the imbalance problem [18,19,34]. (2) Potential bias of smoking information: in our study, the number of smoking patients might be underestimated due to the nature of this patient-reported outcome. This data were reported by patients at the time of their hospital visit and recorded in the nursing consultation notes. (3) The severity of AOM was scaled with standard terminology criteria for adverse events (CTCAE) in v3 or v4.03, almost equivalent to mucositis. Various criteria are available for mucositis grading, such as the those of the Radiation Therapy Oncology Group (RTOG) and the World Health Organization (WHO). These scales have excellent concordance with bundled scores of 3 and 4 to describe severe AOM [51]. The CTCAE is easily conducted by clinicians and nurses and broadly applied in the hospital. (4) The correlations of contributors under AOM for NPC patients are complex. For clinical decision-making, genome information, other clinical information such as fermented-food consumption and EBV infection, and pathological image may also play critical roles. The limited data resources for multimodal data integration are common challenges in the data-mining field. The radiomics data in our study also provided relevant genomic information. Compared with gene test results, the CECT and MRI examination images collected in our research are clinical routines used by clinicians to set the RT plan for NPC patients. These noninvasive examinations could serve as high-throughput screening tools for further application of severe AOM prediction in the future. (5) Other selection of VOIs: for practical consideration, we have not added the VOIs of the oral cavity, tongue, pharyngeal muscles, etc., which may hold potential predictive value for AOM. Further investigation is recommended to incorporate this information to enhance the accuracy of the analysis.

5. Conclusions

AOM is a challenging and distressing complication in NPC patients following RT. Prediction of severe AOM is necessary for timely prevention and intervention, which would further improve the QoL and survival of patients. In this study, we adopted multimodal data (clinical, radiomics, and dosiomics), multimodality of imaging (CECT, cT1WI, and T2WI), and multi-regional information (GTVn, GTVnp, and PTVn) to develop a best-performance model for severe AOM prediction. The simple combination of selected information and data fusion were applied in our work. The results demonstrated that the fusion of radiomics and dosiomics data from the primary tumor could generate the most effective and best-performing model (mean AUC = 0.81 $\pm$ 0.01). The data resources and VOIs selected in this study are routinely used in clinical practice, which has excellent potential for further clinical support. Further validation work on a large cohort is warranted to validate model generalizability.

Author Contributions: Conceptualization, Y.D. and S.L.; methodology, Y.D., J.Z. and J.C. (Jin Cao); software, Y.D. and J.Z.; validation, Y.D. and J.Z.; formal analysis, Y.D.; investigation, Y.D., J.Z., X.Z., X.T. and X.H.; resources, F.K.L., C.W.Y., K.A. and J.C. (Jing Cai); data curation, Y.D. and J.Z.; writing—original draft preparation, Y.D.; writing—review and editing, Y.D., S.L., A.L., Y.Z., X.T., H.L. and J.C. (Jing Cai); visualization, Y.D. and J.Z.; supervision, J.C. (Jing Cai); project administration, S.L. and J.C. (Jing Cai); funding acquisition, J.C. (Jing Cai). All authors have read and agreed to the published version of the manuscript.

Funding: This research was partly supported by research grants of Shenzhen Basic Research Program (JCYJ20210324130209023), Shenzhen-Hong Kong-Macau S&T Program (Category C) (SGDX20201103095002019), Mainland-Hong Kong Joint Funding Scheme (MHKJFS) (MHP/005/20), Project of Strategic Importance Fund (P0035421) and Projects of RISA (P0043001) from The Hong Kong Polytechnic University.

Institutional Review Board Statement: Ethical review and approval were waived for this study due to the natural of retrospective study.

Informed Consent Statement: Patient consent was waived due to the nature of the retrospective study.

Data Availability Statement: For patients' privacy protection, the data availability was not applicable.

Acknowledgments: Thanks to Qi Liu for general support with the initial data analysis.

Conflicts of Interest: The authors declare no conflict of interest.

Appendix A

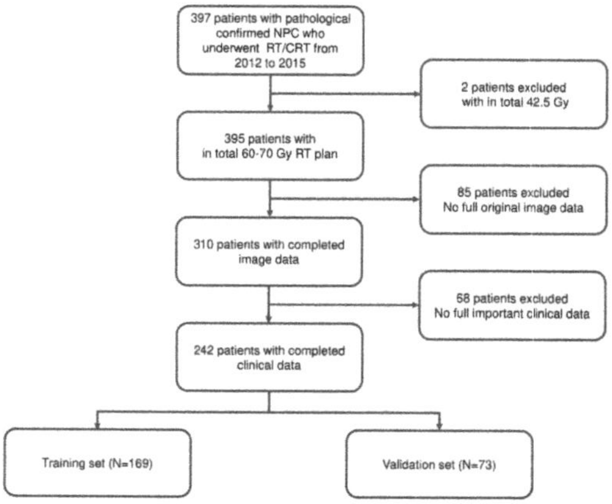

Figure A1. Schematic diagram of patient selection.

Appendix B

CECT. All patients were scanned in the CT stimulator (16-slice Brilliance Big 1Bore CT, Philips Medical Systems, Cleveland, OH) at Queen Elizabeth Hospital in Hong Kong. The scanned regions were from vertex to 5 cm below the sternoclavicular notch. The contrast-enhanced images were acquired at 30 s after intravenous injection of 70 mL iodinated contrast agents. The detailed acquisition parameters are listed in Table A1.

Table A1. Acquisition parameters of CECT.

Parameters	CECT
Pitch	1
Kilovoltage (kV)	120
Current (mAs)	250–350
Slice thickness (mm)	3
Matrix	512 × 512
Scan time (s)	15

MRI. MRI scans were acquired with an MR scanner (1.5 Tesla, Siemens Avanto, Germany). T2 weighted imaging with short-tau-inversion-recovery (STIR) sequence and cT1WI with spin-echo MRI sequence were applied. The detailed acquisition parameters are listed in Table A2.

Table A2. Parameters of T2WI and T1WI.

Parameters	T2-STIR	cT1WI
[TR]/[TE] (ms)	7640/97	739/17
FOV (cm^2)	24 × 24	24 × 24
Number of acquisitions	1	1
Slice thickness (mm×slices)	4 × 25	3 × 28
Spacing (cm^3)	0.75 × 0.75 × 4.4	0.938 × 0.938 × 3.3
Matrix	320 × 320	256 × 256

Appendix C

Table A3. Threshold of RFC selection and feature numbers for further model development.

Modal of Data	Threshold	Number of Features
GTVnp_RD	0.014	5
GTVnp_R_CECTcT1T2	0.01	8
GTVnp_R_CECTcT1	0.0125	5
GTVnp_R_cT1T2	0.125	5
GTVnp_R_cT1	0.015	4
GTVnp_R_CECT	0.01	19
GTVnp_R_T2	0.03	2
GTVnp_D	0.024	6
GTVn_RD	0.02	7
GTVn_R	0.03	7
GTVn_D	0.06	7
PTVn_D	0.03	3
PTVn_60Gy_D	0.03	12
PTVn_70Gy_D	0.042	1
R	0.012	2
D	0.016	13
RD	0.005	13

Appendix D

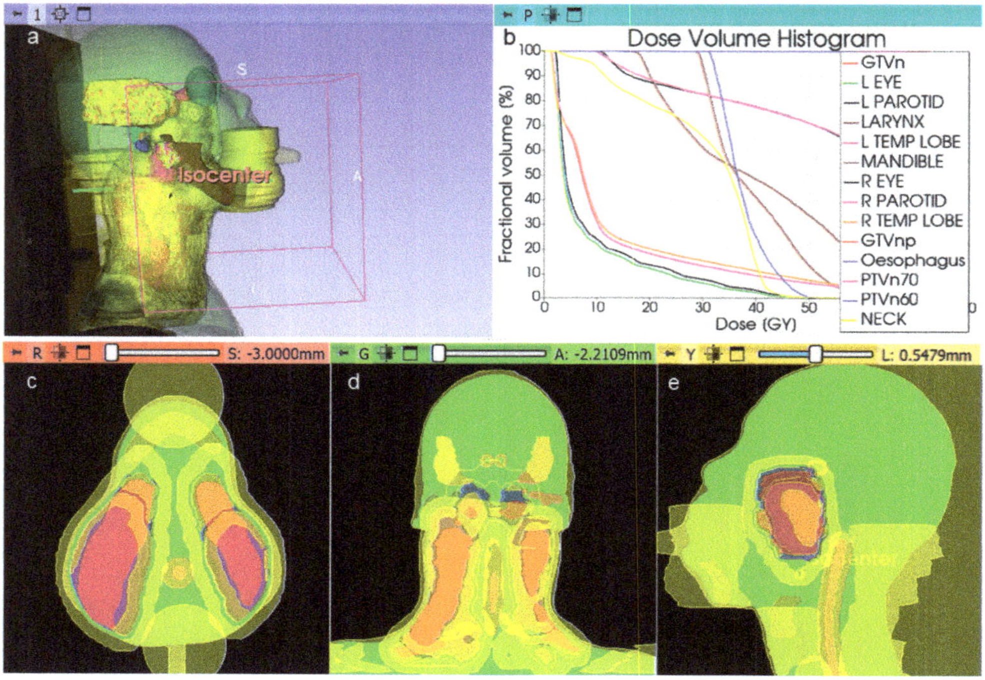

Figure A2. Dose maps of NPC patients undergoing IMRT. (**a**) 3D view of NPC patient. (**b**) DVH of multiple VOIs. (**c**). Axial view of patient in different VOIs. (**d**). Coronal view of patient. (**e**). Sagittal view of patient.

References

1. Chen, Y.-P.; Chan, A.T.C.; Le, Q.-T.; Blanchard, P.; Sun, Y.; Ma, J. Nasopharyngeal carcinoma. *Lancet* **2019**, *394*, 64–80. [CrossRef] [PubMed]
2. Chang, E.T.; Ye, W.; Zeng, Y.-X.; Adami, H.-O. The Evolving Epidemiology of Nasopharyngeal Carcinoma. *Cancer Epidemiol. Biomarkers Prev.* **2021**, *30*, 1035–1047. [CrossRef] [PubMed]
3. Chen, Y.-P.; Ismaila, N.; Chua, M.L.K.; Colevas, A.D.; Haddad, R.; Huang, S.H.; Wee, J.T.S.; Whitley, A.C.; Yi, J.-L.; Yom, S.S.; et al. Chemotherapy in Combination with Radiotherapy for Definitive-Intent Treatment of Stage II-IVA Nasopharyngeal Carcinoma: CSCO and ASCO Guideline. *J. Clin. Oncol.* **2021**, *39*, 840–859. [CrossRef]
4. Bossi, P.; Chan, A.; Licitra, L.; Trama, A.; Orlandi, E.; Hui, E.; Halámková, J.; Mattheis, S.; Baujat, B.; Hardillo, J.; et al. Nasopharyngeal carcinoma: ESMO-EURACAN Clinical Practice Guidelines for diagnosis, treatment and follow-up. *Ann. Oncol.* **2020**, *32*, 452–465. [CrossRef]
5. You, R.; Liu, Y.-P.; Huang, P.-Y.; Zou, X.; Sun, R.; He, Y.-X.; Wu, Y.-S.; Shen, G.-P.; Zhang, H.-D.; Duan, C.-Y.; et al. Efficacy and Safety of Locoregional Radiotherapy with Chemotherapy vs Chemotherapy Alone in De Novo Metastatic Nasopharyngeal Carcinoma: A Multicenter Phase 3 Randomized Clinical Trial. *JAMA Oncol.* **2020**, *6*, 1345. [CrossRef]
6. Tang, L.-L.; Guo, R.; Zhang, N.; Deng, B.; Chen, L.; Cheng, Z.-B.; Huang, J.; Hu, W.-H.; Huang, S.H.; Luo, W.-J.; et al. Effect of Radiotherapy Alone vs Radiotherapy with Concurrent Chemoradiotherapy on Survival Without Disease Relapse in Patients with Low-risk Nasopharyngeal Carcinoma: A Randomized Clinical Trial. *JAMA* **2022**, *328*, 728–736. [CrossRef]
7. Wang, K.; Tepper, J.E. Radiation therapy-associated toxicity: Etiology, management, and prevention. *CA A Cancer J. Clin.* **2021**, *71*, 437–454. [CrossRef]
8. Elad, S.; Yarom, N.; Zadik, Y.; Kuten-Shorrer, M.; Sonis, S.T. The broadening scope of oral mucositis and oral ulcerative mucosal toxicities of anticancer therapies. *CA A Cancer J. Clin.* **2021**, *72*, 57–77. [CrossRef]
9. Raber-Durlacher, J.E.; Elad, S.; Barasch, A. Oral mucositis. *Oral Oncol.* **2010**, *46*, 452–456. [CrossRef] [PubMed]
10. McDowell, L.; Corry, J.; Ringash, J.; Rischin, D. Quality of Life, Toxicity and Unmet Needs in Nasopharyngeal Cancer Survivors. *Front. Oncol.* **2020**, *10*, 930. [CrossRef] [PubMed]

11. Li, H.; You, Y.; Lin, C.; Zheng, M.; Hong, C.; Chen, J.; Li, D.; Au, W.W.; Chen, Z. XRCC1 codon 399Gln polymorphism is associated with radiotherapy-induced acute dermatitis and mucositis in nasopharyngeal carcinoma patients. *Radiat. Oncol.* **2013**, *8*, 31. [CrossRef]
12. Yang, D.-W.; Wang, T.-M.; Zhang, J.-B.; Li, X.-Z.; He, Y.-Q.; Xiao, R.; Xue, W.-Q.; Zheng, X.-H.; Zhang, P.-F.; Zhang, S.-D.; et al. Genome-wide association study identifies genetic susceptibility loci and pathways of radiation-induced acute oral mucositis. *J. Transl. Med.* **2020**, *18*, 1–12. [CrossRef]
13. Orlandi, E.; Iacovelli, N.A.; Rancati, T.; Cicchetti, A.; Bossi, P.; Pignoli, E.; Bergamini, C.; Licitra, L.; Fallai, C.; Valdagni, R.; et al. Multivariable model for predicting acute oral mucositis during combined IMRT and chemotherapy for locally advanced nasopharyngeal cancer patients. *Oral Oncol.* **2018**, *86*, 266–272. [CrossRef]
14. Inada, M.; Nishimura, Y.; Ishikura, S.; Ishikawa, K.; Murakami, N.; Kodaira, T.; Ito, Y.; Tsuchiya, K.; Murakami, Y.; Saito, J.; et al. Organs-at-risk dose constraints in head and neck intensity-modulated radiation therapy using a dataset from a multi-institutional clinical trial (JCOG1015A1). *Radiat. Oncol.* **2022**, *17*, 1–8. [CrossRef] [PubMed]
15. Xu, T.; Liu, Y.; Dou, S.; Li, F.; Guan, X.; Zhu, G. Weekly cetuximab concurrent with IMRT aggravated radiation-induced oral mucositis in locally advanced nasopharyngeal carcinoma: Results of a randomized phase II study. *Oral Oncol.* **2015**, *51*, 875–879. [CrossRef] [PubMed]
16. Köstler, W.J.; Hejna, M.; Wenzel, C.; Zielinski, C.C. Oral Mucositis Complicating Chemotherapy and/or Radiotherapy: Options for Prevention and Treatment. *CA A Cancer J. Clin.* **2001**, *51*, 290–315. [CrossRef]
17. Scully, C.; Epstein, J.; Sonis, S. Oral mucositis: A challenging complication of radiotherapy, chemotherapy, and radiochemotherapy: Part 1, pathogenesis and prophylaxis of mucositis. *Head Neck* **2003**, *25*, 1057–1070. [CrossRef] [PubMed]
18. Shu, Z.; Zeng, Z.; Yu, B.; Huang, S.; Hua, Y.; Jin, T.; Tao, C.; Wang, L.; Cao, C.; Xu, Z.; et al. Nutritional Status and Its Association with Radiation-Induced Oral Mucositis in Patients with Nasopharyngeal Carcinoma During Radiotherapy: A Prospective Study. *Front. Oncol.* **2020**, *10*, 594687. [CrossRef]
19. Elting, L.S.; Cooksley, C.D.; Chambers, M.S.; Garden, A.S. Risk, Outcomes, and Costs of Radiation-Induced Oral Mucositis Among Patients with Head-and-Neck Malignancies. *Int. J. Radiat. Oncol. Biol. Phys.* **2007**, *68*, 1110–1120. [CrossRef]
20. Saito, N.; Imai, Y.; Muto, T.; Sairenchi, T. Low body mass index as a risk factor of moderate to severe oral mucositis in oral cancer patients with radiotherapy. *Support. Care Cancer* **2012**, *20*, 3373–3377. [CrossRef]
21. Rosenthal, D.I.; Trotti, A. Strategies for Managing Radiation-Induced Mucositis in Head and Neck Cancer. *Semin. Radiat. Oncol.* **2009**, *19*, 29–34. [CrossRef]
22. Chong, V.; Ong, C. Nasopharyngeal carcinoma. *Eur. J. Radiol.* **2008**, *66*, 437–447. [CrossRef]
23. Xia, P.; Fu, K.K.; Wong, G.W.; Akazawa, C.; Verhey, L.J. Comparison of treatment plans involving intensity-modulated radiotherapy for nasopharyngeal carcinoma. *Int. J. Radiat. Oncol.* **2000**, *48*, 329–337. [CrossRef] [PubMed]
24. Lahat, D.; Adali, T.; Jutten, C. Multimodal data fusion: An overview of methods, challenges, and prospects. *Proc. IEEE* **2015**, *103*, 1449–1477. [CrossRef]
25. Boehm, K.M.; Khosravi, P.; Vanguri, R.; Gao, J.; Shah, S.P. Harnessing multimodal data integration to advance precision oncology. *Nat. Rev. Cancer* **2021**, *22*, 114–126. [CrossRef] [PubMed]
26. Lipkova, J.; Chen, R.J.; Chen, B.; Lu, M.Y.; Barbieri, M.; Shao, D.; Vaidya, A.J.; Chen, C.; Zhuang, L.; Williamson, D.F.; et al. Artificial intelligence for multimodal data integration in oncology. *Cancer Cell* **2022**, *40*, 1095–1110. [CrossRef]
27. Teng, X.; Zhang, J.; Ma, Z.; Zhang, Y.; Lam, S.; Li, W.; Xiao, H.; Li, T.; Li, B.; Zhou, T.; et al. Improving radiomic model reliability using robust features from perturbations for head-and-neck carcinoma. *Front. Oncol.* **2022**, *12*, 974467. [CrossRef]
28. Lam, S.-K.; Zhang, Y.; Zhang, J.; Li, B.; Sun, J.-C.; Liu, C.Y.-T.; Chou, P.-H.; Teng, X.; Ma, Z.-R.; Ni, R.-Y.; et al. Multi-Organ Omics-Based Prediction for Adaptive Radiation Therapy Eligibility in Nasopharyngeal Carcinoma Patients Undergoing Concurrent Chemoradiotherapy. *Front. Oncol.* **2022**, *11*, 792024. [CrossRef] [PubMed]
29. Zhang, Y.; Lam, S.; Yu, T.; Teng, X.; Zhang, J.; Lee, F.K.-H.; Au, K.-H.; Yip, C.W.-Y.; Wang, S.; Cai, J. Integration of an imbalance framework with novel high-generalizable classifiers for radiomics-based distant metastases prediction of advanced nasopharyngeal carcinoma. *Knowl.-Based Syst.* **2021**, *235*, 107649. [CrossRef]
30. Li, B.; Zheng, X.; Zhang, J.; Lam, S.; Guo, W.; Wang, Y.; Cui, S.; Teng, X.; Zhang, Y.; Ma, Z.; et al. Lung Subregion Partitioning by Incremental Dose Intervals Improves Omics-Based Prediction for Acute Radiation Pneumonitis in Non-Small-Cell Lung Cancer Patients. *Cancers* **2022**, *14*, 4889. [CrossRef]
31. Li, B.; Ren, G.; Guo, W.; Zhang, J.; Lam, S.-K.; Zheng, X.; Teng, X.; Wang, Y.; Yang, Y.; Dan, Q.; et al. Function-Wise Dual-Omics analysis for radiation pneumonitis prediction in lung cancer patients. *Front Pharmacol.* **2022**, *13*, 971849. [CrossRef]
32. Hansen, C.; Bertelsen, A.; Zukauskaite, R.; Johnsen, L.; Bernchou, U.; Thwaites, D.; Eriksen, J.; Johansen, J.; Brink, C. Prediction of radiation-induced mucositis of H&N cancer patients based on a large patient cohort. *Radiother. Oncol.* **2020**, *147*, 15–21. [CrossRef]
33. Li, P.-J.; Li, K.-X.; Jin, T.; Lin, H.-M.; Fang, J.-B.; Yang, S.-Y.; Shen, W.; Chen, J.; Zhang, J.; Chen, X.-Z.; et al. Predictive Model and Precaution for Oral Mucositis During Chemo-Radiotherapy in Nasopharyngeal Carcinoma Patients. *Front. Oncol.* **2020**, *10*, 596822. [CrossRef] [PubMed]
34. Dodd, M.J.; Miaskowski, C.; Shiba, G.H.; Dibble, S.L.; Greenspan, D.; MacPhail, L.; Paul, S.M.; Larson, P. Risk factors for chemotherapy-induced oral mucositis: Dental appliances, oral hygiene, previous oral lesions, and history of smoking. *Cancer Investig.* **1999**, *17*, 278–284. [CrossRef]

35. Zwanenburg, A.; Vallières, M.; Abdalah, M.A.; Aerts, H.J.W.L.; Andrearczyk, V.; Apte, A.; Ashrafinia, S.; Bakas, S.; Beukinga, R.J.; Boellaard, R.; et al. The Image Biomarker Standardization Initiative: Standardized Quantitative Radiomics for High-Throughput Image-based Phenotyping. *Radiology* **2020**, *295*, 328–338. [CrossRef]
36. van Griethuysen, J.J.M.; Fedorov, A.; Parmar, C.; Hosny, A.; Aucoin, N.; Narayan, V.; Beets-Tan, R.G.H.; Fillion-Robin, J.-C.; Pieper, S.; Aerts, H.J.W.L. Computational Radiomics System to Decode the Radiographic Phenotype. *Cancer Res.* **2017**, *77*, e104–e107. [CrossRef]
37. Lowekamp, B.C.; Chen, D.T.; Eibanez, L.; Eblezek, D. The Design of SimpleITK. *Front. Neuroinform.* **2013**, *7*, 45. [CrossRef] [PubMed]
38. Grégoire, V.; Levendag, P.; Ang, K.K.; Bernier, J.; Braaksma, M.; Budach, V.; Chao, C.; Coche, E.; Cooper, J.S.; Cosnard, G.; et al. CT-based delineation of lymph node levels and related CTVs in the node-negative neck: DAHANCA, EORTC, GORTEC, NCIC, RTOG consensus guidelines. *Radiother. Oncol.* **2003**, *69*, 227–236. [CrossRef]
39. Gabryś, H.S.; Buettner, F.; Sterzing, F.; Hauswald, H.; Bangert, M. Design and Selection of Machine Learning Methods Using Radiomics and Dosiomics for Normal Tissue Complication Probability Modeling of Xerostomia. *Front. Oncol.* **2018**, *8*, 35. [CrossRef]
40. Buettner, F.; Miah, A.B.; Gulliford, S.L.; Hall, E.; Harrington, K.J.; Webb, S.; Partridge, M.; Nutting, C.M. Novel approaches to improve the therapeutic index of head and neck radiotherapy: An analysis of data from the PARSPORT randomised phase III trial. *Radiother. Oncol.* **2012**, *103*, 82–87. [CrossRef] [PubMed]
41. Lundberg, S.M.; Lee, S.I. A unified approach to interpreting model predictions. *Adv. Neural Inf. Process. Syst.* **2017**, 30. [CrossRef]
42. Satheeshkumar, P.S.; El-Dallal, M.; Mohan, M.P. Feature selection and predicting chemotherapy-induced ulcerative mucositis using machine learning methods. *Int. J. Med. Inform.* **2021**, *154*, 104563. [CrossRef] [PubMed]
43. Rossi, L.; Bijman, R.; Schillemans, W.; Aluwini, S.; Cavedon, C.; Witte, M.; Incrocci, L.; Heijmen, B. Texture analysis of 3D dose distributions for predictive modelling of toxicity rates in radiotherapy. *Radiother. Oncol.* **2018**, *129*, 548–553. [CrossRef] [PubMed]
44. Sanguineti, G.; Sormani, M.P.; Marur, S.; Gunn, G.B.; Rao, N.; Cianchetti, M.; Ricchetti, F.; McNutt, T.; Wu, B.; Forastiere, A. Effect of Radiotherapy and Chemotherapy on the Risk of Mucositis During Intensity-Modulated Radiation Therapy for Oropharyngeal Cancer. *Int. J. Radiat. Oncol.* **2012**, *83*, 235–242. [CrossRef] [PubMed]
45. Dean, J.A.; Wong, K.H.; Welsh, L.C.; Jones, A.-B.; Schick, U.; Newbold, K.L.; Bhide, S.A.; Harrington, K.J.; Nutting, C.M.; Gulliford, S.L. Normal tissue complication probability (NTCP) modelling using spatial dose metrics and machine learning methods for severe acute oral mucositis resulting from head and neck radiotherapy. *Radiother. Oncol.* **2016**, *120*, 21–27. [CrossRef]
46. Giraud, P.; Giraud, P.; Gasnier, A.; El Ayachy, R.; Kreps, S.; Foy, J.-P.; Durdux, C.; Huguet, F.; Burgun, A.; Bibault, J.-E. Radiomics and Machine Learning for Radiotherapy in Head and Neck Cancers. *Front. Oncol.* **2019**, *9*, 174. [CrossRef]
47. Moreno, A.C.; Frank, S.J.; Garden, A.S.; Rosenthal, D.I.; Fuller, C.D.; Gunn, G.B.; Reddy, J.P.; Morrison, W.H.; Williamson, T.D.; Holliday, E.B.; et al. Intensity modulated proton therapy (IMPT)–The future of IMRT for head and neck cancer. *Oral Oncol.* **2019**, *88*, 66–74. [CrossRef]
48. Plevová, P. Prevention and treatment of chemotherapy- and radiotherapy-induced oral mucositis: A review. *Oral Oncol.* **1999**, *35*, 453–470. [CrossRef]
49. Melo, W.W.P.; Aragão, W.A.B.; Baia-Da-Silva, D.C.; Nascimento, P.C.; Lima, R.R.; de Souza-Rodrigues, R.D. Effects of Photobiomodulation on Oral Mucositis: Visualization and Analysis of Knowledge. *Life* **2022**, *12*, 1940. [CrossRef]
50. Araújo, S.N.M.; Luz, M.H.B.A.; da Silva, G.R.F.; Andrade, E.M.L.R.; Nunes, L.C.C.; Moura, R.O. Cancer patients with oral mucositis: Challenges for nursing care. *Rev. Latino-Am. Enferm.* **2015**, *23*, 267–274. [CrossRef]
51. Villa, A.; Vollemans, M.; De Moraes, A.; Sonis, S. Concordance of the WHO, RTOG, and CTCAE v4.0 grading scales for the evaluation of oral mucositis associated with chemoradiation therapy for the treatment of oral and oropharyngeal cancers. *Support. Care Cancer* **2021**, *29*, 6061–6068. [CrossRef] [PubMed]

cancers

MDPI

Article

Ensemble Deep Learning Model to Predict Lymphovascular Invasion in Gastric Cancer

Jonghyun Lee [1], Seunghyun Cha [2], Jiwon Kim [3], Jung Joo Kim [4], Namkug Kim [5], Seong Gyu Jae Gal [5], Ju Han Kim [6], Jeong Hoon Lee [7], Yoo-Duk Choi [8], Sae-Ryung Kang [9], Ga-Young Song [10], Deok-Hwan Yang [10], Jae-Hyuk Lee [11], Kyung-Hwa Lee [11], Sangjeong Ahn [12], Kyoung Min Moon [13,14,*] and Myung-Giun Noh [11,*]

1 Department of Medical and Digital Engineering, Hanyang University College of Engineering, Seoul 04763, Republic of Korea; jonghyunlee1993@gmail.com
2 Department of Pre-Medicine, Chonnam National University Medical School, 322 Seoyang-ro, Hwasun-eup, Hwasun-gun, Gwangju 58128, Republic of Korea; robird@naver.com
3 NetTargets, 495 Sinseong-dong, Yuseong, Daejeon 34109, Republic of Korea
4 AMGINE, Inc., Jeongui-ro 8-gil 13, Seoul 05836, Republic of Korea; jkim325@aucklanduni.ac.nz
5 Department of Convergence Medicine, Asan Medical Center, University of Ulsan College of Medicine, Seoul 25440, Republic of Korea; namkugkim@gmail.com (N.K.); tobeor3009@gmail.com (S.G.J.G.)
6 Division of Biomedical Informatics, Seoul National University Biomedical Informatics (SNUBI), Seoul National University College of Medicine, Seoul 03080, Republic of Korea; juhan@snu.ac.kr
7 Department of Radiology, Stanford University School of Medicine, Stanford, CA 94305-5101, USA; sosal@stanford.edu
8 Department of Pathology, Chonnam National University Medical School, Gwangju 61469, Republic of Korea; drydchoi@hanmail.net
9 Department of Nuclear Medicine, Clinical Medicine Research Center, Chonnam National University Hospital, 671 Jebongno, Gwangju 61469, Republic of Korea; srkang@jnu.ac.kr
10 Departments of Hematology-Oncology, Chonnam National University Hwasun Hospital, 322 Seoyangro, Hwasun 58128, Republic of Korea; drgyssong@gmail.com (G.-Y.S.); drydh1685@hotmail.com (D.-H.Y.)
11 Department of Pathology, Chonnam National University Hwasun Hospital and Medical School, 322 Seoyang-ro, Hwasun-eup, Hwasun-gun, Hwasun 58128, Republic of Korea; mdkaylee@gmail.com (K.-H.L.)
12 Department of Pathology, Korea University Anam Hospital, Korea University College of Medicine, 73 Goryeodae-ro, Seongbuk-gu, Seoul 02841, Republic of Korea; vanitasahn@gmail.com
13 Division of Pulmonary and Allergy Medicine, Department of Internal Medicine, Chung-Ang University Hospital, Chung-Ang University College of Medicine, Seoul 06973, Republic of Korea
14 Artificial Intelligence, ZIOVISION Co., Ltd., Chuncheon 24341, Republic of Korea
* Correspondence: ml.pulmogicu@gmail.com (K.M.M.); mdmgnoh@gmail.com (M.-G.N.); Tel.: +82-33-610-3058 (K.M.M.); +82-61-379-7069 (M.-G.N.); Fax: +82-33-610-4760 (K.M.M.); +82-61-379-7099 (M.-G.N.)

Citation: Lee, J.; Cha, S.; Kim, J.; Kim, J.J.; Kim, N.; Jae Gal, S.G.; Kim, J.H.; Lee, J.H.; Choi, Y.-D.; Kang, S.-R.; et al. Ensemble Deep Learning Model to Predict Lymphovascular Invasion in Gastric Cancer. *Cancers* **2024**, *16*, 430. https://doi.org/10.3390/cancers16020430

Received: 13 December 2023
Revised: 11 January 2024
Accepted: 16 January 2024
Published: 19 January 2024

Simple Summary: Lymphovascular invasion (LVI) serves as a crucial predictor in gastric cancer, indicating an increased likelihood of lymph node spread and poorer patient outcomes. Detecting LVI(+) within gastric cancer histopathology presents challenges due to its elusive nature, leading to the proposal of a deep learning-based detection method using H&E-stained whole-slide images. Remarkably, both the classification and detection models demonstrated superior performance, and their ensemble exhibited outstanding predictive capabilities in identifying LVI areas. This innovative approach holds promise in precision medicine, potentially streamlining examinations and reducing discrepancies among pathologists.

Abstract: Lymphovascular invasion (LVI) is one of the most important prognostic factors in gastric cancer as it indicates a higher likelihood of lymph node metastasis and poorer overall outcome for the patient. Despite its importance, the detection of LVI(+) in histopathology specimens of gastric cancer can be a challenging task for pathologists as invasion can be subtle and difficult to discern. Herein, we propose a deep learning-based LVI(+) detection method using H&E-stained whole-slide images. The ConViT model showed the best performance in terms of both AUROC and AURPC among the classification models (AUROC: 0.9796; AUPRC: 0.9648). The AUROC and AUPRC of YOLOX computed based on the augmented patch-level confidence score were slightly lower (AUROC: −0.0094; AUPRC: −0.0225) than those of the ConViT classification model. With weighted averaging of the patch-level confidence scores, the ensemble model exhibited the best AUROC, AUPRC, and F1 scores of 0.9880, 0.9769, and 0.9280, respectively. The proposed model is expected to contribute to precision medicine by potentially saving examination-related time and labor and reducing disagreements among pathologists.

Keywords: digital pathology; artificial intelligence; gastric cancer; lymphovascular invasion

1. Introduction

Gastric cancer is the most common type of cancer, accounting for 12% of all cancer cases in Korea according to data from the National Cancer Center in 2018 [1]. In 2020, more than 1 million (1,089,103) new cases of gastric cancer were estimated worldwide, resulting in 768,793 deaths [2]. Lymph node metastasis is the most significant prognostic factor for patients with gastric cancer, and the presence of lymphovascular invasion (LVI) is the most significant risk factor for lymph node metastasis [3–6]. LVI is defined as the invasion of vessel walls by tumor cells and/or the presence of tumor emboli within an endothelial-lined space [7]. Predictive value and prevalence of LVI are highly dependent on the type of cancer, and the presence of LVI is a recognized prognostic factor in a variety of solid malignancies, including breast cancer, urothelial carcinoma, and colorectal cancer [8]. Since the proclamation of LVI as an important factor in the prognosis of gastric cancer by Talamonti et al. [9], the American Joint Committee on Cancer has recommended the evaluation of LVI [10]. According to the current Japanese guidelines, LVI in gastric cancer is not clinically useful information except for predicting the possibility of curative endoscopic resection. LVI is the most significant risk factor associated with lymph node metastases in individuals with early gastric cancer [6,11–13]. The rate of lymph node metastasis observed in patients exhibiting LVI (25.7–32.1%) was much higher compared to that in those without LVI (1.5–2.3%) [6,11,13,14]. In addition, Fusikawa et al. showed that a significant difference was observed between the values of 79.8% in the LVI(−) group and 67.2% in the LVI(+) group in advanced cancer [7]. The five-year survival rate of advanced cancers with nodal metastases is 76.7% in the LVI(−) group and 60.9% in the LVI(+) group [7]. Therefore, LVI is an independent prognostic marker in gastric cancer and tends to worsen the prognosis, particularly in cases of advanced malignancy with lymph node metastasis.

The recognition of lymphatic tumor emboli in microscopic sections is dependent on the pathologist [15]. There is potential for significant inter-observer variations in the diagnosis of LVI amongst pathologists [16]. Inter-observer disagreement can be expected in the diagnosis of LVI as retraction artifacts that isolate tumor aggregates can be caused by tissue shrinkage during fixation, which are easily confused with true tumor emboli during routine examination of hematoxylin and eosin (H&E) stained sections [17,18]. Tumors may be artefactually displaced into vessels during specimen cut up or processing [19]. For instance, Gilchrist et al. noted that when three surgical pathologists were told to assess for LVI in a pT1-2 N0 M0 histological mastectomy case, all three concurred in only 12 of 35 breast cancer cases [15,16]. Several attempts have been made to overcome these limitations. The monoclonal D2-40 antibody can selectively detect lymphatic vessels as it is expressed in the lymphatic endothelium but not in blood vessels, and D2-40 staining is reportedly more sensitive than H&E staining for detecting lymphatic invasion (LI) [17,20,21]. Elastin staining may also be used for a clearer recognition of blood vessels as it identifies the elastic fibers of blood vessels [22–25]. Inter-observer agreement in the diagnoses of LVI was improved by adding ancillary D2-40 and elastin staining, regardless of the experience of the pathologists [4]. However, the assessment of LVI by pathologists is inherently limited owing to human errors. Examining large areas of tumors for LVI is time-consuming and challenging because the foci of LVI can be small and subjective. Nonetheless, the presence of LVI can have a marked impact on disease management, and the identification of a genuine single focus is sufficient to label a case as LVI(+). This automated identification of possible LVI(−)indicating lesions may have significant clinical utility [19].

Digital pathology defines the creation of whole-slide images (WSI) from a histology slide that can be viewed on a screen to form a diagnostic report [26]. Traditionally, histological diagnosis and pathological staging by pathologists have been evaluated using glass slides and microscopes [26]. Digital pathology is now increasingly being implemented in laboratories around the world, and digital support management is seen as a key component of health service planning aimed at improving efficiency, network operation, and quality [26]. There is great potential for using artificial intelligence (AI) to assist pathologists and derive new biological insights into disease biology, even in areas imperceptible to human observers [27]. However, the majority of AI medical devices that have received FDA approval and have been introduced to the market thus far are primarily focused on radiology. In contrast, only a limited number of devices have been approved for use in the field of pathology [28]. Moreover, it is important to explore the potential of these AI technologies as many pathology departments do not have enough pathologists.

AI algorithms that utilize convolutional neural networks (CNNs) for image analysis have already shown significant promise in the pathological evaluation of various solid tumors, including prostate cancer screening in prostate biopsies [29,30], leading to new evaluations of clinical outcomes, providing [31,32] or predicting the presence of mutations [33] or molecular subtypes [34] in H&E-stained sections. The usefulness of these algorithms in identifying small regions of prognostic significance in digital WSI has previously been demonstrated in the context of identifying metastatic breast cancer within lymph nodes [35,36]. In addition, the AI model can automatically find LVI in the WSI of testicular cancer [19]. AI model can identify LVI foci better than a human expert (recall score: 0.68 vs. 0.56).

In this study, we developed an algorithm to identify LVI foci related to the prognosis of gastric cancer. The image classification and detection models were trained and validated at both the patch and WSI levels. The ensemble approach was used to combine the predictions of these sub-models to improve the overall performance of the model. The sub-models were trained on a dataset of WSI of gastric cancer, with annotations of vascular and lymphatic vascular invasion. A conceptual diagram of the LVI prediction model is shown in Figure 1.

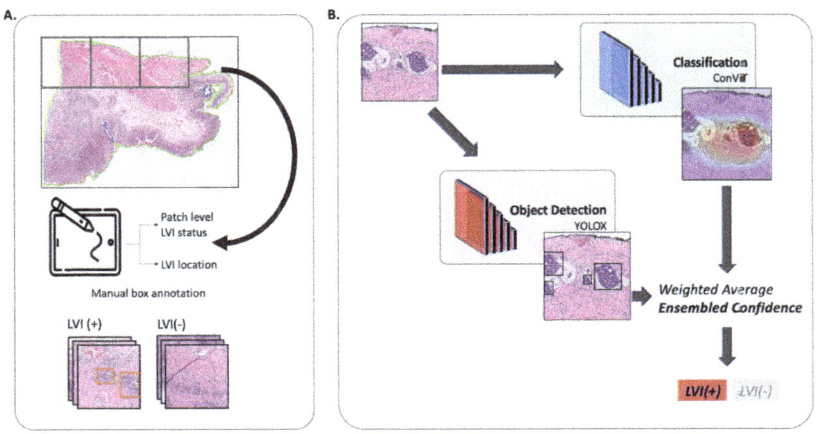

Figure 1. Schematic of the LVI Net. Panel (**A**) portrays the preprocessing step and annotations, while Panel (**B**) illustrates the workflow of the LVI Net. The patch image is input into both the classification and detection models. Subsequently, the prediction outcomes from these models conducted weighted averaging, resulting in the computation of the final confidence level (referred to as the ensemble confidence). This ensemble confidence is then utilized to predict the ultimate diagnosis of LVI(+) or LVI(−).

2. Methods

2.1. Patients and Tumor Samples

Gastric adenocarcinoma slides were obtained from 88 patients who underwent endoscopic submucosal dissection, subtotal gastrectomy, or total gastrectomy at the Chonnam National University Hwasun Hospital from 2018 to 2021. The availability of adequate tissue and the histological diagnosis of gastric cancer were the inclusion criteria. One hundred WSI were collected from these patient samples. Clinical information was collected from the electronic medical records maintained in the electronic database of the hospital. This study was approved by the Institutional Review Board (IRB) of the Chonnam National University Hwasun Hospital (CNUHH-2021-197) and conducted in accordance with the Declaration of Helsinki. Informed consent from patients was waived with IRB approval.

2.2. Datasets

The slides were scanned using a Leica-Aperio GT450 Scanner (Leica Biosystems) using an 40× objective. Using QuPath 0.3.0 tools, the LVI(+) regions were annotated by two board-certified pathologists. The examples of LVI(+) and LVI(−) are depicted in Figure 2. We performed CD34 and D2-40 immunohistochemical staining on all slides to confirm LVI(+) foci and to increase the accuracy of marking LVI(+) foci. For training, validation, and test splitting, we randomly selected WSI with a 6:2:2 ratio. We patchified WSIs using conventional digital pathology image analysis (Figure 1, preprocessing panel). LVI(+) foci were generated based on LVI(+) annotations. The sliding windowing approach generated LVI(−) patches from the remaining WSI. Without any overlap, we visited all WSI regions that did not include LVI(+) foci. To handle class imbalances and remove redundancy in LVI(−) patches, one-third was sampled from all LVI(−) patches. The LVI(+) and LVI(−) patches were generated at 20×-level (0.5 µm/pixel) with 512 × 512 pixels.

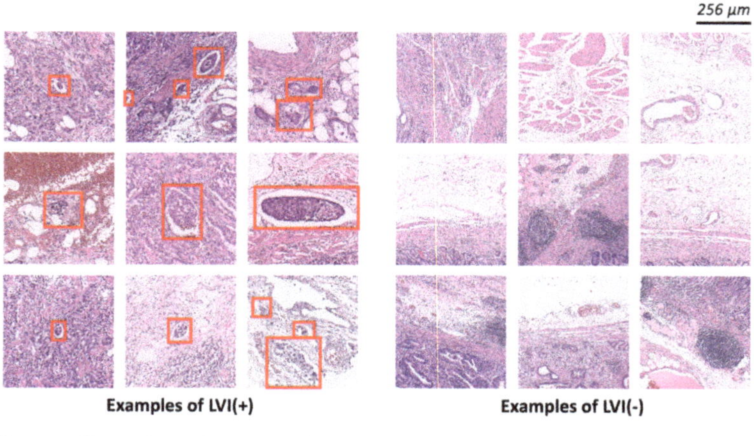

Figure 2. Example patch images of LVI(+) and LVI(−). The left panel displays a patch associated with LVI(+) classification, while the right panel represents LVI(−). LVI foci refer to tumors located within identifiable white, rounded structures that align anatomically with blood vessels and lymph nodes. A patch is classified as positive if it contains one or more regions indicating the presence of LVI. The LVI areas are marked as red boxes.

To conduct external validation, we utilized a publicly accessible classification dataset that contained patch images pertaining to lymphatic invasion [37]. Comprising 48 WSIs sourced from 27 patients, this external validation dataset comprised 302 positive instances and 671 negative instances. The patch images were captured at a 5×-level magnification (2 µm/pixel) with dimensions of 512 × 512 pixels. Notably, this external validation dataset was acquired using a distinct scanner (Leica-Aperio AT2) and originated from a different hospital setting.

2.3. Model Development

We fine-tuned the image classification and detection models to identify the LVI foci in a given patch image. The following analysis was conducted using Python 3.8, Pytorch 1.13.1, and a single A100 GPU.

2.4. Classification Models

We defined the classification problem as a binary classification. The ResNet 50 [38], EfficeientNet B3 [39], and ConViT (Small) [40] models were fine-tuned on the LVI datasets. The parameters of the selected image classification models ranged from 20 to 30 M. In an empirical study, we found that large parameters converged into overfitting because of our limited dataset volume. We utilized ImageNet [41] pretrained weights with entire layers that can be updated by considering the modality gap between a conventional RGB and digital pathology images. Image augmentations were applied, including affine transform, elastic transform, blurring, brightness, and color jittering. Balanced weight-sampling methods were applied during training to alleviate data imbalance. The image classification models were trained using the Adam optimizer (learning rate: 1×10^{-4}), cosine annealing learning rate scheduler, and automated mixed precision.

2.5. Detection Models

The detection model was utilized to classify and localize the desired object in the entire image simultaneously. A regression operation was applied to localize the object using a bounding box. We utilized a one-stage object detection model called the YOLO model [42]. YOLO detection uses the concept of an anchor box. The anchor box has a predefined shape and ratio of the bounding box that is utilized in the bounding box location prediction. For example, human objects commonly exhibited square shapes with long heights and short widths. In contrast, the dog objects had square shapes with short heights and log widths. Anchor-based methods have been actively utilized to ease the prediction performance. However, in terms of LVI, the shape of LVI was arbitrary; several LVI foci assumed a square shape, and the others assumed a rectangular shape with variants of size. To compare the impact of the anchor box assumption on LVI foci detection, we trained both an anchor box assumption-based detection model (YOLO v3) [43] and detection model without the anchor box assumption (YOLOX) [44]. To match the number of parameters, the medium size of YOLOX was selected. The hyperparameters and data augmentations followed the recommendations of each framework. The detection model could detect as many LVI(+) regions as possible. Therefore, unlike a classification model, a single-patch image can have multiple prediction confidence scores. To aggregate multiple confidence scores, we computed the augmented confidence score of each patch image using the maximum operator.

2.6. Ensemble Model

The ensembled confidence score (C_{ens}) is calculated as the weighted average of the confidence score of the classification model (C_{clf}) and the augmented confidence score of the detection model (C_{det}), according to Equation (1):

$$C_{ens} = \frac{\left(w_{clf} \times C_{clf} + w_{det} \times C_{det}\right)}{2}, \quad (1)$$

where w_{clf} and w_{det} denote weighted factors of classification and detection models, respectively. Considering the performances of each model, we empirically set the w_{clf} to 1.0 and w_{det} to 1.0, respectively. The ensembled confidence score was treated as a final confidence score.

2.7. Evaluation Metrics

Generally, to evaluate the classification performance, the true positive (TP), false positive (FP), false negative (FN), and true negative (TN) are computed by comparing the prediction confidence that a model returns and the ground truth. Furthermore, the TP, FP, FN, and TN, accuracy score, recall (sensitivity), precision (positive predicted value, PPV), F1 score, AUROC, AUPRC are obtained. The detection performance was evaluated based on the intersection over union (IOU) of the bounding box predicted by the model and ground truth bounding box. With the IOU threshold, we could determine whether the model prediction was true or false. Using the precision and recall scores, we can summarize the detection performance as an average precision (AP) score [45]. The AP_{50} score corresponded to the AP score at the IOU threshold of 50%. The classification performance of the detection model was computed based on the augmented confidence score that aggregated multiple prediction outputs.

3. Results

3.1. Patient Characteristics

All the patients were LVI(+). The mean age of the patients was 69.6 years (±10.2), and the majority were men (73.0%) (Table 1). Poorly differentiated tumors comprised 46.0% of the cases. Despite being LVI(+), 10 patients (18.2%) did not exhibit LNM. The number of lymph node involvement was 12.0 (±13.7). Perineural invasion was observed in 39 (61.9%) patients. The clinicopathological features of the cases are summarized in Table 1.

Table 1. Baseline characteristics of the study population.

Variable	Total (N = 63)
Age [1]	69.6 (10.2)
Sex [2]	
Male	46 (73.0%)
Female	17 (27.0%)
Lauren Classification [2]	
Intestinal	36 (57.1%)
Diffuse	12 (19.0%)
Mixed	15 (23.8%)
Grade [2]	
Well differentiated	3 (4.8%)
Moderately differentiated	31 (49.2%)
Poorly differentiated	29 (46.0%)
T Staging [2]	
pT1a	2 (3.2%)
pT1b	17 (27.0%)
pT2	5 (7.9%)
pT3	13 (20.6%)
pT4a	23 (36.5%)
pT4b	3 (4.8%)
N Staging [2]	
pN0	10 (18.2%)
pN1	6 (10.9%)
pN2	13 (23.6%)
pN3a	8 (14.5%)
pN3b	18 (32.7%)
LN Involvement [1]	12.0 (13.7)
Perineural Invasion [2]	
Present	39 (61.9%)
Not identified	24 (38.1%)
IHC Expression of C-erb B2 [2]	
0	36 (57.1%)
1+	11 (17.5%)
2+	6 (9.5%)
3+	7 (11.1%)
Not available	3 (4.8%)

LN, Lymph Node; IHC, Immunohistochemistry. [1] Mean (S.D.); [2] Number of items (Percentage).

3.2. Patch-Level Analysis

With WSI-level splitting, each WSI was randomly allocated as a training, valid, or test dataset. Each WSI image had a different prognosis for LVI. Therefore, the number of LVI foci and patch images was heterogeneous. The dataset configurations are presented in Table 2. The patch-level analysis was components: classification, detection, and an ensemble of both classification and detection. Figure 3 illustrated the example outputs of ground truths, classification focused areas, and detection outputs.

Table 2. Dataset configuration.

	# of WSI	# of Positive Per WSI [1]	# of Negative Per WSI [1]
Train set	64	68.77 (90.04)	159.23 (87.54)
Valid set	16	28.50 (25.61)	161.12 (60.69)
Test set	20	105.3 (91.73)	201.4 (102.48)

WSI, Whole slide image; [1] Mean (S.D.).

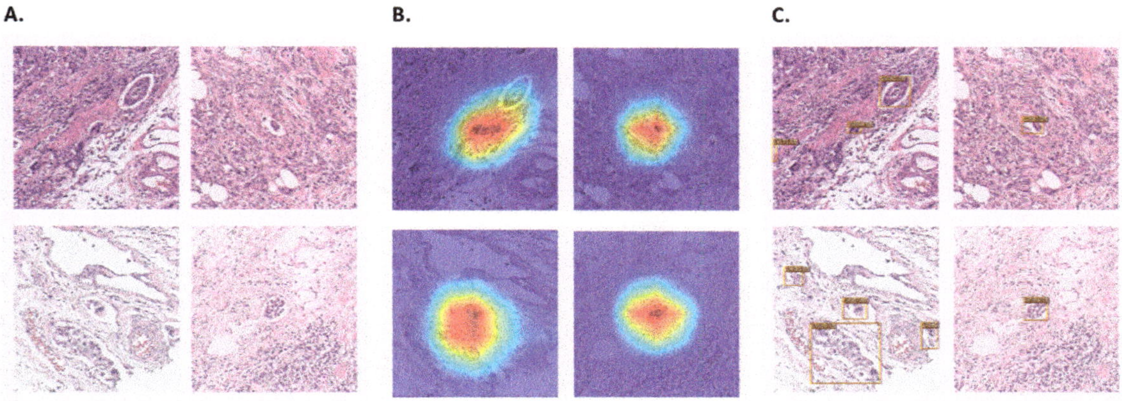

Figure 3. The output of classification and detection model. The summarization of the classification and detection results for the same patch image is presented. Panel (**A**) displays the original image, while panels (**B**,**C**) showcase the classification and detection results, respectively. The heatmap generated using Grad-CAM highlights the areas of focus by the classification model, with red areas indicating greater attention. This visual representation indicates that the classification model exhibits a relatively focused perspective. Conversely, the detection model predicts the object's location by enclosing it within a bounding box and provides the confidence level for each prediction. It is evident that the detection model successfully identifies various dispersed regions within the image.

3.3. Patch-Level Analysis: Classification Models

The patch classification results were outstanding for all classification models without any considerable performance gap. The ConViT model showed the best performance in terms of both the area under the receiver operating characteristics (AUROC) and area under the precision-recall curve (AUPRC) in the classification models (AUROC: 0.9796; AUPRC: 0.9648). The accuracy, precision, recall score, and F1 score were computed with a confidence score threshold of 0.5.

3.4. Patch-Level Analysis: Detection Models

In detection, the YOLOX model outperformed the YOLO v3 model in both detection (AP_{50}) and classification metrics. The AP_{50} of YOLOX and YOLO v3 were 0.55 and 0.66, respectively. The AUROC and AUPRC values of YOLOX were higher than those for YOLO v3 (0.9666 vs. 0.9702 for the AUROC and 0.9423 vs. 0.9302 for the AUPRC). However, the

AUROC and AUPRC of YOLOX computed based on the augmented patch-level confidence score were slightly lower (AUROC: −0.0094; AUPRC: −0.0225) than those of the ConViT classification model. In the detection models, the accuracy, precision, recall score, and F1 score were computed with an augmented patch-level confidence score threshold of 0.7. The threshold was adjusted to be stricter than the value utilized in the image classification model to mitigate the heavy false positives that could occur during detection.

3.5. Patch-Level Analysis: Ensemble Model

Notably, the YOLOX model exhibited an outstanding F1 score (+0.0039 points compared with that of ConViT) in all benchmark models. Considering the AUROC, AUPRC, and F1 scores, we attempted to mix the best-performing models in an ensemble approach. With simple averaging of the patch-level confidence scores, the ensemble model showed the best AUROC, AUPRC, and F1 scores of 0.9880, 0.9769, and 0.9280, respectively. The performances are summarized in Table 3.

Table 3. Performance of trained model using the patch images.

Method	Model	AUROC	AUPRC	Accuracy	F1 Score
Classification	ResNet50	0.9762 (0.9726–0.9798)	0.9593 (0.9447–0.9739)	0.9319 (0.9254–0.9384)	0.8992 (0.8895–0.9089)
	EfficientNetB3	0.9731 (0.9693–0.9769)	0.9551 (0.935–0.9752)	0.9281 (0.9217–0.9345)	0.8929 (0.8827–0.9031)
	ConViT	0.9796 (0.9765–0.9827)	0.9648 (0.9592–0.9704)	0.9348 (0.9288–0.9408)	0.9025 (0.8935–0.9115)
Detection	YOLOv3	0.9666 (0.9623–0.9709)	0.9302 (0.9203–0.9401)	0.927 (0.9196–0.9344)	0.8977 (0.8868–0.9086)
	YOLOX	0.9702 (0.9648–0.9756)	0.9423 (0.9323–0.9523)	0.9353 (0.9278–0.9428)	0.9064 (0.8962–0.9166)
Ensemble		0.988 (0.9852–0.9908)	0.9769 (0.9717–0.9821)	0.9514 (0.9459–0.9569)	0.928 (0.9198–0.9362)

Mean (95% confidence interval).

3.6. WSI-Level Analysis

The WSI consists of multiple patch images, allowing for aggregation of these patches at the WSI level. The conceptual diagram of WSI-level analysis is shown in Figure 4. Each patch prediction result was aggregated at the WSI-level, and the WSI-level prediction result was aggregated once more in the entire test dataset. The WSI-level prediction performance is summarized in Table 4. The performance was consistent with the results of the patch-level prediction (Table 3). We adjusted the threshold such that the positive and negative could be determined as the medium points (0.5); however, this threshold could be rescaled depending on the interests of the researcher. In our dataset, WSIs generally included multiple LVI regions. Therefore, we concluded that the benefit of reducing false positives was more significant. If the LVI region is small, such as in patients with early-stage cancer, a strategy can be adopted to reduce false negatives by lowering the threshold.

Table 4. Performance of trained model using the whole slide images.

Method	True Negative Rate [1]	False Positive Rate [1]	False Negative Rate [1]	True Positive Rate [1]
ConViT	96.63 (0.03)	3.37 (0.03)	11.88 (0.09)	88.12 (0.09)
YOLOX	94.99 (0.03)	5.01 (0.03)	10.88 (0.12)	89.12 (0.12)
Ensemble	97.56 (0.02)	2.44 (0.02)	10.21 (0.07)	89.79 (0.07)

[1] Mean (S.D.).

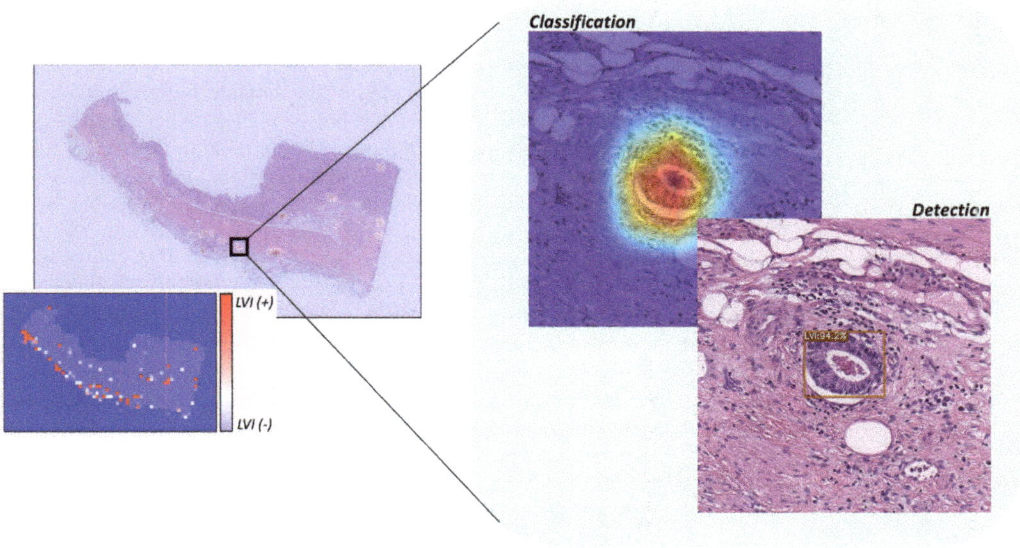

Figure 4. The example WSI-level analysis. A WSI-level analysis can be visualized by combining the results of patch-level analysis. Additionally, it can be illustrated as a WSI-level heatmap, which utilizes the spatial information of the patch images. In the heatmap representation, red points indicate regions that exhibit high confidence in being classified as LVI(+) cases, whereas blue points indicate regions with high confidence in being classified as LVI(−) cases. The image on the right showcases a magnified view of the area identified as LVI(+), presenting the respective judgments made by both the classification model and the detection model.

3.7. External Validation

To measure the efficacy of the ensemble approach, we conducted an external validation using a preexisting dataset. Employing this classification dataset facilitated the application of our model to ascertain positive or negative LVIs [37]. The ensemble model demonstrated superior performance compared to both classification and object detection models (Table 5). Specifically, the AUROC of the ensemble model exhibited improvements of 0.025 (2.8%) and 0.052 (5.9%) in contrast to the classification and detection models, respectively. Furthermore, the AUPRC of the ensemble model saw enhancements of 0.044 (5.1%) and 0.081 (9.8%), respectively. Analogous to the internal validation dataset, the ensemble model exhibited robustness when compared to the classification and detection-only models.

Table 5. Performance of trained model using the external validation dataset.

Method	Model	AUROC	AUPRC	Accuracy	F1 Score
Classification	ConViT	0.9184 (0.8975–0.9393)	0.869 (0.8338–0.9041)	0.8674 (0.8465–0.8883)	0.7896 (0.7543–0.8248)
Detection	YOLOX	0.8915 (0.8638–0.9192)	0.8319 (0.7876–0.8763)	0.8592 (0.8364–0.882)	0.7934 (0.7577–0.8291)
Ensemble		0.9438 (0.9258–0.9619)	0.9132 (0.8875–0.939)	0.8983 (0.879–0.9175)	0.8358 (0.8035–0.8681)

Mean (95% confidence interval).

4. Discussion

In this study, we present a deep-learning model for predicting gastric LVI from the patch images from WSI. Two models were developed: image classification and detection. The ConViT (classification) and YOLOX (detection) models showed comparable performances. The final ensemble model showed outstanding performance in predicting gastric LVI.

In a previous study, Ghosh et al. demonstrated that a deep-learning model could predict LVI foci in testicular LVI [36]. They applied the semantic segmentation-based model (DeeplabV3) [46] to predict the mask of LVI foci; however, the number of LVI(+) foci to train and evaluate a semantic segmentation model were small. Therefore, the model performance can be further improved. With few samples of LVI foci included in the test dataset (34 foci), it could be difficult to determine the generalized performance of LVI prediction.

One of the primary tasks of digital pathology is the detection of mitosis, which often employs a two-stage framework comprising object detection and classification [47]. This approach is preferred due to the small size of mitotic objects, which makes the model predictions highly susceptible to false positives and false negatives. Initially, candidate regions are identified through object detection, and subsequently refined using classification techniques. While the sequential application of this two-stage framework may not pose significant challenges in studies based on limited benchmark datasets, it can prove time-consuming in typical medical scenarios. Therefore, to address this issue, we propose an ensemble approach that combines the advantages of the two-stage model while enabling parallel processing.

In our experimental setting, the classification model ConViT exhibited an outstanding performance among the candidate classification models. The ConViT model attempted to fuse the outstanding performance of transformer-based architectures with the advantages of CNN. The ability of the transformer to focus on global information and the ability of CNNs to focus on local patterns boosted the prediction performance. The LVI foci had heterogeneous shape and size characteristics. In addition, it is essential to determine whether the LVI is located in the lymph node site or blood vessels. The most common false positives occurred in detachment artifacts owing to the failure to interpret peripheral contexts.

The detection model also showed comparable performance in detecting LVI foci. The anchor-free assumption-based model YOLOX was more appropriate because of the varying sizes and shapes of the LVI foci. The YOLOX model exhibited a comparable performance with regard to the AUROC and AUPRC than the ConViT model. However, it exhibited a slightly better performance with regard to the F1 score. The ensemble model exhibited improved AUROC, AUPRC, and F1 scores compared with the classification and detection-only model (improved gain: 0.0084, AUROC; 0.012, AUPRC; 0.022, F1 score). Additionally, the improvement of the ensemble model was also found in the external validation (AUROC: 2.8%; AUPRC: 5.1%).

Our model predicted LVI foci in WSI; in other words, it identified whether LVI foci existed. However, LVI is essentially a histological finding that suggests the possibility of metastasis to the lymph nodes. Previous studies have reported models to predict LNM from pathological slide images of solid tumors, such as breast, colorectal, bladder, and prostate cancers. Although LNM is one of the most important prognostic factors, a model for predicting LNM in gastric cancer has not yet been reported. Wang et al. reported a model for predicting the prognosis of gastric cancer using the histopathology of resected lymph nodes; however, this was not a model for predicting metastasis to the lymph nodes. This algorithm, which detects LVI(+) foci, is expected to significantly help pathologists at the actual reading site, However, predicting LVI(+) in the clinical field is not sufficient to predict the prognosis of a patient. It is necessary to conduct additional studies on the association of the LVI(+) foci identified by this algorithm with the number of lymph node metastases and patient survival prognosis, and thus, further investigation into this is anticipated.

In addition, semi-supervised and active learning pipelines for generating LVI focal labeling more easily need to be further developed. Our YOLOX model can predict the LVI foci using a bounding box. Therefore, we can assume that the prediction results of YOLOX are newly annotated LVI foci in the other datasets. With the supervision of human experts who reject or accept newly annotated LVI foci (active learning), the labeled dataset expands rapidly. Additionally, in this study, we hypothesized that detection would be sufficient to predict LVI foci. However, a previous study utilized semantic-segmentation-based modeling for testicular LVI foci detection. LVI foci share similar patterns despite differences in organs, such as tumors surrounded by blood vessels or lymph nodes. Therefore, in the future, we aim to expand our work to compare semantic segmentation, object detection, and classification models to predict the LVI foci.

Our study has several limitations. First, a number of LVI(+) foci imbalances may exist for each slide. This data imbalance problem may cause distortion in the learning process. We applied WSI-level data splitting to resolve the LVI(+) foci imbalance problem. The best option for data splitting involves splitting the patient-level data. However, we encountered varying LVI(+) foci depending on the patient status. Furthermore, of the multiple sections of slides that may be present in a single gastric cancer tissue, we selected no more than five slides from the same patient. Therefore, patient-level data splitting can be coupled with a heavy class imbalance that is harmful to supervised learning procedures. To mitigate this issue, we alternately selected WSI-level splitting. Second, LVI(+) foci always contain the possibility of false positives or negatives. To reduce false-positive or false-negative foci marks at the annotation step, we confirmed CD34 and D2-40 immunohistochemical staining on all slides. In addition, LVI(+) confirmation was performed by two pathologists. However, annotation marking for foci may be missed because LVI(+) is a relatively small lesion within the WSI. This results in missed marking annotations for some LVI(+) foci and marked LVI(+) foci for some artifacts. Similarly, when the trained algorithmic model predicts LVI(+) positive foci, it may be a false positive. To discriminate false positives, all areas predicted to be LVI(+) positive foci were individually checked by two pathologists. Through this process, we were able to improve the accuracy of the model in predicting LVI(+) foci. Spatial heterogeneity is a crucial factor that must be taken into account in studies on artificial intelligence learning in digital pathology. Stomach cancer is specifically recognized as a type of cancerous tissue that exhibits significant and pronounced spatial heterogeneity within the tissue. Nevertheless, spatial heterogeneity was not a significant factor that needed to be taken into account for this project. LVI is histopathologically defined by the presence of tumor emboli within lymphatic/vascular channels and exhibits morphological features that are rather homogeneous. For instance, the presence of LVI is not exclusive to stomach cancer but is also observed in various other forms of cancer. These findings indicate that the scope of this research extends beyond stomach cancer and has potential for further application to other types of cancer.

5. Conclusions

This research presents an ensemble deep-learning model for detecting vascular and lymphatic vascular invasion in WSI of histopathology of gastric cancer. The ensemble deep-learning model has been demonstrated as more robust and accurate than single models, and it can be used as a valuable tool for pathologists in diagnosing gastric cancer and may help improve the accuracy of diagnosis and prognosis of the disease. This approach can be considered an alternative to traditional methods and as a step toward computer-aided diagnotic systems in histopathology.

Author Contributions: Conceptualization, K.M.M. and M.-G.N.; Methodology, J.L., S.C. and K.M.M.; Validation, S.A.; Investigation, J.L., S.C., J.K., J.J.K., K.M.M. and M.-G.N.; Resources, Y.-D.C., S.-R.K., G.-Y.S., D.-H.Y., J.-H.L., K.-H.L., S.A. and M.-G.N.; Data curation, M.-G.N.; Writing—original draft, J.L. and M.-G.N.; Writing—review and editing, J.H.L., S.G.J.G. and M.-G.N.; Visualization, J.L.; Supervision, N.K., S.G.J.G., J.H.K., J.H.L. and S.A.; Project administration, D.-H.Y., K.M.M. and M.-G.N.; Funding acquisition, D.-H.Y. and K.M.M. All authors have read and agreed to the published version of the manuscript.

Funding: This research was supported by following research grants: the Korea Health Technology R & D Project through the Korea Health Industry Development Institute (KHIDI) funded by the Ministry of Health & Welfare, Republic of Korea (grant number: HR20C0021, HI23C1494); the Bio & Medical Technology Development Program of the National Research Foundation (NRF) &

funded by the Korean government (MSIT) (NRF-2019M3E5D1A02067961); the Chonnam National University Hwasun Hospital Institute for Biomedical Science (HCRU23001, HCGI23009); the Medical Research Promotion Program through the Gangneung Asan Hospital funded by the Asan Foundation (2021 B002); the National IT Industry Promotion Agency (NIPA) of KOREA grant funded by the Korea government (MSIT) (S1402-23-1001, Development of Medical AI Software to Improve Patient Safety Through the Detection and Evaluation of the Type and Position of Tubes and Catheters in Chest Radiograph).

Institutional Review Board Statement: This study was approved by the Institutional Review Board (IRB) of the Chonnam National University Hwasun Hospital (protocol code: CNUHH-2021-197 and date of approval: 28 September 2021) and conducted in accordance with the Declaration of Helsinki.

Informed Consent Statement: Informed consent from patients was waived with IRB approval.

Data Availability Statement: Data cannot be shared publicly because of sensitive personal medical information. The datasets generated and/or analyzed during the current study are available from the corresponding author upon reasonable request. The external validation dataset is available at https://zenodo.org/records/10020633 (accessed on 3 January 2024). The source code of this project available at https://github.com/jonghyunlee1993/LVINet (accessed on 9 January 2024).

Acknowledgments: This study was conducted as a team project of the Korea Human Resources Development Institute (KOHI) Medical Imaging Artificial Intelligence Expert Training Project (2021).

Conflicts of Interest: The authors declare that the research was conducted in the absence of any commercial or financial relationships that could be considered potential conflicts of interest.

References

1. Hong, S.; Won, Y.J.; Lee, J.J.; Jung, K.W.; Kong, H.J.; Im, J.S.; Seo, H.G.; The Community of Population-Based Regional Cancer Registries. Cancer statistics in Korea: Incidence, mortality, survival, and prevalence in 2018. *Cancer Res. Treat.* **2021**, *53*, 301–315. [CrossRef] [PubMed]
2. Lordick, F.; Carneiro, F.; Cascinu, S.; Fleitas, T.; Haustermans, K.; Piessen, G.; Vogel, A.; Smyth, E.C. Gastric cancer: ESMO clinical practice guideline for diagnosis, treatment and follow-up. *Ann. Oncol.* **2022**, *33*, 1005–1020. [CrossRef] [PubMed]
3. Ferlay, J.; Ervik, M.; Lam, F.; Colombet, M.; Mery, L.; Piñeros, M. Global Cancer Observatory: Cancer Today. International Agency for Research on Cancer. Available online: https://gco.iarc.fr/today (accessed on 14 June 2023).
4. Takada, K.; Yoshida, M.; Aizawa, D.; Sato, J.; Ono, H.; Sugino, T. Lymphovascular invasion in early gastric cancer: Impact of ancillary D2-40 and elastin staining on interobserver agreement. *Histopathology* **2020**, *76*, 888–897. [CrossRef] [PubMed]
5. Nitti, D.; Marchet, A.; Olivieri, M.; Ambros, A.; Mencarelli, R.; Belluco, C.; Lise, M. Ratio between metastatic and examined lymph nodes is an independent prognostic factor after D2 resection for gastric cancer: Analysis of a large European monoinstitutional experience. *Ann. Surg. Oncol.* **2003**, *10*, 1077–1085. [CrossRef] [PubMed]
6. Sekiguchi, M.; Oda, I.; Taniguchi, H.; Suzuki, H.; Morita, S.; Fukagawa, T.; Sekine, S.; Kushima, R.; Katai, H. Risk stratification and predictive risk-scoring model for lymph node metastasis in early gastric cancer. *J. Gastroenterol.* **2016**, *51*, 961–970. [CrossRef] [PubMed]
7. Gotoda, T.; Yanagisawa, A.; Sasako, M.; Ono, H.; Nakanishi, Y.; Shimoda, T.; Kato, Y. Incidence of lymph node metastasis from early gastric cancer: Estimation with a large number of cases at two large centers. *Gastric Cancer* **2000**, *3*, 219–225. [CrossRef] [PubMed]
8. Fujikawa, H.; Koumori, K.; Watanabe, H.; Kano, K.; Shimoda, Y.; Aoyama, T.; Yamada, T.; Hiroshi, T.; Yamamoto, N.; Cho, H.; et al. The clinical significance of lymphovascular invasion in gastric cancer. *In Vivo* **2020**, *34*, 1533–1539. [CrossRef]
9. Song, Y.J.; Shin, S.H.; Cho, J.S.; Park, M.H.; Yoon, J.H.; Jegal, Y.J. The role of lymphovascular invasion as a prognostic factor in patients with lymph node-positive operable invasive breast cancer. *J. Breast Cancer* **2011**, *14*, 198–203. [CrossRef]
10. Talamonti, M.S.; Kim, S.P.; Yao, K.A.; Wayne, J.D.; Feinglass, J.; Bennett, C.L.; Rao, S. Surgical outcomes of patients with gastric carcinoma: The importance of primary tumor location and microvessel invasion. *Surgery* **2003**, *134*, 720–727, Discussion 727–729. [CrossRef]
11. Amin, M.B.; Greene, F.L.; Edge, S.B.; Compton, C.C.; Gershenwald, J.E.; Brookland, R.K.; Meyer, L.; Gress, D.M.; Byrd, D.R.; Winchester, D.P. The eighth edition AJCC cancer staging manual: Continuing to build a bridge from a population-based to a more "personalized" approach to cancer staging. *CA Cancer J. Clin.* **2017**, *67*, 93–99. [CrossRef]
12. Kim, Y.I.; Kook, M.C.; Choi, J.E.; Lee, J.Y.; Kim, C.G.; Eom, B.W.; Yoon, H.M.; Ryu, K.W.; Kim, Y.W.; Choi, I.J. Evaluation of submucosal or lymphovascular invasion detection rates in early gastric cancer based on pathology section interval. *J. Gastric Cancer* **2020**, *20*, 165–175. [CrossRef] [PubMed]
13. Kwee, R.M.; Kwee, T.C. Predicting lymph node status in early gastric cancer. *Gastric Cancer* **2008**, *11*, 134–148. [CrossRef]
14. Kim, H.; Kim, J.H.; Park, J.C.; Lee, Y.C.; Noh, S.H.; Kim, H. Lymphovascular invasion is an important predictor of lymph node metastasis in endoscopically resected early gastric cancers. *Oncol. Rep.* **2011**, *25*, 1589–1595. [CrossRef]

15. Lee, S.Y.; Yoshida, N.; Dohi, O.; Lee, S.P.; Ichikawa, D.; Kim, J.H.; Sung, I.K.; Park, H.S.; Otsuji, E.; Itoh, Y.; et al. Differences in prevalence of lymphovascular invasion among early gastric cancers between Korea and Japan. *Gut Liver* **2017**, *11*, 383–391. [CrossRef] [PubMed]
16. Zaorsky, N.G.; Patil, N.; Freedman, G.M.; Tuluc, M. Differentiating lymphovascular invasion from retraction artifact on histological specimen of breast carcinoma and their implications on prognosis. *J. Breast Cancer* **2012**, *15*, 478–480. [CrossRef] [PubMed]
17. Gilchrist, K.W.; Gould, V.E.; Hirschl, S.; Imbriglia, J.E.; Patchefsky, A.S.; Penner, D.W.; Pickren, J.; Schwartz, I.S.; Wheeler, J.E.; Barnes, J.M.; et al. Interobserver variation in the identification of breast carcinoma in intramammary lymphatics. *Hum. Pathol.* **1982**, *13*, 170–172. [CrossRef] [PubMed]
18. Gresta, L.T.; Rodrigues-Junior, I.A.; de Castro, L.P.; Cassali, G.D.; Cabral, M.M. Assessment of vascular invasion in gastric cancer: A comparative study. *World J. Gastroenterol.* **2013**, *19*, 3761–3769. [CrossRef] [PubMed]
19. Ghosh, A.; Sirinukunwattana, K.; Khalid Alham, N.; Browning, L.; Colling, R.; Protheroe, A.; Protheroe, E.; Jones, S.; Aberdeen, A.; Rittscher, J.; et al. The potential of artificial intelligence to detect lymphovascular invasion in testicular cancer. *Cancers* **2021**, *13*, 1325. [CrossRef]
20. Yonemura, Y.; Endou, Y.; Tabachi, K.; Kawamura, T.; Yun, H.Y.; Kameya, T.; Hayashi, I.; Bandou, E.; Sasaki, T.; Miura, M. Evaluation of lymphatic invasion in primary gastric cancer by a new monoclonal antibody, D2-40. *Hum. Pathol.* **2006**, *37*, 1193–1199. [CrossRef]
21. Arigami, T.; Natsugoe, S.; Uenosono, Y.; Arima, H.; Mataki, Y.; Ehi, K.; Yanagida, S.; Ishigami, S.; Hokita, S.; Aikou, T. Lymphatic invasion using D2-40 monoclonal antibody and its relationship to lymph node micrometastasis in pN0 gastric cancer. *Br. J. Cancer* **2005**, *93*, 688–693. [CrossRef]
22. Sako, A.; Kitayama, J.; Ishikawa, M.; Yamashita, H.; Nagawa, H. Impact of immunohistochemically identified lymphatic invasion on nodal metastasis in early gastric cancer. *Gastric Cancer* **2006**, *9*, 295–302. [CrossRef] [PubMed]
23. Araki, I.; Hosoda, K.; Yamashita, K.; Katada, N.; Sakuramoto, S.; Moriya, H.; Mieno, H.; Ema, A.; Kikuchi, S.; Mikami, T.; et al. Prognostic impact of venous invasion in stage IB node-negative gastric cancer. *Gastric Cancer* **2015**, *18*, 297–305. [CrossRef] [PubMed]
24. Harris, E.I.; Lewin, D.N.; Wang, H.L.; Lauwers, G.Y.; Srivastava, A.; Shyr, Y.; Shakhtour, B.; Revetta, F.; Washington, M.K. Lymphovascular invasion in colorectal cancer: An interobserver variability study. *Am. J. Surg. Pathol.* **2008**, *32*, 1816–1821. [CrossRef] [PubMed]
25. Kirsch, R.; Messenger, D.E.; Riddell, R.H.; Pollett, A.; Cook, M.; Al-Haddad, S.; Streutker, C.J.; Divaris, D.X.; Pandit, R.; Newell, K.J.; et al. Venous invasion in colorectal cancer impact of an elastin stain on detection and interobserver agreement among gastrointestinal and nongastrointestinal pathologists. *Am. J. Surg. Pathol.* **2013**, *37*, 200–210. [CrossRef] [PubMed]
26. Nam, S.; Chong, Y.; Jung, C.K.; Kwak, T.Y.; Lee, J.Y.; Park, J.; Rho, M.J.; Go, H. Introduction to digital pathology and computer-aided pathology. *J. Pathol. Transl. Med.* **2020**, *54*, 125–134. [CrossRef]
27. Ahmad, Z.; Rahim, S.; Zubair, M.; Abdul-Ghafar, J. Artificial intelligence (AI) in medicine, current applications and future role with special emphasis on its potential and promise in pathology: Present and future impact, obstacles including costs and acceptance among pathologists, practical and philosophical considerations. A comprehensive review. *Diagn. Pathol.* **2021**, *16*, 24. [CrossRef]
28. Joshi, G.; Jain, A.; Araveeti, S.R.; Adhikari, S.; Garg, H.; Bhandari, M. FDA approved Artificial Intelligence and Machine Learning (AI/ML)-Enabled Medical Devices: An updated landscape. *medRxiv* **2022**. [CrossRef]
29. Pantanowitz, L.; Quiroga-Garza, G.M.; Bien, L.; Heled, R.; Laifenfeld, D.; Linhart, C.; Sandbank, J.; Albrecht Shach, A.; Shalev, V.; Vecsler, M.; et al. An artificial intelligence algorithm for prostate cancer diagnosis in whole slide images of core needle biopsies: A blinded clinical validation and deployment study. *Lancet Digit. Health* **2020**, *2*, e407–e416. [CrossRef]
30. Campanella, G.; Hanna, M.G.; Geneslaw, L.; Miraflor, A.; Werneck Krauss Silva, V.; Busam, K.J.; Brogi, E.; Reuter, V.E.; Klimstra, D.S.; Fuchs, T.J. Clinical-grade computational pathology using weakly supervised deep learning on whole slide images. *Nat. Med.* **2019**, *25*, 1301–1309. [CrossRef]
31. Turkki, R.; Byckhov, D.; Lundin, M.; Isola, J.; Nordling, S.; Kovanen, P.E.; Verrill, C.; von Smitten, K.; Joensuu, H.; Lundin, J.; et al. Breast cancer outcome prediction with tumour tissue images and machine learning. *Breast Cancer Res. Treat.* **2019**, *177*, 41–52. [CrossRef]
32. Bychkov, D.; Linder, N.; Turkki, R.; Nordling, S.; Kovanen, P.E.; Verrill, C.; Walliander, M.; Lundin, M.; Haglund, C.; Lundin, J. Deep learning based tissue analysis predicts outcome in colorectal cancer. *Sci. Rep.* **2018**, *8*, 3395. [CrossRef]
33. Wang, S.; Shi, J.; Ye, Z.; Dong, D.; Yu, D.; Zhou, M.; Liu, Y.; Gevaert, O.; Wang, K.; Zhu, Y.; et al. Predicting EGFR mutation status in lung adenocarcinoma on computed tomography image using deep learning. *Eur. Respir. J.* **2019**, *53*, 1800986. [CrossRef] [PubMed]
34. Hinata, M.; Ushiku, T. Detecting immunotherapy-sensitive subtype in gastric cancer using histologic image-based deep learning. *Sci. Rep.* **2021**, *11*, 22636. [CrossRef] [PubMed]
35. Bejnordi, B.E.; Veta, M.; van Diest, P.J.; van Ginneken, B.; Karssemeijer, N.; Litjens, G.; van der Laak, J.A.W.M.; Consortium, C. Diagnostic assessment of deep learning algorithms for detection of lymph node metastases in women with breast cancer. *JAMA* **2017**, *318*, 2199–2210. [CrossRef]

36. Steiner, D.F.; MacDonald, R.; Liu, Y.; Truszkowski, P.; Hipp, J.D.; Gammage, C.; Thng, F.; Peng, L.; Stumpe, M.C. Impact of deep learning assistance on the histopathologic review of lymph nodes for metastatic breast cancer. *Am. J. Surg. Pathol.* **2018**, *42*, 1636–1646. [CrossRef] [PubMed]
37. Lee, J.; Ahn, S.; Kim, H.S.; An, J.; Sim, J. A robust model training strategy using hard negative mining in a weakly labeled dataset for lymphatic invasion in gastric cancer. *J. Pathol. Clin. Res.* **2024**, *10*, e355. [CrossRef] [PubMed]
38. He, K.; Zhang, X.; Ren, S.; Sun, J. Deep residual learning for image recognition. In Proceedings of the IEEE Conference on Computer Vision and Pattern Recognition (CVPR), Las Vegas, NV, USA, 27–30 June 2016; pp. 770–778.
39. Tan, M.; Le, Q. EfficientNet: Rethinking model scaling for convolutional neural networks. In Proceedings of the 36th International Conference on Machine Learning, Long Beach, CA, USA, 9–15 June 2019; pp. 6105–6114.
40. d'Ascoli, S.; Touvron, H.; Leavitt, M.L.; Morcos, A.S.; Biroli, G.; Sagun, L. Convit: Improving vision transformers with soft convolutional inductive biases. In Proceedings of the 38th International Conference on Machine Learning, Virtual, 18–24 July 2021; pp. 2286–2296.
41. Deng, J.; Dong, W.; Socher, R.; Li, L.-J.; Li, K.; Fei-Fei, L. ImageNet: A large-scale hierarchical image database. In Proceedings of the 2009 IEEE Conference on Computer Vision and Pattern Recognition (CVPR), Miami, FL, USA, 20–25 June 2009; pp. 248–255.
42. Redmon, J.; Divvala, S.; Girshick, R.; Farhadi, A. You only look once: Unified, real-time object detection. In Proceedings of the IEEE Conference on Computer Vision and Pattern Recognition (CVPR), Las Vegas, NV, USA, 27–30 June 2016; pp. 779–788.
43. Redmon, J.; Farhadi, A. YOLOv3: An incremental improvement. *arXiv* **2018**. [CrossRef]
44. Ge, Z.; Liu, S.; Wang, F.; Li, Z.; Sun, J. YOLOX: Exceeding YOLO series in 2021. *arXiv* **2021**. [CrossRef]
45. Lin, T.-Y.; Maire, M.; Belongie, S.; Hays, J.; Perona, P.; Ramanan, D.; Dollár, P.; Zitnick, C.L. Microsoft COCO: Common objects in context. In *Computer Vision—ECCV 2014*; Fleet, D., Pajdla, T., Schiele, B., Tuytelaars, T., Eds.; Springer: Cham, Switzerland, 2014; Volume 8693, pp. 740–755.
46. Chen, L.-C.; Papandreou, G.; Schroff, F.; Adam, H. Rethinking atrous convolution for semantic image segmentation. *arXiv* **2017**. [CrossRef]
47. Piansaddhayanaon, C.; Santisukwongchote, S.; Shuangshoti, S.; Tao, Q.; Sriswasdi, S.; Chuangsuwanich, E. ReCasNet: Improving consistency within the two-stage mitosis detection framework. *Artif. Intell. Med.* **2023**, *135*, 102462. [CrossRef]

cancers

MDPI

Article

Quantitative Spatial Characterization of Lymph Node Tumor for N Stage Improvement of Nasopharyngeal Carcinoma Patients

Jiang Zhang [1], Xinzhi Teng [1], Saikit Lam [2,3], Jiachen Sun [1], Andy Lai-Yin Cheung [4], Sherry Chor-Yi Ng [4], Francis Kar-Ho Lee [5], Kwok-Hung Au [5], Celia Wai-Yi Yip [5], Victor Ho-Fun Lee [4,6], Zhongshi Lin [7], Yongyi Liang [7], Ruijie Yang [8], Ying Han [9], Yuanpeng Zhang [10], Feng-Ming (Spring) Kong [6,9] and Jing Cai [1,3,*]

1 Department of Health Technology and Informatics, The Hong Kong Polytechnic University, Hong Kong SAR, China
2 Department of Biomedical Engineering, Faculty of Engineering, The Hong Kong Polytechnic University, Hong Kong SAR, China
3 Research Institute for Smart Ageing, The Hong Kong Polytechnic University, Hong Kong SAR, China
4 Department of Clinical Oncology, Queen Mary Hospital, Hong Kong SAR, China
5 Department of Clinical Oncology, Queen Elizabeth Hospital, Hong Kong SAR, China
6 Department of Clinical Oncology, The University of Hong Kong, Hong Kong SAR, China
7 Shenzhen Institute for Drug Control (Shenzhen Testing Center of Medical Devices), Shenzhen 518057, China
8 Department of Radiation Oncology, Peking University Third Hospital, Beijing 100191, China
9 Department of Clinical Oncology, The University of Hong Kong-Shenzhen Hospital, Shenzhen 518009, China
10 Department of Medical Informatics, Nantong University, Nantong 226007, China
* Correspondence: jing.cai@polyu.edu.hk; Tel.: +852-3400-8645

Simple Summary: The N staging system for Nasopharyngeal Carcinoma (NPC) is constantly improving for better survival risk stratification with accumulating clinical evidence. Discovering new prognostic factors often depends on clinical observations, which often lack comprehensiveness and precision. This study aimed to propose new quantitative spatial characterizations of LN tumor and demonstrate their feasibility of improving N stage. Independent anatomical prognostic factors were discovered and achieved superior risk stratification performance when combined with N stage. This quantitative approach could be applied to other cancer sites to discover new prognostic or predictive factors and ultimately benefit precision medicine.

Abstract: This study aims to investigate the feasibility of improving the prognosis stratification of the N staging system of Nasopharyngeal Carcinoma (NPC) from quantitative spatial characterizations of metastatic lymph node (LN) for NPC in a multi-institutional setting. A total of 194 and 284 NPC patients were included from two local hospitals as the discovery and validation cohort. Spatial relationships between LN and the surrounding organs were quantified by both distance and angle histograms, followed by principal component analysis. Independent prognostic factors were identified and combined with the N stage into a new prognostic index by univariate and multivariate Cox regressions on disease-free survival (DFS). The new three-class risk stratification based on the constructed prognostic index demonstrated superior cross-institutional performance in DFS. The hazard ratios of the high-risk to low-risk group were 9.07 ($p < 0.001$) and 4.02 ($p < 0.001$) on training and validation, respectively, compared with 5.19 ($p < 0.001$) and 1.82 ($p = 0.171$) of N3 to N1. Our spatial characterizations of lymph node tumor anatomy improved the existing N-stage in NPC prognosis. Our quantitative approach may facilitate the discovery of new anatomical characteristics to improve patient staging in other diseases.

Keywords: nasopharyngeal carcinoma; N stage; lymph node tumor; tumor geometry

Citation: Zhang, J.; Teng, X.; Lam, S.; Sun, J.; Cheung, A.L.-Y.; Ng, S.C.-Y.; Lee, F.K.-H.; Au, K.-H.; Yip, C.W.-Y.; Lee, V.H.-F.; et al. Quantitative Spatial Characterization of Lymph Node Tumor for N Stage Improvement of Nasopharyngeal Carcinoma Patients. *Cancers* **2023**, *15*, 230. https://doi.org/10.3390/cancers15010230

Academic Editors: Wei Wu and Trever G. Bivona

Received: 5 December 2022
Revised: 28 December 2022
Accepted: 29 December 2022
Published: 30 December 2022

1. Introduction

Nasopharyngeal carcinoma (NPC) has a high prevalence in southeast Asia [1,2]. With the development of the intensity modulated radiation therapy (IMRT) technique, better survival patterns can be achieved for patients with early and late stage NPC, especially local and regional tumor control [3,4]. However, distant metastasis remained the primary failure pattern with a high occurrence rate in five years for patients with advanced lymph node (LN) metastasis [5,6]. In addition, nodal metastasis is associated with poor prognosis in other head-and-neck cancer (HNC) subtypes, such as paranasal squamous cell carcinoma [7]. Thus, effective prognosis stratification, especially for the LN tumor, is necessary to guide more accurate clinical decision-making for personalized treatments [8,9].

N stage, which belongs to the tumor–node–metastasis (TNM) staging system jointly proposed by the American Joint Committee on Cancer (AJCC) and the Union for International Cancer Control (UICC), is one of the most robust and widely used LN classifications [10]. The current edition (8th) for NPC is based on anatomical characterization, including size, laterality, and location. However, N stage has been suggested to be less comprehensive and precise due to the qualitative definitions [11].

Over the past decades, various new LN anatomical descriptors have been proposed to improve the current N staging system [12]. For instance, parotid lymph node (PLN) involvement was found to be associated with a poor prognosis in distant metastasis, and an upgrade to the N3 classification was recommended [13,14]. Besides, the current N-staging system categorizes retropharyngeal lymph node (RLN) involvement ($\leq$6 cm) as N1 disease. However, Huang et al. suggested an upgrade of patients with bilateral retropharyngeal lymph node involvement to N2 due to the distinctive prognostic performance within N1 [15]. Other anatomical characteristics of LN, such as extra-nodal extension [16–18] and positive LN numbers [11,19] have been proposed to improve the existing N stage classification system for NPC.

Despite the tremendous efforts made, the development of a more accurate N staging system was still hindered by the rather complex LN anatomical environment. In the era of IMRT, detailed tumor and normal tissue delineations have become the standard procedure for treatment planning with the increasing availability of advanced imaging techniques such as MRI and PET [20–22]. Quantitative spatial characterization of metastatic LN may provide more accurate descriptions of its anatomy, enabling the holistic discovery of anatomical prognostic factors by a data-driven approach.

Therefore, this study aims to investigate the feasibility of improving the prognosis stratification of N staging system from quantitative spatial characterizations of metastatic LN. We designed two types of geometric histograms based on the distances and angles of LN tumor volume to surrounding normal tissues. Independent prognostic factors were extracted by principal component analysis and combined into one prognostic index. A new risk stratification from the combined index was proposed and evaluated on multiple survival endpoints, including disease-free survival (DFS), overall survival (OS), relapse-free survival (RFS) and distant metastasis-free survival (DMFS) both internally and externally. Our methodology may promote accelerated improvement of the LN classification for NPC and can be potentially generalized to other cancer sites.

2. Materials and Methods

Two cohorts of biopsy-proven NPC patients receiving chemoradiotherapy were retrospectively recruited from Hong Kong Queen Mary Hospital (QMH) between 2013 and 2019 and Hong Kong Queen Elizabeth Hospital (QEH) between 2012 and 2015, respectively. Informed consents from patients were waived due to the retrospective nature of this study. The total number of included patients was 194 from QMH and 284 from QEH after excluding patients with (1) co-existing cancer or distance metastasis before treatment, (2) radiation therapy only without concurrent chemoradiotherapy, (3) patients in stage N0 who do not have visible tumor in the lymph node region and (4) incomplete clinical record and missing segmentations. Patients from the QMH cohort were used for deriving independent prognostic factors and development of prognostic index, while the QEH cohort was used solely for external validation.

Clinical factors, including age, sex, T stage, N stage, M stage, overall stage, chemotherapy strategy, and survival information were collected from patient folders. The time of OS, RFS, DMFS, and DFS is defined from the date of treatment to the earliest occurrence of death from any cause, local or regional tumor recurrence, distant metastasis, and the combination of above all, respectively. The TNM stage was administered according to the 7th edition of the AJCC protocol for the QEH cohort and switched to the 8th edition after 2017 for the QMH cohort. Treatment planning structure sets were retrieved from the Picture Archiving and Communication System (PACS) in Digital Imaging and Communications in Medicine (DICOM) format. The gross tumor volume in LN (GTVn) was contoured from contrast-enhanced CT fused with MRI in QEH and an extra imaging modality of PET/CT in QMH by oncologists with at least five years of experience.

Distance and angle histograms were designed to describe the spatial configuration of GTVn relative to the surrounding organs at risk (OARs). OARs that were consistently delineated across the two institutions, including SpinalCord, Parotids (combined Left and Right Parotid), Mandible, Larynx, and Brainstem, were included in this study. Overlap volume histogram (OVH) was first proposed by Kazhdan et al. for quantifying patient geometries [23] and successfully applied by Wu et al. to predict the optimal dose-volume histogram for knowledge-based treatment planning [24]. It summarizes the distances between OAR and the target volume by recording the fractional OAR volume as a function of the maximum distance from the PTV surface:

$$OVH(d) = \frac{count_i(r(v^i_{OAR}, S_{GTVn}))}{V_{OAR}}, \quad (1)$$

where $r(v^i_{OAR}, S_{GTVn})$ is the surface distance defined as the minimum Euclidean distance from OAR voxel v^i_{OAR} to all the LN tumor surface points v^k_{GTVn}:

$$r(v^i_{OAR}, S_{GTVn}) = \min_k\{\|v^i_{OAR} - v^k_{GTVn}\| | v^k_{GTVn} \in S_{GTVn}\}. \quad (2)$$

The surface distance is positive for an OAR voxel outside GTVn and negative when inside. We used the signed Euclidean distance transform algorithm [25] provided by the Python package SimpleITK (version 2.1.1) [26] to calculate the surface distance map and acquired the OVH as the cumulative histogram within the OAR mask. An example GTVn surface distance map is visualized by the heat map in Figure 1c where the left parotid (Parotid_L) is drawn as a red contour.

Spatial configuration of the lymph node tumor could not be precisely determined by distance alone due to the complex organ structures in the head-and-neck region. We designed the projection overlap volume (POV) histogram to quantify the angular relationships between GTVn and the surrounding OARs. POV is defined as the relative OAR volume that overlaps with the parallel projection of GTVn:

$$POV(\alpha) = \frac{\sum_i \chi_{\alpha i}}{V}, \chi_{\alpha i} = f(x) = \begin{cases} 1, & \text{if } \min_j \theta_{ij} < \alpha < \max_j \theta_{ij} \\ 0, & \text{otherwise} \end{cases}, \quad (3)$$

where V is the voxel volume of the OAR, and θ_{ij} is the angle from GTVn surface point v_j to OAR voxel point v_i on the axial plane. POV histogram is calculated by summing up the masked OAR sinogram along the angle direction. The masked OAR sinogram is the modified radon transform of the OAR mask volume around the axial axis; only the voxels located before GTVn are counted for each OAR mask volume projection. One Parotid_L masked sinogram is shown in Figure 1d.

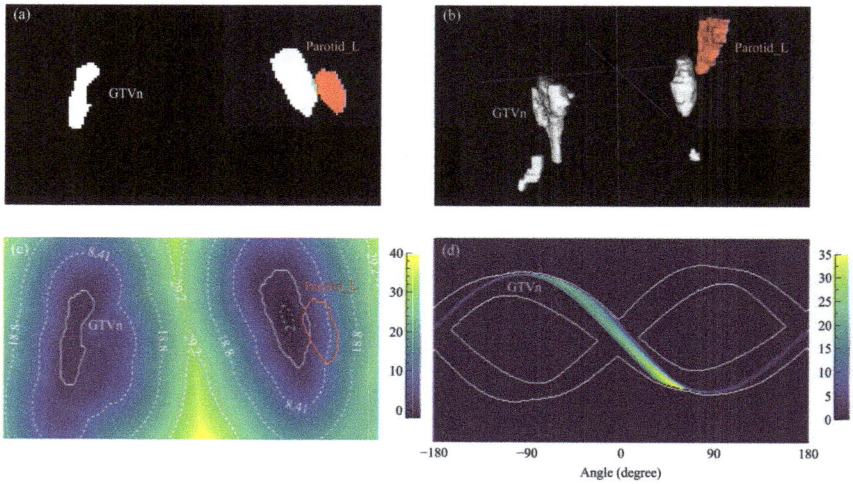

Figure 1. Distance and angle maps based on example GTVn and Parotid_L structures. (**a**) One axial slice of the structure masks (white: GTVn, red: Parotid_L) with the overlap region highlighted by blue. (**b**) The rendered three-dimensional structures. (**c**) One axial slice of the GTVn distance map with annotated contour lines and the Parotid_L contour. (**d**) One slice of the Parotid_L angle map masked by the GTVn sinogram edges (white contours).

Dimensions of the OVH and POV histograms were further reduced by principal component analysis (PCA), where the components that explained the greatest variance across patients were highlighted. This study included the smallest number of principal components (PCs) of OVH and POV that explained 75% of the cumulative variance for each OAR. The coefficients of the principal components (PCs) were extracted as the potential prognostic factors.

Independent prognostic factors were identified from the selected PCs by univariate Cox regression on DFS followed by the covariate independency test with N stage through multivariate Cox regression. The final prognostic index was built by combining the independent prognostic factors with N stage through multivariate Cox regression and evaluated by concordance index (C-index). The confidence interval and p-values for baseline N stage comparison were determined by 1000-iteration bootstrapping. Risk stratification performance was assessed by Kaplan–Meier (KM) analysis, where patients were equally stratified into high (G1), median (G2), and low (G3) risk groups based on the prognostic index in the discovery cohort. The stratification thresholds were applied to the testing cohort as well for the three-grade stratification. Hazard ratios (HRs) with 95% confidence interval (95CI) and the log-rank p-values between risk groups were acquired from univariate Cox regression. All Cox regressions and KM analysis were implemented by the Python package lifelines (version 0.27.0) [27], and the p-value of 0.05 was considered significant.

3. Results

3.1. Baseline Patient Characteristics

Distributions of the baseline patient characteristics for the two cohorts were listed in Table 1. Consistent distributions of age, sex, overall stage, chemotherapy strategy, and World Health Organization (WHO) histology were found between the discovery and validation cohort. The T stage and N stage were significantly different ($p \leq 0.05$) between the two institutions. The median follow-up time of the discovery cohort is 2.5 years and 4.6 years for the validation cohort. Of the 194 discovery patients within the follow-up period, 22 developed local recurrence, 17 with regional recurrence, 29 with distant metastases, and 25 died. The three-year DFS, OS, RFS, and DMFS rates were 72.1%, 90.0%, 82.4%, and 82.4%, respectively. In the validation cohort, 34, 25, 44, and 40 patients of

284 developed local recurrence, regional recurrence, distant metastasis, and death, and the five-year DFS, OS, RFS, DMFS are 74.3%, 94.0%, 85.0%, and 86.2%.

Table 1. Baseline patient characteristics of the discovery and validation cohort.

	Discovery Cohort	Validation Cohort	*p*-Value
Age			
Mean	53.39	52.16	0.249
Sex			
Female	41	70	0.667
Male	153	214	
N stage			
N1	62	17	0.035
N2	93	228	
N3	39	39	
Chemotherapy			
CCRT	33	178	0.330
CCRT + ACT	78	61	
CCRT + ICT	83	43	
WHO histology			
Type 2	27	74	0.142
Type 3	167	210	

Note: Staging was performed according to the 7th edition of the AJCC protocol for the validation cohort and switched to the 8th edition after 2017 for the discovery cohort. Abbreviations: CCRT, concurrent chemoradiotherapy; ACT, adjuvant chemotherapy; ICT, induction chemotherapy; WHO, World Health Organization.

3.2. Prognostic LN Spatial Factors

Thirty-one PCs were extracted from the OVH and POV histograms in total, including four OVH PC and three POV PC of SpinalCord, five OVH PC and three POV PC of Parotids, two OVH PC and two POV PC of Brainstem, three OVH OC and three POV PC of Larynx, and four OVH PC and two POV OC of Mandible. After univariate and multivariate Cox regressions, two spatial factors including the first PC of spinal cord OVH ($OVH_{SC,PC1}$) and the third PC of spinal cord POV ($POV_{SC,PC3}$) were selected as independently prognostic to DFS. Between the two spatial factors, $OVH_{SC,PC1}$ demonstrated a higher discriminability to DFS with C-index of 0.66 at discovery and 0.56 at external validation, while 0.57 at discovery and 0.54 at external validation for $POV_{SC,PC3}$.

As listed in Table 2, $POV_{SC,PC3}$ contributed the highest positive hazard (HR = 3.35, 95CI: 1.41–7.99), followed by the N stage (HR = 2.26, 95CI: 1.46–3.49). On the other hand, $OVH_{SC,PC1}$ had the negative impact of survival hazard (HR = 0.63, 95CI: 0.48–0.83). Figure 2a presents the distributions of the two spatial factors of the 3-year disease and non-disease progressed patients at both discovery and validation. Patients who developed disease progression within three years had significantly lower $OVH_{SC,PC1}$ (mean: −0.80 vs. −0.07, $p = 0.007$) and higher $POV_{SC,PC3}$ (mean: 0.082 vs. 0.057, $p = 0.012$) at discovery, but smaller differences were found on the validation cohort ($OVH_{SC,PC1}$: 0.46 vs. 0.74, $p = 0.032$; $POV_{SC,PC3}$: −0.18 vs. −0.25, $p = 0.089$). Moreover, the spinal cord OVH appeared to be overall larger in the validation but smaller for the POV. After binarizing the two spatial factors by the median values in the discovery cohort, more patients in the validation cohort fell into the low-risk groups, as indicated by Figure 2b. The odds ratios were 0.30 ($p = 0.006$) for $OVH_{SC,PC1}$ and 2.21 ($p = 0.052$) for $POV_{SC,PC3}$ in the discovery cohort. They were less significant for $OVH_{SC,PC1}$ (odds ratio = 0.60, $p = 0.275$) but more significant for $POV_{SC,PC3}$ (odds ratio = 2.83, $p = 0.004$) in the validation cohort.

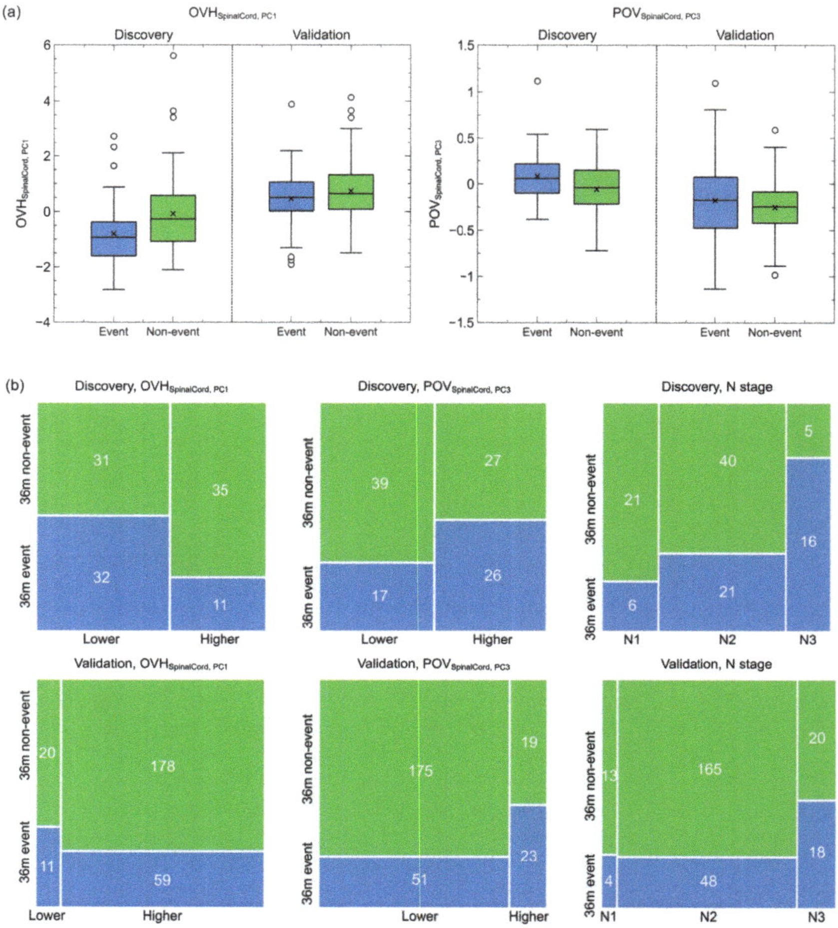

Figure 2. Continuous and binarized spatial factor distributions and N stage distributions for 3-year disease progressed and non-disease progressed patients in the discovery and validation cohort. (**a**) Box plots of continuous spatial factor distributions. Patients with disease progression within three years had lower mean OVH principle values and higher mean POV principle values at both discovery and validation. (**b**) Mosaic plots of the binarized spatial factor and N stage distributions of patients with and without 3-year disease progression.

Table 2. Hazard ratios and *p*-values of the selected spatial factors and N stage from multivariate Cox regression on disease-free survival.

Covariant	HR (95CI)	*p*-Value
$OVH_{SC,PC1}$	0.63 (0.48–0.83)	<0.001
$POV_{SC,PC3}$	3.35 (1.40–7.99)	0.006
N stage	2.26 (1.46–3.49)	<0.001

3.3. Combined Prognostic Index

The combined prognostic index had better discriminability than N stage on all the survival endpoints but showed statistical significance mainly in DFS and RFS, as reported in Table 3. C-index in DFS increased from 0.654 (training) and 0.568 (external validation) to 0.722 (training) and 0.603 (external validation) when combining the two new spatial factors with N stage. Such improvement was significant in training (p-value = 0.020) while much less in external validation (0.086). On the other hand, the training and validation improvements were both significant in RFS with C-index reaching 0.723 (p-value = 0.020) and 0.603 (p-value = 0.019), respectively.

Table 3. Concordance index of the proposed geometric prognostic index and N stage in the training and validation cohort.

Survival Endpoint	Training Cohort			Validation Cohort		
	Prognostic Index (95CI)	N Stage (95CI)	p-Value	Prognostic Index (95CI)	N Stage (95CI)	p-Value
DFS	0.72 (0.65–0.79)	0.65 (0.57–0.73)	0.020	0.60 (0.54–0.67)	0.57 (0.52–0.62)	0.086
OS	0.75 (0.63–0.84)	0.72 (0.64–0.80)	0.245	0.60 (0.48–0.71)	0.58 (0.50–0.67)	0.395
RFS	0.72 (0.62–0.82)	0.64 (0.54–0.73)	0.020	0.60 (0.52–0.69)	0.53 (0.47–0.60)	0.019
DMFS	0.72 (0.63–0.81)	0.65 (0.54–0.76)	0.062	0.57 (0.47–0.67)	0.57 (0.50–0.65)	0.536

Better risk stratifications were achieved by the combined prognostic index in DFS and DMFS than N stage itself, as shown by the KM curves in Figure 3. Table 4 reports the hazard ratios and the corresponding p-values between different risk groups as well as the three-year survival rates in DFS, OS, RFS, and DMFS. On the discovery cohort, the DFS survivals of the three new risk groups were statistically different ($p \leq 0.05$) whereas much lower statistical significance was found between the N1 and N2 groups ($p = 0.139$). Higher hazard ratios were observed between G2 (4.49) and G3 (9.07) to G1 compared to the N stage (N1 vs. N2: 1.83, N1 vs. N3: 5.19). However, the HR was less between G3 to G2 (1.913) compared to the one between N3 to N2 (2.988). A similar trend was found in DMFS where G2 (4.11) and G3 (10.41) were better separated from G1 but worse between G2 and G3 (2.26). In the validation cohort, the HRs between G2 (DFS: 1.71, $p = 0.021$; DMFS: 1.72, $p = 0.101$) and G3 (DFS: 4.02, $p < 0.01$; DMFS: 2.93, $p = 0.014$) to G1 also increased significantly compared to that between N2 (DFS: 0.772, $p = 0.518$; DMFS: 0.552, $p = 0.271$) and N3 to N1 (DFS: 1.821, $p = 0.171$; DMFS: 1.876, $p = 0.216$) in both DFS and DMFS. Similarly, a less HR was found between G2 and G3 (DFS: 2.44, $p = 0.006$; DMFS: 1.74, $p = 0.219$) than between N2 and N3 (DFS: 2.66, $p \leq 0.001$; DMFS: 3.17, $p = 0.001$).

The remaining survival endpoints showed heterogeneous patterns under the new risk stratification (Table 4). Significant HR improvements were observed in OS, but marginal in RFS for the discovery cohort. On the other hand, RFS showed significantly higher stratification performance in the validation cohort, but no improvement in OS was observed. Moreover, the validation cohort demonstrated higher 3-year survival rates on G1 and lower on G2 for RFS and DMFS, whereas marginal improvement of 3-year survival rates was found in the discovery cohort.

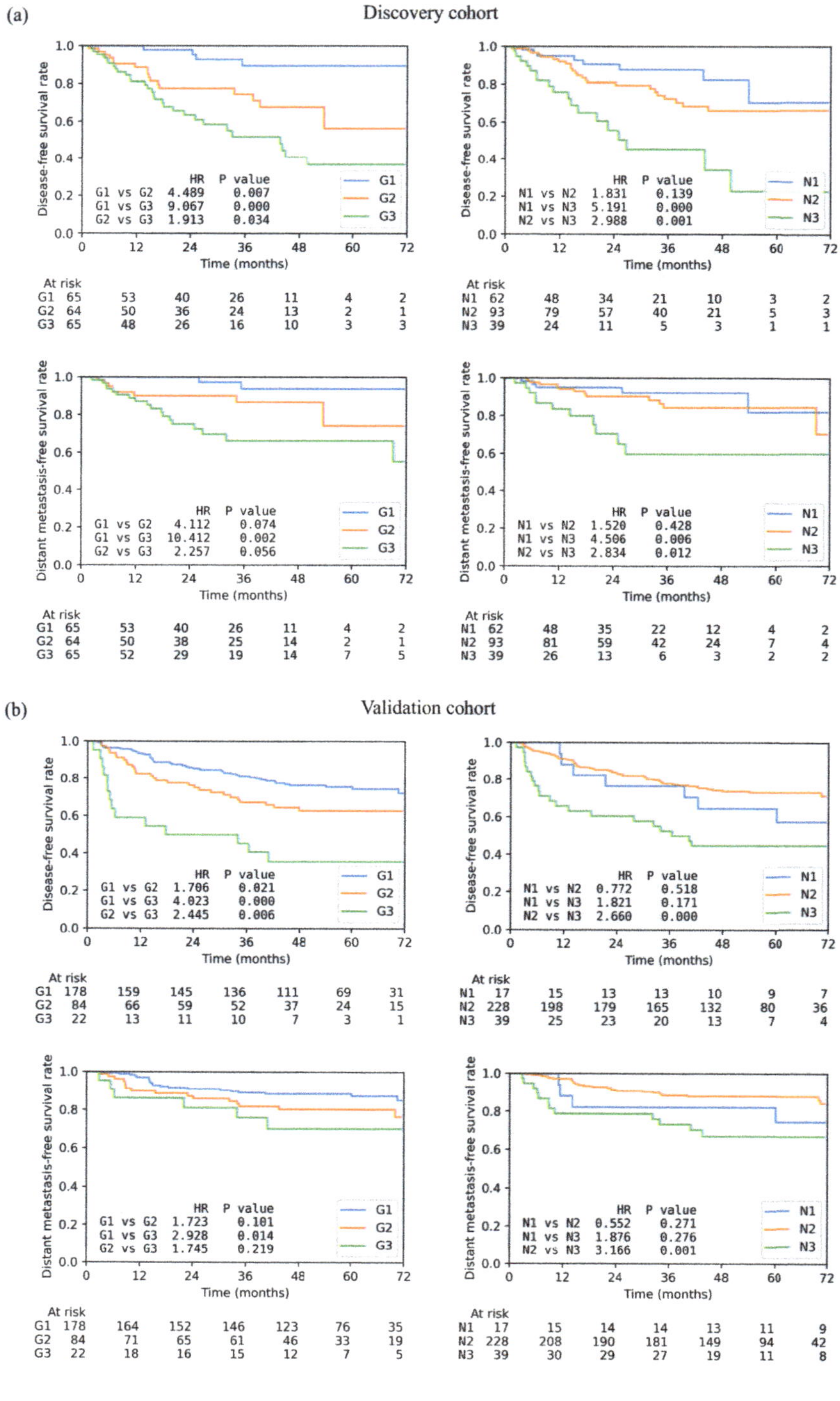

Figure 3. Kaplan–Meier curves of the low-(G1), median-(G2), and high-risk (G3) patient groups based on the new spatial index and the three N stages on (**a**) disease-free survival and (**b**) distant metastasis-free survival. Each plot also contains the hazard ratio (HR) and the corresponding *p*-value between each two groups.

Table 4. Risk stratification performance of the proposed risk groups and N stage in multiple survival endpoints and discovery and validation cohort.

Survival Endpoint	Proposed Risk Stratification				N Stage			
	Group	HR	*p*-Value	3y SR	Group	HR	*p*-Value	3y SR
Discovery cohort								
DFS	G1	—	—	89.6%	N1	—	—	87.9%
	G2	4.49	0.007	74.6%	N2	1.83	0.139	72.6%
	G3	9.07	<0.001	52.1%	N3	5.19	<0.001	45.6%
OS	G1	—	—	97.3%	N1	—	—	100.0%
	G2	7.66	0.055	92.7%	N2	3.33	0.115	89.4%
	G3	13.98	0.011	79.7%	N3	11.62	0.002	72.6%
RFS	G1	—	—	89.6%	N1	—	—	93.0%
	G2	2.23	0.181	85.3%	N2	2.64	0.079	79.5%
	G3	4.76	0.005	72.2%	N3	4.59	0.014	74.9%
DMFS	G1	—	—	94.0%	N1	—	—	92.2%
	G2	4.11	0.074	86.8%	N2	1.52	0.428	84.5%
	G3	10.41	0.002	66.5%	N3	4.51	0.006	59.8%
Validation cohort						r		
DFS	G1	—	—	81.2%	N1	—	—	76.5%
	G2	1.71	0.021	67.2%	N2	0.77	0.518	77.8%
	G3	4.02	<0.001	45.5%	N3	1.82	0.171	52.7%
OS	G1	—	—	95.2%	N1	—	—	87.8%
	G2	1.36	0.384	93.5%	N2	1.56	0.548	95.3%
	G3	2.28	0.076	85.9%	N3	2.57	0.223	89.0%
RFS	G1	—	—	88.7%	N1	—	—	87.8%
	G2	1.46	0.219	82.9%	N2	0.84	0.736	85.9%
	G3	3.69	0.001	62.7%	N3	1.20	0.764	78.2%
DMFS	G1	—	—	89.3%	N1	—	—	82.4%
	G2	1.72	0.101	82.0%	N2	0.55	0.271	88.7%
	G3	2.93	0.014	76.2%	N3	1.88	0.276	73.5%

Note: HR and *p*-value were relative to the low-risk group (G1) or N1. Abbreviations: HR, hazard ratio; 3y SR: 3-year survival rate; DFS, disease-free survival; OS, overall survival; RFS, relapse-free survival; DMFS, distant metastasis-free survival; 95CI: 95% confidence interval.

3.4. Representative Cases

To further explain the contribution of the two anatomical factors in better identifying the risk of disease progression, we selected two representative cases from the discovery cohort with the same N stage but distinct risks based on the spatial index. The high-risk patient was classified as G1 and the low-risk one as G3, both having the same N stage (N2) and chemotherapy strategy (CCRT + ACT). The high-risk patient developed distant metastases at 32.3 months, while the low-risk patient showed no signs of disease progression for at least 34.3 months. Figure 4a presents the 2D axial masks and the 3D volumes of GTVn and three OARs for the high and low-risk patient. Anatomically, both patients had metastatic retropharyngeal LN, but a significantly larger extent of the right cervical LN tumor was observed in the high-risk patient. Meanwhile, distinct patterns of the spinal OVH and POV curves were found, as drawn in Figure 4b, where the selected PC vectors were also included. The OVH curve of the high-risk patient was significantly higher than that of the low-risk patient with the largest overlap volume difference emphasized at around the global minimum (~75 mm) of the first PC vector. The POV at the first local maximum (~25 degrees) of the PC vector was much higher in the high-risk patient, exceeding the higher POV of the low-risk patient at the second local maximum (~125 degrees).

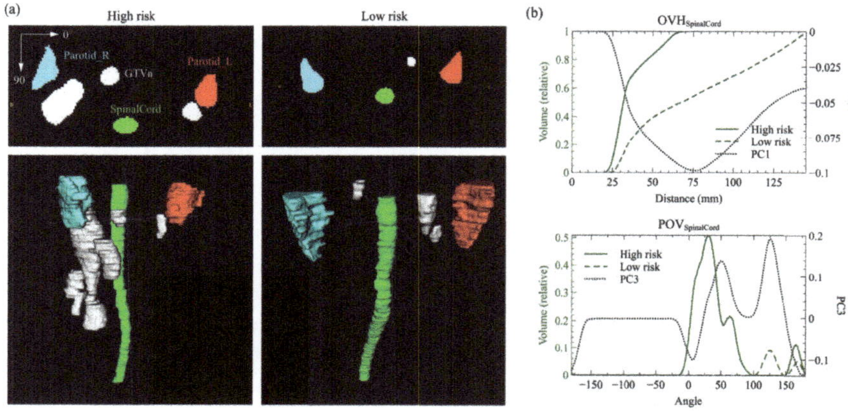

Figure 4. Quantitative anatomical characterizations of the high-risk and low-risk patient. (**a**) The axial slice masks and rendered 3D volumes of GTVn (lymph node tumor), Parotid_L, Parotid_R, and SpinalCord structures. (**b**) The SpinalCord overlap volume histogram (POV) and projection overlap volume (POV) of the two patients and the corresponding selected principal component (PC) vector. Significant differences in lymph node anatomy were captured by the large variations in the histograms and highlighted by the PCs.

4. Discussion

This study demonstrated the feasibility of discovering new prognostic factors from quantitative spatial characterization of LN tumor for better LN risk stratification with high cross-site generalizability. Two histograms precisely characterized the LN tumor anatomy by distances (OVH) and angles (POV). PCA effectively reduced the high-dimensional histograms into several informative and independent anatomical factors, and two final independent prognostic factors were discovered by Cox regressions in DFS. The prognostic index that combines the independent prognostic spatial factors and the N stage achieved better new three-level risk stratifications than the N stage itself in DFS and DMFS at both discovery and external validation.

Only the spinal cord spatial factor $OVH_{SC,PC1}$ and $POV_{SC,PC3}$ were identified as the independent prognostic factors to DFS. $OVH_{SC,PC1}$ highlights the overlap of the lower spinal cord with the expansion of isotropic LN tumor by approximately 75 mm (Figure 4b), indicating a smaller axial expansion of LN. The PC vector of $POV_{SC,PC3}$ has two peaks at around 25 and 125 degrees and reaches local minimums at 0 and 180 degrees (Figure 4b). Higher projection overlaps at the peak angles indicate more volume of LN tumor in the anterior direction of the spinal cord, whereas the valley angles suggest less involvement of the LN tumor on the lateral sides. Additionally, both factors are correlated with the axial extent of the LN tumor due to the thin cylindrical structure of the spinal cord. Such correlation was also demonstrated by the two example patients in Figure 4a where the high-risk patient with lower $OVH_{SC,PC1}$ and higher $POV_{SC,PC3}$ had a significantly larger axial extent of cervical LN.

Previous clinical observations on the prognostic power of the anatomy of LN tumors were highly correlated with our quantitative findings. The results of our survival analysis suggest an increased risk of disease progression with lower $OVH_{SC,PC1}$ (adjusted HR = 0.63, 95CI: 0.48–0.83; $p \leq 0.001$) and higher $POV_{SC,PC3}$ (adjusted HR = 3.35, 95CI: 1.41–7.99), regardless of the N stage. Their independent prognostic power could be explained by the two example patients in whom the high-risk one developed early distant metastases despite their identical N stage. As discussed in the previous paragraph, a higher prognostic index value suggests a higher axial expansion and extent of the LN tumor, which supports the ongoing discussion of the high prognostic value of the quantitative LN burden. Previous clinical studies reported the number of metastatic LN regions as an independent predictor of DMFS [11,19]. For $POV_{SC,PC3}$, a higher value may also indicate retropharyngeal LN metastasis with a larger size or bilateral involvement. Retropharyngeal LN has also been suggested to indicate worse in DFS and DMFS [28,29]. Specifically, the size of the metastatic retropharyngeal LN with a cutoff axial diameter of 6mm has been identified as a significant prognostic factor for OS and DMFS [30,31]. It was also suggested that the bilateral involvement of the retropharyngeal lymph nodes should be upgraded to N2 disease due to the worse 5-year OS and DMFS [15]. These anatomical characteristics have been partially included in the definition of the N1 classification of the 7th and 8th N staging system [32], where metastasis is limited above the caudal border of cricoid cartilage and/or retropharyngeal lymph node(s) does not exceed 6mm in greatest dimension. Our quantitative anatomical factors may provide more precise descriptions of various LN anatomy characterizations, thus independent of the existing N stage classifications.

The two final spatial factors were predictive of three-year DFS and DMFS at both discovery and validation. However, the binarization thresholds were less generalizable from discovery to validation due to the overall different magnitudes of the spatial factor values. As a result, much higher low-risk patients were classified in the validation cohort when using the median values in the discovery as the binarization thresholds. The systematic cross-institutional variations in the spatial factor magnitudes could be attributed to the inconsistent spinal cord volume definitions, especially the starting and ending point. A higher spinal cord extent may lead to a lower relative overlap volume for both OVH and POV at the same absolute distance and angle, and the resulting PC coefficients are expected to be smaller. For clinical utility, consistent organ and tumor segmentations are important to ensure a reliable quantitative spatial characterization. Further adjustments in the spatial factor definitions for enhanced robustness are needed in future studies.

Despite the promising performance of the spatial characterization of lymph node tumors in survival prognosis, the analysis involves standardized tumor and OAR segmentations [33] as well as complex computations of distance and angle histograms for thorough characterization, which often require specific training. The potential long learning curve for clinicians may hinder the clinical application of the proposed predictors. Integration of AI-based systems for auto-segmentation [34] and dedicated calculation scripts into the existing treatment planning system could be one solution for fast implementation in daily clinical practice. On the other hand, other types of biomarkers, which are easier to implement in clinics, have been proposed as strong survival predictors for patients with NPC and other HNC diseases. Systematic inflammation indicators, which can be directly measured from blood test results, have been reported to be prognostic in multiple HNC subtypes. For example, pre-treatment neutrophil-to-lymphocyte ratio (NLR) has been investigated, and a strong statistical correlation was observed with positive neck occult metastasis in laryngeal squamous cell carcinoma [35]. Another study by Orabona et al. confirmed the independent prognostic power of the systemic immune-inflammation index (SII) and the systemic inflammation response index (SIRI) on OS of patients who received malignant salivary gland tumor surgery [36].

The constructed prognostic index results in improved risk stratifications in DFS and DMFS compared to the existing N stage both internally and externally. It is consistent with previous findings on the improved DMFS prognostication of the LN tumor region number [11] and the involvement of the retropharyngeal LN tumor [15]. Better risk stratifications on OS were only observed on the discovery cohort and RFS on the external validation cohort. Several reasons could contribute to the heterogeneous results. First, the thresholds of the prognostic index for the three-class risk classification could be suboptimal and less generalizable. The threshold optimization method for risk stratification requires a more careful design and wide validation for clinical practice. As discussed in the previous paragraph, the overall magnitudes of the spatial factors were inconsistent, which may contribute to the reduced generalizability of the prognostic index and the resulting risk groups. Second, some patient characteristics, such as stages and chemotherapy treatments, are rather different between discovery and external validation. They may affect the generalizability of the risk stratification performance due to the different baseline performances. Third, the sample sizes and follow-up durations are limited, especially in the discovery cohort. Less patients remained as uncensored samples, resulting in less reliable results. Increasing the sample size with more complete follow-up information is needed in future studies to enhance the clinical evidence of our findings.

5. Conclusions

This study used the distance histogram OVH and the newly proposed angle histogram POV to quantitatively characterize the anatomy of the LN tumor in relation to the surrounding spinal cord and parotids. Independent prognostic factors on DFS were discovered from the principal components of the anatomical histograms and combined with the N stage into an spatial index. It surpassed the N stage itself in risk discrimination and stratification. The proposed quantitative approach may facilitate the discovery of new anatomical characteristics in a more holistic and precise way to improve patient staging in other diseases.

Author Contributions: Conceptualization, J.Z., X.T. and J.C.; Data curation, S.L. and J.S.; Formal analysis, J.Z.; Funding acquisition, J.C.; Investigation, J.Z.; Methodology, J.Z. and X.T.; Project administration, J.C.; Resources, S.L., A.L.-Y.C., S.C.-Y.N., F.K.-H.L., K.-H.A., C.W.-Y.Y., V.H.-F.L., Z.L., Y.L., R.Y., Y.H. and F.-M.K.; Software, J.Z.; Supervision, J.C.; Validation, J.Z., X.T. and Y.Z.; Visualization, J.Z. and X.T.; Writing—original draft, J.Z.; Writing—review & editing, J.Z., S.L. and J.C. All authors have read and agreed to the published version of the manuscript.

Funding: This study has partly supported by research grants of Shenzhen-Hong Kong-Macau S&T Program (Category C) (SGDX20201103095002019), Shenzhen Basic Research Program (JCYJ20210324130209023), Innovation and Technology Fund-Mainland-Hong Kong Joint Funding Scheme (ITF-MHKJFS) (MHP/005/20), and Project of Strategic Importance Fund (P0035421) and Projects of RISA (P0043001) of The Hong Kong Polytechnic University.

Institutional Review Board Statement: The use of data was approved by the Institutional Review Board of the University of Hong Kong/Hospital Authority Hong Kong West Cluster (HKU/HA HKW IRB), reference number UW21-412, and the Research Ethics Committee (Kowloon Central/Kowloon East), reference number KC/KE-18-0085/ER-1.

Informed Consent Statement: Patient consent was waived due to the retrospective nature of this study.

Data Availability Statement: The data presented in this study are available on request from the corresponding author. The data are not publicly available due to patient privacy protection.

Conflicts of Interest: The authors declare no conflict of interest.

References

1. Ferlay, J.; Soerjomataram, I.; Dikshit, R.; Eser, S.; Mathers, C.; Rebelo, M.; Parkin, D.M.; Forman, D.; Bray, F. Cancer incidence and mortality worldwide: Sources, methods and major patterns in GLOBOCAN 2012. *Int. J. Cancer* **2015**, *136*, E359–E386. [CrossRef] [PubMed]
2. Chua, M.L.; Wee, J.T.; Hui, E.P.; Chan, A.T. Nasopharyngeal carcinoma. *Lancet* **2016**, *387*, 1012–1024. . [CrossRef] [PubMed]
3. Zhang, M.X.; Li, J.; Shen, G.P.; Zou, X.; Xu, J.J.; Jiang, R.; You, R.; Hua, Y.J.; Sun, Y.; Ma, J. Intensity-modulated radiotherapy prolongs the survival of patients with nasopharyngeal carcinoma compared with conventional two-dimensional radiotherapy: A 10-year experience with a large cohort and long follow-up. *Eur. J. Cancer* **2015**, *51*, 2587–2595. [CrossRef] [PubMed]
4. Lee, A.W.; Ng, W.T.; Chan, L.L.; Hung, W.M.; Chan, C.C.; Sze, H.C.; Chan, O.S.; Chang, A.T.; Yeung, R.M. Evolution of treatment for nasopharyngeal cancer–success and setback in the intensity-modulated radiotherapy era. *Radiother. Oncol.* **2014**, *110*, 377–384. [CrossRef]
5. Sun, X.; Su, S.; Chen, C.; Han, F.; Zhao, C.; Xiao, W.; Deng, X.; Huang, S.; Lin, C.; Lu, T. Long-term outcomes of intensity-modulated radiotherapy for 868 patients with nasopharyngeal carcinoma: An analysis of survival and treatment toxicities. *Radiother. Oncol.* **2014**, *110*, 398–403. [CrossRef]
6. Qu, W.; Li, S.; Zhang, M.; Qiao, Q. Pattern and prognosis of distant metastases in nasopharyngeal carcinoma: A large-population retrospective analysis. *Cancer Med.* **2020**, *9*, 6147–6158. [CrossRef]
7. Sireci, F.; Dispenza, F.; Lorusso, F.; Immordino, A.; Immordino, P.; Gallina, S.; Peretti, G.; Canevari, F. Tumours of Nasal Septum: A Retrospective Study of 32 Patients. *Int. J. Environ. Res. Public Health* **2022**, *19*, 1713. [CrossRef]

8. Kam, M.K.; Teo, P.M.; Chau, R.M.; Cheung, K.; Choi, P.H.; Kwan, W.; Leung, S.; Zee, B.; Chan, A.T. Treatment of nasopharyngeal carcinoma with intensity-modulated radiotherapy: The Hong Kong experience. *Int. J. Radiat. Oncol.* **2004**, *60*, 1440–1450. [CrossRef]
9. Xu, Y.; Huang, T.; Fan, L.; Jin, W.; Chen, X.; Chen, J. Patterns and prognostic value of lymph node metastasis on distant metastasis and survival in nasopharyngeal carcinoma: A surveillance, epidemiology, and end results study, 2006–2015. *J. Oncol.* **2019**, *2019*, 4094395 . [CrossRef]
10. Egner, J.R. AJCC cancer staging manual. *JAMA* **2010**, *304*, 1726–1727. [CrossRef]
11. Zhou, X.; Ou, X.; Yang, Y.; Xu, T.; Shen, C.; Ding, J.; Hu, C. Quantitative metastatic lymph node regions on magnetic resonance imaging are superior to AJCC N classification for the prognosis of nasopharyngeal carcinoma. *J. Oncol.* **2018**, *2018*, 9172585. [CrossRef] [PubMed]
12. Chiang, C.L.; Guo, Q.; Ng, W.T.; Lin, S.; Ma, T.S.W.; Xu, Z.; Xiao, Y.; Li, J.; Lu, T.; Choi, H.C.W. Prognostic Factors for Overall Survival in Nasopharyngeal Cancer and Implication for TNM Staging by UICC: A Systematic Review of the Literature. *Front. Oncol.* **2021**, *11*, 703995. [CrossRef] [PubMed]
13. Xu, Y.; Chen, X.; Zhang, M.; Xiao, Y.; Zong, J.; Guo, Q.; Qiu, S.; Zheng, W.; Lin, S.; Pan, J. Prognostic effect of parotid area lymph node metastases after preliminary diagnosis of nasopharyngeal carcinoma: A propensity score matching study. *Cancer Med.* **2017**, *6*, 2213–2221. [CrossRef] [PubMed]
14. Zhang, Y.; Zhang, Z.C.; Li, W.F.; Liu, X.; Liu, Q.; Ma, J. Prognosis and staging of parotid lymph node metastasis in nasopharyngeal carcinoma: An analysis in 10,126 patients. *Oral Oncol.* **2019**, *95*, 150–156. [CrossRef] [PubMed]
15. Huang, L.; Zhang, Y.; Liu, Y.; Li, H.; Wang, S.; Liang, S.; Zhou, J.; Cui, C.; Sun, Y.; Chen, M. Prognostic value of retropharyngeal lymph node metastasis laterality in nasopharyngeal carcinoma and a proposed modification to the UICC/AJCC N staging system. *Radiother. Oncol.* **2019**, *140*, 90–97. [CrossRef]
16. Ai, Q.Y.; King, A.D.; Poon, D.M.; Mo, F.K.; Hui, E.P.; Tong, M.; Ahuja, A.T.; Ma, B.B.; Chan, A.T. Extranodal extension is a criterion for poor outcome in patients with metastatic nodes from cancer of the nasopharynx. *Oral Oncol.* **2019**, *88*, 124–130. [CrossRef]
17. Lu, T.; Hu, Y.; Xiao, Y.; Guo, Q.; Huang, S.H.; O'Sullivan, B.; Fang, Y.; Zong, J.; Chen, Y.; Lin, S. Prognostic value of radiologic extranodal extension and its potential role in future N classification for nasopharyngeal carcinoma. *Oral Oncol.* **2019**, *99*, 104438. [CrossRef]
18. Mao, Y.; Wang, S.; Lydiatt, W.; Shah, J.P.; Colevas, A.D.; Lee, A.W.; O'Sullivan, B.; Guo, R.; Luo, W.; Chen, Y.; et al. Unambiguous advanced radiologic extranodal extension determined by MRI predicts worse outcomes in nasopharyngeal carcinoma: Potential improvement for future editions of N category systems. *Radiother. Oncol.* **2021**, *157*, 114–121. [CrossRef]
19. Ma, H.; Liang, S.; Cui, C.; Zhang, Y.; Xie, F.; Zhou, J.; Dong, A.; Chen, M.; Xie, C.; Li, H. Prognostic significance of quantitative metastatic lymph node burden on magnetic resonance imaging in nasopharyngeal carcinoma: A retrospective study of 1224 patients from two centers. *Radiother. Oncol.* **2020**, *151*, 40–46. [CrossRef]
20. Sun, Y.; Yu, X.L.; Luo, W.; Lee, A.W.; Wee, J.T.S.; Lee, N.; Zhou, G.Q.; Tang, L.L.; Tao, C.J.; Guo, R.; et al. Recommendation for a contouring method and atlas of organs at risk in nasopharyngeal carcinoma patients receiving intensity-modulated radiotherapy. *Radiother. Oncol.* **2014**, *110*, 390–397. [CrossRef]
21. Peng, Y.; Chen, S.; Qin, A.; Chen, M.; Gao, X.; Liu, Y.; Miao, J.; Gu, H.; Zhao, C.; Deng, X.; et al. Magnetic resonance-based synthetic computed tomography images generated using generative adversarial networks for nasopharyngeal carcinoma radiotherapy treatment planning. *Radiother. Oncol.* **2020**, *150*, 217–224. [CrossRef] [PubMed]
22. Luo, X.; Liao, W.; Chen, J.; Song, T.; Chen, Y.; Zhang, S.; Chen, N.; Wang, G.; Zhang, S. Efficient semi-supervised gross target volume of nasopharyngeal carcinoma segmentation via uncertainty rectified pyramid consistency. In Proceedings of the International Conference on Medical Image Computing and Computer-Assisted Intervention, Strasbourg, France, 27 September–1 October 2021; Springer: Berlin/Heidelberg, Germany, 2021; pp. 318–329. ._30. [CrossRef]
23. Kazhdan, M.; Simari, P.; McNutt, T.; Wu, B.; Jacques, R.; Chuang, M.; Taylor, R. A Shape Relationship Descriptor for Radiation Therapy Planning. In *Lecture Notes in Computer Science, Proceedings of the Medical Image Computing and Computer-Assisted Intervention—MICCAI 2009, London, UK, 20–24 September 2009*; Yang, G.Z., Hawkes, D., Rueckert, D., Noble, A., Taylor, C., Eds.; Springer: Berlin/Heidelberg, Germany, 2009; pp. 100–108. ._13. [CrossRef]
24. Wu, B.; Ricchetti, F.; Sanguineti, G.; Kazhdan, M.; Simari, P.; Chuang, M.; Taylor, R.; Jacques, R.; McNutt, T. Patient geometry-driven information retrieval for IMRT treatment plan quality control. *Med Phys.* **2009**, *36*, 5497–5505. [CrossRef] [PubMed]
25. Maurer, C.; Qi, R.; Raghavan, V. A linear time algorithm for computing exact Euclidean distance transforms of binary images in arbitrary dimensions. *IEEE Trans. Pattern Anal. Mach. Intell.* **2003**, *25*, 265–270. [CrossRef]
26. Lowekamp, B.; Chen, D.; Ibáñez, L.; Blezek, D. The Design of SimpleITK. *Front. Neuroinform.* **2013**, *7*, 45. . [CrossRef] [PubMed]
27. Davidson-Pilon, C. lifelines: Survival analysis in Python. *J. Open Source Softw.* **2019**, *4*, 1317. [CrossRef]
28. Ma, J.; Liu, L.; Tang, L.; Zong, J.; Lin, A.; Lu, T.; Cui, N.; Cui, C.; Li, L. Retropharyngeal lymph node metastasis in nasopharyngeal carcinoma: Prognostic value and staging categories. *Clin. Cancer Res.* **2007**, *13*, 1445–1452. [CrossRef]
29. Tang, L.L.; Guo, R.; Zhou, G.; Sun, Y.; Liu, L.Z.; Lin, A.H.; Mai, H.; Shao, J.; Li, L.; Ma, J. Prognostic value and staging classification of retropharyngeal lymph node metastasis in nasopharyngeal carcinoma patients treated with intensity-modulated radiotherapy. *PLoS ONE* **2014**, *9*, e108375. [CrossRef]

30. Li, Y.Z.; Xie, C.M.; Wu, Y.P.; Cui, C.Y.; Huang, Z.L.; Lu, C.Y.; Wu, P.H. Nasopharyngeal carcinoma patients with retropharyngeal lymph node metastases: A minimum axial diameter of 6 mm is a more accurate prognostic predictor than 5 mm. *Am. J. Roentgenol.* **2015**, *204*, 20–23. [CrossRef]
31. Chen, J.; Luo, J.; He, X.; Zhu, C. Evaluation of contrast-enhanced computed tomography (CT) and magnetic resonance imaging (MRI) in the detection of retropharyngeal lymph node metastases in nasopharyngeal carcinoma patients. *Cancer Manag. Res.* **2020**, *12*, 1733. [CrossRef]
32. Guo, R.; Mao, Y.P.; Tang, L.L.; Chen, L.; Sun, Y.; Ma, J. The evolution of nasopharyngeal carcinoma staging. *Br. J. Radiol.* **2019**, *92*, 20190244. [CrossRef]
33. Lee, A.W.; Ng, W.T.; Pan, J.J.; Poh, S.S.; Ahn, Y.C.; AlHussain, H.; Corry, J.; Grau, C.; Grégoire, V.; Harrington, K.J.; et al. International guideline for the delineation of the clinical target volumes (CTV) for nasopharyngeal carcinoma. *Radiother. Oncol.* **2018**, *126*, 25–36. .: 10.1016/j.radonc.2017.10.032. [CrossRef] [PubMed]
34. Lin, L.; Dou, Q.; Jin, Y.M.; Zhou, G.Q.; Tang, Y.Q.; Chen, W.L.; Su, B.A.; Liu, F.; Tao, C.J.; Jiang, N.; et al. Deep Learning for Automated Contouring of Primary Tumor Volumes by MRI for Nasopharyngeal Carcinoma. *Radiology* **2019**, *291*, 677–686. [CrossRef] [PubMed]
35. Salzano, G.; Perri, F.; Maglitto, F.; Togo, G.; De Fazio, G.R.; Apolito, M.; Calabria, F.; Laface, C.; Vaira, L.A.; Committeri, U.; et al. Pre-Treatment Neutrophil-to-Lymphocyte and Platelet-to-Lymphocyte Ratios as Predictors of Occult Cervical Metastasis in Clinically Negative Neck Supraglottic and Glottic Cancer. *J. Pers. Med.* **2021**, *11*, 1252. [CrossRef] [PubMed]
36. Abbate, V.; Barone, S.; Troise, S.; Laface, C.; Bonavolontà, P.; Pacella, D.; Salzano, G.; Iaconetta, G.; Califano, L.; Dell'Aversana Orabona, G. The Combination of Inflammatory Biomarkers as Prognostic Indicator in Salivary Gland Malignancy. *Cancers* **2022**, *14*, 5934. [CrossRef] [PubMed]

cancers

MDPI

Article

Preoperative Prediction of Perineural Invasion and Prognosis in Gastric Cancer Based on Machine Learning through a Radiomics–Clinicopathological Nomogram

Heng Jia [1,†], Ruzhi Li [2,†], Yawei Liu [3], Tian Zhan [1], Yuan Li [4,*] and Jianping Zhang [1,*]

1 Department of General Surgery, The Second Affiliated Hospital of Nanjing Medical University, Nanjing 210011, China; lbhengjia@163.com (H.J.); 15851863223@163.com (T.Z.)
2 Department of Endoscopic Center, The Fourth Affiliated Hospital of Nanjing Medical University, Nanjing 210031, China; zhizhi1012@njmu.edu.cn
3 Department of General Surgery, Nanjing Drum Tower Hospital Clinical College of Nanjing Medical University, Nanjing 210008, China; 18262635889@163.com
4 Key Laboratory of Modern Toxicology, Ministry of Education, School of Public Health, Nanjing Medical University, Nanjing 211166, China
* Correspondence: liyuan@njmu.edu.cn (Y.L.); drzhangjp@njmu.edu.cn (J.Z.)
† These authors contributed equally to this work.

Simple Summary: Gastric cancer remains the world's fifth most lethal malignancy. Perineural invasion (PNI) is a common growth pattern of gastric cancer. Currently, the diagnosis of PNI relies on postoperative pathology, which is an invasive approach. In this study, we built a radiomics–clinicopathological model based on logistic regression analysis to preoperatively predict PNI. The radiomics–clinicopathological model yielded AUC values of 0.851 (95%CI: 0.769–0.933) in the training set, 0.842 (95%CI: 0.713–0.970) in the testing set and 0.813 (95%CI: 0.672–0.954) in the validation set. This proposed model may help clinicians make clinical decisions and provide personalized treatment to gastric cancer patients. In this research, the value of perineural invasion (PNI) in predicting prognoses for gastric cancer patients was also studied.

Abstract: Purpose: The aim of this study was to construct and validate a nomogram for preoperatively predicting perineural invasion (PNI) in gastric cancer based on machine learning, and to investigate the impact of PNI on the overall survival (OS) of gastric cancer patients. Methods: Data were collected from 162 gastric patients and analyzed retrospectively, and radiomics features were extracted from contrast-enhanced computed tomography (CECT) scans. A group of 42 patients from the Cancer Imaging Archive (TCIA) were selected as the validation set. Univariable and multivariable analyses were used to analyze the risk factors for PNI. The *t*-test, Max-Relevance and Min-Redundancy (mRMR) and the least absolute shrinkage and selection operator (LASSO) were used to select radiomics features. Radscores were calculated and logistic regression was applied to construct predictive models. A nomogram was developed by combining clinicopathological risk factors and the radscore. The area under the curve (AUC) values of receiver operating characteristic (ROC) curves, calibration curves and clinical decision curves were employed to evaluate the performance of the models. Kaplan–Meier analysis was used to study the impact of PNI on OS. Results: The univariable and multivariable analyses showed that the T stage, N stage and radscore were independent risk factors for PNI ($p < 0.05$). A nomogram based on the T stage, N stage and radscore was developed. The AUC of the combined model yielded 0.851 in the training set, 0.842 in the testing set and 0.813 in the validation set. The Kaplan–Meier analysis showed a statistically significant difference in OS between the PNI group and the non-PNI group ($p < 0.05$). Conclusions: A machine learning-based radiomics–clinicopathological model could effectively predict PNI in gastric cancer preoperatively through a non-invasive approach, and gastric cancer patients with PNI had relatively poor prognoses.

Keywords: machine learning; radiomics; gastric cancer; perineural invasion

Citation: Jia, H.; Li, R.; Liu, Y.; Zhan, T.; Li, Y.; Zhang, J. Preoperative Prediction of Perineural Invasion and Prognosis in Gastric Cancer Based on Machine Learning through a Radiomics–Clinicopathological Nomogram. *Cancers* **2024**, *16*, 614. https://doi.org/10.3390/cancers16030614

Academic Editor: Hajime Isomoto

Received: 28 December 2023
Revised: 26 January 2024
Accepted: 27 January 2024
Published: 31 January 2024

1. Introduction

Gastric cancer is one of the most lethal cancers, ranking fifth among all cancers in the world [1]. The current standard treatment for gastric cancer is radical gastrectomy, supplemented by radiotherapy and chemotherapy. However, despite this standard treatment, the overall survival (OS) of gastric cancer patients remains low [2]. Although significant progress has been made in chemotherapy and radiotherapy in recent years, radical gastrectomy remains the most effective therapy. Nevertheless, a large number of gastric cancer patients still experience recurrence and metastasis after radical gastrectomy [3]. Therefore, it is crucial to identify new prognostic factors that contribute to the poor survival of gastric cancer patients, enabling the development of customized treatment plans to enhance patients' OS.

The TNM staging system is currently recognized as a robust prognostic indicator for gastric cancer [4]. With the continuous study of the clinicopathology and other aspects of gastric cancer patients, some pathological factors and molecular indicators have been confirmed to be related to the prognoses of gastric cancer patients [5,6]. Previous studies suggested that in addition to the TNM stage, perineural invasion (PNI) is an independent factor that affects the prognosis of patients with gastric cancer [7]. PNI is a tumor growth pattern that can lead to poor prognoses such as tumor metastasis and peritoneal recurrence [8]. In gastric cancer, PNI is primarily defined as the circumstance where at least 33% of a nerve is surrounded by tumor cells [9]. The incidence of PNI in gastric cancer varies from 6.9% to 75.6% [10]. Studies have shown that patients with PNI have shorter survival times compared to those without PNI [11]. Consequently, treatments such as surgery or neoadjuvant chemoradiotherapy may be affected if PNI occurs. Currently, the assessment of PNI in gastric cancer relies on post-operative pathology, which is an invasive approach. Therefore, it would be highly significant to preoperatively predict PNI in gastric cancer through a non-invasive approach.

Radiomics has played a crucial role in the diagnosis and treatment of diseases in recent years [12,13]. In gastrointestinal diseases, radiomics is mainly used for tumor staging, differential diagnosis and prognosis analysis [14,15]. Previous studies have explored non-invasive methods for preoperatively assessing PNI in gastric cancer. Zheng et al. investigated the efficacy of radiomics and clinical models based on machine learning for preoperatively predicting PNI in gastric cancer. They found that their approach performed well when identifying PNI before surgery [16]. However, their study did not include enough clinical and pathological factors, and there was no analysis of the impact of PNI on prognosis. Here, we explored the role of a combined radiomics–clinicopathological model in predicting PNI in gastric cancer using machine learning, and analyzed the impact of PNI on prognosis.

2. Materials and Methods

2.1. Patients

This study analyzed clinicopathological and contrast-enhanced computed tomography (CECT) data from 162 patients treated at the Second Affiliated Hospital of Nanjing Medical University from January 2019 to December 2022. The inclusion criteria were as follows: (1) CECT was performed less than 2 weeks before surgery and (2) the patients had a pathological diagnosis of gastric cancer. The exclusion criteria were as follows: (1) patients with unfilled stomachs; (2) patients with poor CECT quality; (3) patients with combined severe primary disease; and (4) patients with incomplete clinical data or follow-up information. The workflow of the inclusion and exclusion is shown in Figure 1. Our study was approved by the ethics committee of the Second Affiliated Hospital of Nanjing Medical University (2023-KY-162-01), and the patients' or their family members' informed consent was acquired. Our study adhered to the Image Biomarker Standardization Initiative (IBSI) guidelines and the Declaration of Helsinki [17].

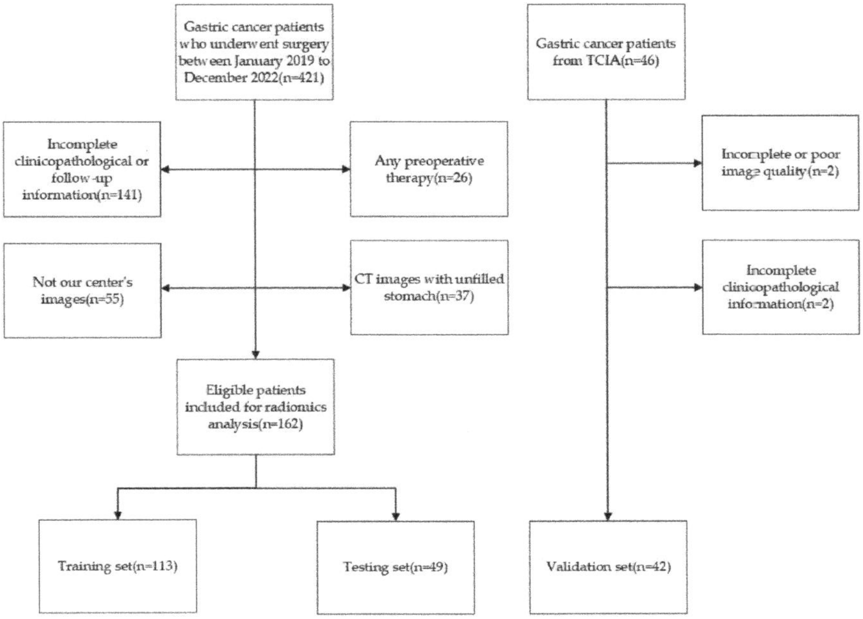

Figure 1. Flow diagram of inclusion and exclusion criteria.

Patients' baseline information was collected, including age, gender, tumor location, differentiation type, Lauren type, lymphovascular invasion (LVI) status, smoking status, drinking status, carcinoembryonic antigen (CEA) levels, cancer antigen 125 (CA125) levels, cancer antigen 199 (CA199) levels and human epidermal growth factor receptor 2 (HER-2) levels. Pathological diagnoses were confirmed by two pathologists, and when there was a dispute between the two pathologists, the diagnosis was confirmed by a third pathologist. These patients were randomly divided into a training set and a testing set with a ratio of 7:3, and 42 patients from the Cancer Imaging Archive (TCIA) with complete pathological data were also enrolled as a validation set.

The follow-up information of the 162 patients was also collected via telephone or from outpatient records. The endpoint event was the OS time, which was defined as the time from the day of the surgery to the day of death due to any cause, or 1 October 2023.

2.2. CT Image Acquisition

We included the arterial phase in this study because the arterial phase of CECT has better diagnostic performance [18]. All of the patients signed informed consent forms before enhanced CT examinations were carried out. The patients were required to fast for at least 6 h before the CT examination, and drank 1000 mL of water to keep their stomachs dilated. All of the patients underwent 64-slice dual source CT. Patients received 1.5 mL/kg of an iodinated contrast agent (Ioversol Injection 320 mg I/mL, Jiangsu Hengrui Pharmaceuticals Co. Ltd., Lianyungang, China) at a flow rate of 3.0 mL/sec using an automatic syringe pump. The arterial-phase imaging and venous-phase imaging followed a 30-s and a 60-s delay after the intravenous injection, respectively. The scan parameters were as follows: tube voltage, 120 kV; tube current, 150–300 mA; field of view, 30–50 cm; matrix, 512 × 512; rotation time, 0.5 s; and pitch, 1.0. The images were reconstructed with section thicknesses of 2 mm.

2.3. Regions of Interest and Extraction of Radiomics Features

We obtained Digital Imaging and Communications in Medicine (DICOM)-format images from the Picture Archiving and Communication System (PACS), and two experienced radiologists manually drew regions of interest (ROIs), slice by slice, using ITK-SNAP software (version 3.8.0) [19]. One radiologist manually drew ROIs for all of the patients, while the other draw ROIs for 30 randomly selected patients. The Python package PyRadiomics (version 2.1.2) was used to extract radiomics features from the ROIs [20]. All private patient information was removed. A Laplace of Gaussian filter with sigma values of 3 mm, 4 mm and 5 mm was used to reconstruct the images. The Pingouin package was utilized to calculate the intraclass coefficients (ICCs), and features with ICCs > 0.75 were considered effective [21].

2.4. Feature Selection and Calculation of Radscores

A total of 1595 features were extracted from ROIs delineated using ITK-SNAP software. The Synthetic Minority Over-Sampling Technique (SMOTE) was used to manage the imbalanced data in the training set. A Z-score transformation was used to normalize the features to the same level. First, a *t*-test was conducted to identify features that distinguish the PNI group from the non-PNI group. Second, the Max-Relevance and Min-Redundancy (mRMR) approach was utilized to filter irrelevant features, and the top 15 features were selected for further consideration. Finally, the least absolute shrinkage and selection operator (LASSO) algorithm was used to select features, and a 10-fold cross-validation was performed to obtain the optimal lambda. The features' radscores were calculated based on their coefficients and their values via the following: Radscore = $\sum_{i=0}^{n} C_i \times X_i + \mathrm{b}$, where C_i is the coefficient of the *i*-th selected feature, X_i is the value of the selected feature and b is the intercept.

2.5. Model Construction and Evaluation

Univariable and multivariable analyses were conducted, combining the radscore and clinicopathological variables. A nomogram was developed to predict PNI by combining clinicopathological and radiomics features. The performance of the model was evaluated based on area under receiver operating characteristics (ROC) curves (AUCs). The calibration curve measured the consistency between the predicted and actual probabilities of PNI. Clinical decision curves were used to analyze and evaluate the clinical practicality of the model. Overall, calibration curves and clinical decision curves were used to evaluate the model's efficacy. The workflow of our study is shown in Figure 2.

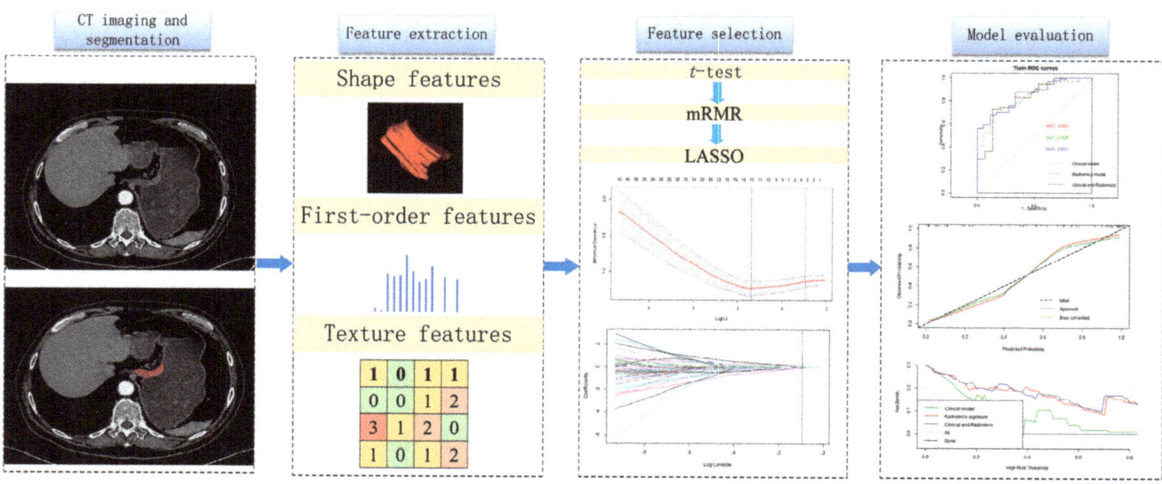

Figure 2. Workflow of perineural invasion prediction. A total of 1595 features were extracted from arterial phases of each patient. The *t*-test, Max-Relevance and Min-Redundancy (mRMR) and the least absolute shrinkage and selection operator (LASSO) were used to select features. Receiver operating characteristics (ROCs), calibration curves and clinical decision curves were used to evaluate the models.

2.6. Statistical Analysis

Statistical analyses were performed using Python (version 3.7.0), R software (version 4.2.3) and SPSS (version 22.0) software. Kaplan–Meier analyses were performed to assess OS. The *t*-test was used to test normally distributed continuous variables, and the Mann–Whitney U test was used to test non-normally distributed continuous variables. The chi-square test and Fisher test were used to test categorical variables. The R package "glmnet" was used to perform a LASSO algorithm analysis. The R package "rms" was used to draw the nomogram and the calibration curves. The R package "pROC" was used to draw ROC curves, and R package "dca.R" was used to analyze clinical decision curves. The R package "shiny" and "DynNom" were utilized to construct the online nomogram.

3. Results

3.1. Clinical Characteristics

A total of 162 patients were included in this study. The patients were randomly split into a training set and a testing set with a ratio of 7:3, with 113 patients in the training set and 49 patients in the testing set. Among the 113 patients, 95 patients were pathologically diagnosed with PNI and 18 patients were not. In the testing set, 33 patients were diagnosed with PNI and 16 patients were not. The differences in the clinical and pathological factors are shown in Table 1. According to the univariable analysis, the tumors' T stage ($p < 0.01$), N stage ($p < 0.01$) and LVI ($p = 0.042$) were closely related to PNI. We included the T stage, N stage and LVI to build clinical models based on a logistic regression. Additionally, 42 patients from TCIA were recruited as the validation set. As shown in Figure 3, the AUCs of the clinical models with the logistic regression were 0.820 (95%CI: 0.695–0.944) in the training set and 0.768 (95%CI: 0.596–0.491) in the testing set. The clinical model was validated using the validation set, and the AUC was 0.669 (95%CI: 0.497–0.842).

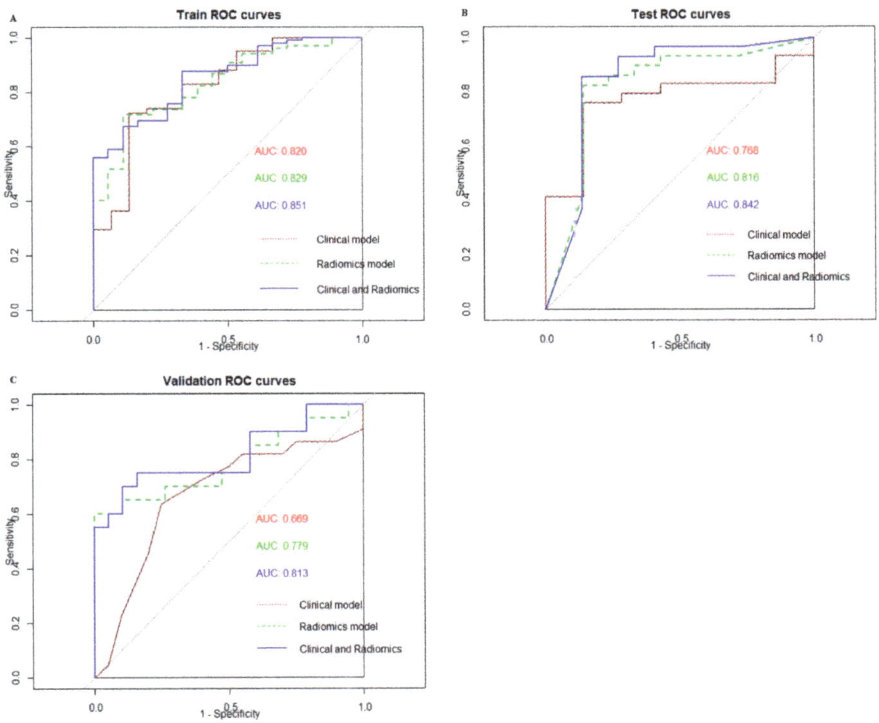

Figure 3. (**A**) ROC curves of three models in the training set. (**B**) ROC curves of three models in the testing set. (**C**) ROC curves of three models in the validation set.

Table 1. Clinical and pathological characteristics of recruited patients.

		Training Set			Testing Set		
		PNI (+) N = 95	PNI (−) N = 18	*p*	PNI (+) N = 33	PNI (−) N = 16	*p*
Age (mean ± SD [1])		66.11 ± 10.85	65.72 ± 8.85	0.194	62.91 ± 11.38	67.8 ± 3.42	0.494
Gender, No. (%)							0.218
	Male	70 (73.7%)	10 (55.6%)	0.350	26 (78.8%)	5 (31.3%)	
	Female	25 (26.3%)	8 (44.4%)		7 (21.2%)	11 (68.7%)	
Size, No. (%)				0.191			0.189
	<5 cm	45 (47.4%)	12 (66.7%)		18 (54.5%)	13 (81.2%)	
	≥5 cm	50 (52.6%)	6 (33.3%)		15 (45.5%)	3 (18.8%)	
Location, No. (%)							
	Antrum	37 (38.9%)	9 (50%)	0.315	7 (21.2%)	4 (25.0%)	0.205
	Body	27 (28.4%)	2 (11.1%)		5 (15.2%)	5 (31.3%)	
	Fundus	31 (32.6%)	7 (38.9%)		21 (63.6%)	7 (43.8%)	
Tissue differentiation, No. (%)							
	High	51 (53.7%)	5 (27.8%)	0.119	18 (54.5%)	9 (56.3%)	0.075
	Middle	28 (29.5%)	9 (50%)		10 (30.3%)	6 (37.5%)	
	Low	16 (16.8%)	4 (22.2%)		5 (15.2%)	1 (6.3%)	
Lauren type, No. (%)							
	Intestinal	20 (21.5%)	9 (50.0%)	>0.9	5 (15.2%)	2 (12.5%)	0.051
	Diffuse	35 (37.6%)	3 (16.7%)		11 (33.3%)	1 (6.3%)	
	Mixed	38 (40.9%)	6 (33.3%)		17 (51.5%)	13 (81.3%)	
T stage, No. (%)							
	1	11 (11.6%)	5 (27.8%)	<0.001	8 (24.2%)	12 (75.0%)	0.008
	2	7 (7.37%)	7 (38.9%)		2 (6.1%)	3 (18.8%)	
	3	43 (45.3%)	2 (11.1%)		13 (39.4%)	1 (6.3%)	
	4	34 (35.8%)	4 (22.2%)		10 (30.3%)	0	
N stage, No. (%)							
	0	7 (7.37%)	10 (55.6%)	<0.001	5 (15.2%)	3 (18.8%)	0.003
	1	24 (25.3%)	1 (5.56%)		3 (9.1%)	12 (75%)	
	2	18 (18.9%)	5 (27.8%)		9 (27.3%)	1 (6.3%)	
	3	46 (48.4%)	2 (11.1%)		16 (48.5%)	0	
LVI [2], No. (%)							
	Yes	58 (61.1%)	8 (44.4%)	0.014	25 (75.8%)	12 (75%)	0.001
	No	37 (38.9%)	10 (55.6%)		8 (24.2%)	4 (25%)	
HER-2 [3], No. (%)							
	(0–1+)	54 (74.0%)	11 (61.1%)	0.847	19 (57.6%)	5 (31.3%)	0.881
	(++–+++)	19 (26.0%)	7 (38.9%)		14 (42.4%)	11 (68.7%)	
Neutrophils, median (IQR)		4.09 (2.79, 4.69)	3.56 (2.73, 4.27)	0.417	3.63 (3.03, 4.74)	3.19 (2.83, 5.35)	0.706
Lymphocytes, median (IQR)		1.46 (1.14, 1.80)	1.33 (1.19, 1.87)	0.812	1.51 (1.03, 1.96)	1.39 (1.22, 1.82)	1.000
Albumin, median (IQR)		39.80 (36.75, 41.60)	42.00 (35.85, 44.35)	0.460	41.5 (35.9, 45.5)	38.1 (36.45, 40.4)	0.448
CEA [4], No. (%)							
	≤10	90(95.2%)	16(88.9%)	0.374	27 (81.9%)	14 (87.5%)	0.545
	>10	5(4.8%)	2(11.1%)		6 (18.2%)	2 (12.5%)	
CA125 [5], No. (%)							
	≤35	83 (87.3%)	16 (88.9%)	0.133	19 (57.6%)	10 (62.5%)	0.628
	>35	8 (12.7%)	2 (11.1%)		14 (42.4%)	6 (37.5%)	
CA199 [6], No. (%)							
	≤37.0 U/mL	71 (74.6%)	15 (83.3%)	0.096	27 (81.8%)	10 (62.5%)	0.545
	>37.0 U/mL	24 (25.4%)	3 (16.7%)		6 (18.2%)	6 (37.5%)	
Smoking history, No. (%)							
	Yes	8 (12.7%)	2 (11.1%)	0.983	29 (87.9%)	4 (25.0%)	0.245
	No	83 (87.3%)	16 (88.9%)		4 (12.1%)	12 (75.0%)	
Drinking history, No. (%)							
	Yes	11 (11.6%)	4 (22.2%)	0.876	30 (90.9%)	1 (6.3%)	0.262
	No	84 (88.4%)	14 (77.8%)		3 (9.1%)	15 (93.7%)	
Radscore, median (IQR)		1.74 (1.58, 1.99)	1.43 (1.14, 1.60)	<0.001	1.90 (1.69, 2.07)	1.50 (1.33, 1.64)	<0.001

[1] standard deviation; [2] lymphovascular invasion; [3] human epidermal growth factor receptor 2; [4] carcinoembryonic antigen; [5] cancer antigen 125; [6] cancer antigen 199.

3.2. Radiomics Features Selection and Model Construction

A total of 1595 features were extracted using PyRadiomics, including 14 shape features, 306 first-order features and 1275 texture features. The texture features consisted of 408 gray-level co-occurrence matrix (glcm) features, 238 gray-level dependence matrix (gldm) features, 272 gray-level run-length matrix (glrlm) features, 272 gray-level size-zone matrix (glszm) features and 85 neighboring gray-tone difference matrix (ngtdm) features. After performing the *t*-test, 224 features were retained. Subsequently, the Max-Relevance and Min-Redundancy (mRMR) approach was utilized to select features, and the top fifteen features were chosen for the least absolute shrinkage and selection operator (LASSO) analysis. A 10-fold cross validation was used to select the optimal "Lambda" value, and the best "Lambda" value was used to select features (Figure 4). After three rounds of feature selection, five features (wavelet.HLL_glcm_InverseVariance, original_firstorder_90Percentile, wavelet.HHH_firstorder_Minimum, wavelet.LLL_firstorder_Median and gradient_gldm_DependenceNonUniformityNormalized) were ultimately used to build the radiomics models. The coefficients of the selected features are shown in Table 2. The radiomics model, based on the five features, exhibited AUC values of 0.829 (95%CI: 0.738–0.921) in the training set, 0.816 (95%CI: 0.683–0.950) in the testing set and 0.779 (95%CI: 0.625–0.933) in the validation set (Figure 3).

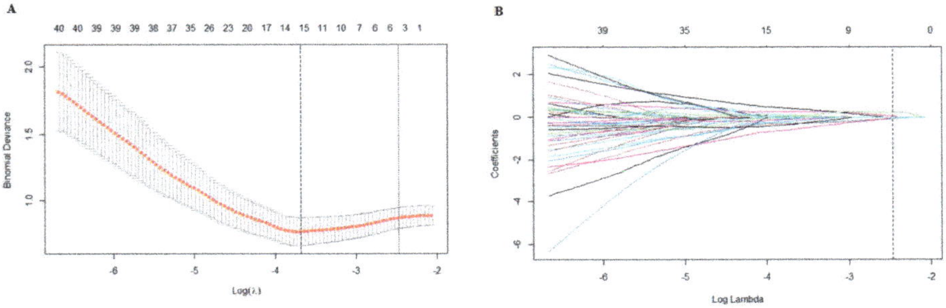

Figure 4. (**A**) A 10-fold cross validation was used to select the optimal "Lambda" value. (**B**) The different colors referred to different features. The best "Lambda" value was used to select features.

Table 2. The coefficients of the selected features.

Features	Coefficients
wavelet.HLL_glcm_InverseVariance	0.17618597
original_firstorder_90Percentile	0.06464257
wavelet.HHH_firstorder_Minimum	−2.96234129
wavelet.LLL_firstorder_Median	0.84062825
gradient_gldm_DependenceNonUniformityNormalized	1.20886446

3.3. *Construction and Evaluation of Combined Radiomics–Clinicopathological Model*

According to the univariable analysis, the T stage, N stage, LVI and radscore were associated with PNI ($p < 0.05$). According to the multivariable analysis, the T stage, N stage and radscore were independent risk factors for PNI ($p < 0.05$), and the OR values of the T stage, N stage and radscore were 8.013 (95%CI: 2.604–24.660), 2.382 (95%CI: 1.266–6.564) and 3.040 (95%CI: 1.250–7.397), respectively (Table 3). The combined radiomics–clinicopathological model based on the T stage, N stage and radscore showed robust efficacy in predicting PNI. The combined radiomics–clinicopathological model was superior in predicting PNI compared to both the clinical model and the radiomics model; the AUC values of the radiomics–clinicopathological model were 0.851 (95%CI: 0.769–0.933) and 0.842 (95%CI: 0.713–0.970) in the training set and testing set, respectively. The AUC value of the radiomics–clinicopathological model in the validation set was 0.813 (95%CI: 0.672–0.954). The ROC curves of the training set, testing set and validation set are shown in Figure 3. The detailed model performance is shown in Table 4.

Table 3. Results of predicted factors and their odds ratio (OR) values in the multivariate analysis.

Predicted Factors	OR [1]	95%CI [2]	*p*-Value
Age	1.008	0.961–1.057	0.754
Gender	0.786	0.272–2.270	0.656
Size	1.031	0.846–1.258	0.760
Location	1.086	0.638–1.847	0.762
Tissue differentiation	0.646	0.352–1.185	0.158
Lauren type	1.261	0.693–2.296	0.447
T stage	8.013	2.604–24.660	0.001
N stage	2.882	1.266–6.564	0.012
LVI [3]	1.344	0.162–11.130	0.784
HER-2 [4]	0.831	0.302–2.286	0.720
Neutrophils	0.981	0.840–1.146	0.981
Lymphocytes	1.028	0.432–2.448	0.950
Albumin	1.007	0.923–1.100	0.870
CEA [5]	1.091	0.925–1.288	0.300
CA-125 [6]	1.046	0.980–1.116	0.181
CA-199 [7]	1.027	0.990–1.065	0.154
Smoking history	0.432	0.074–2.531	0.352
Drinking history	2.169	0.260–18.120	0.475
Radscore	3.040	1.250–7.397	0.014

[1] odds ratio; [2] confidence interval; [3] lymphovascular invasion; [4] human epidermal growth factor receptor 2; [5] carcinoembryonic antigen; [6] cancer antigen 125; [7] cancer antigen 199.

Table 4. The detailed information for the three models.

	Training Set (n = 113)				Testing Set (n = 49)				Validation Set (n = 42)			
	AUC (95%CI)	ACC	SEN	SPE	AUC (95%CI)	ACC	SEN	SPE	AUC (95%CI)	ACC	SEN	SPE
Model1	0.820 (0.695–0.944)	0.832	0.851	0.400	0.768 (0.596–0.941)	0.837	0.851	0.500	0.669 (0.497–0.842)	0.619	0.600	0.667
Model2	0.829 (0.738–0.921)	0.876	0.879	0.833	0.816 (0.683–0.950)	0.836	0.851	0.772	0.779 (0.625–0.933)	0.718	0.737	0.700
Model3	0.851 (0.769–0.933)	0.929	0.886	0.750	0.842 (0.713–0.970)	0.837	0.851	0.818	0.813 (0.672–0.954)	0.744	0.750	0.737

Model1: clinicopathological model; Model2: radiomics model; Model3: combined model; AUC: area under curve; ACC: accuracy; SEN: sensitivity; SPE: specificity.

As shown in Figure 5, a nomogram of the radiomics–clinicopathological model was developed. In order to improve the generalizability and clinical application value of the model, we designed an online nomogram, and the website is as follows: https://lbhengjia.shinyapps.io/PNI_Predict/ (accessed on 25 January 2024). The calibration curve showed a good balance between the observed and predicted probability scores. The decision curves showed that the combined model provided a greater benefit than the clinical and radiomics models (Figures 6 and 7).

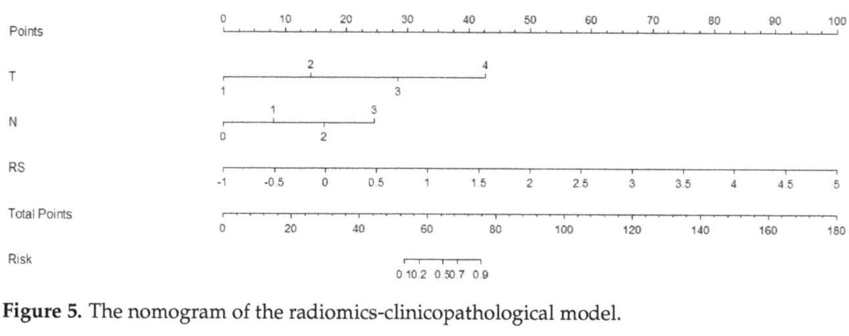

Figure 5. The nomogram of the radiomics-clinicopathological model.

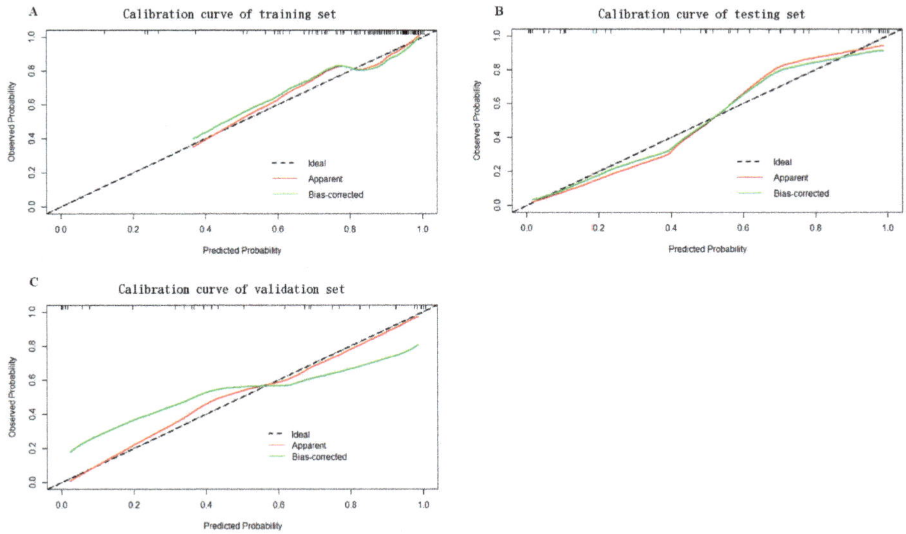

Figure 6. The calibration curves of the training set (**A**), testing set (**B**) and validation set (**C**).

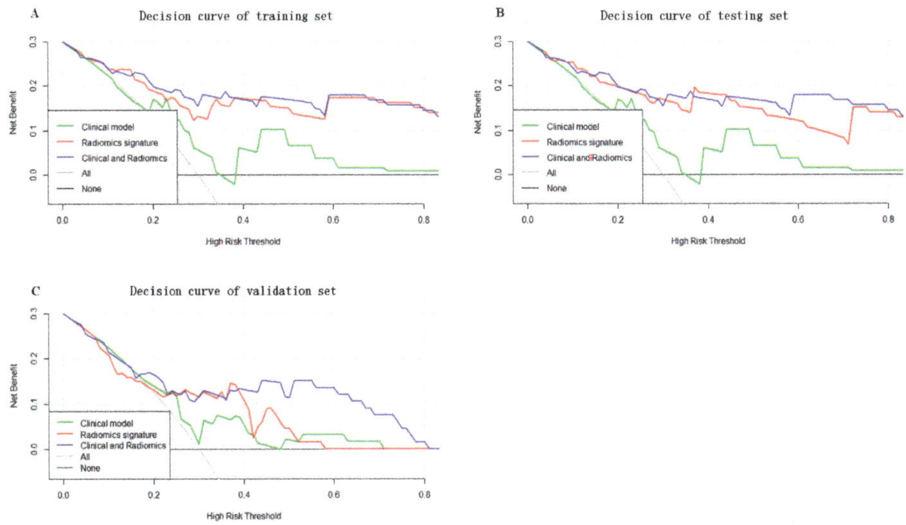

Figure 7. The decision curves of the training set (**A**), testing set (**B**) and validation set (**C**).

3.4. Survival Analysis

The median OS time of the 162 patients was 542 days (range: 7–1038 days). Of the 162 patients, 34 patients had no PNI, and 3 among them died. In total, 128 patients had PNI, and there were 33 deaths in this group. As shown in Figure 8, there was a statistically significant difference in OS between the PNI group and the non-PNI group ($p < 0.05$).

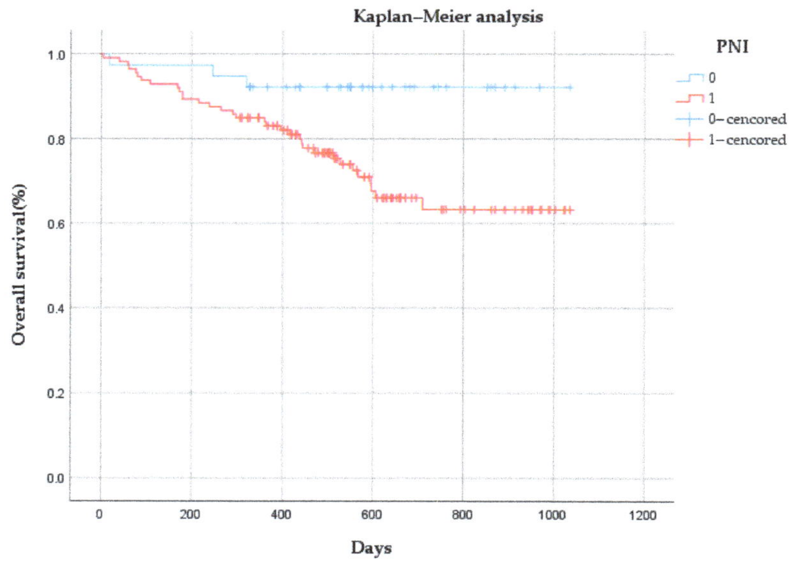

Figure 8. Kaplan–Meier analysis of overall survival based on perineural invasion (PNI) status among 162 patients.

4. Discussion

Early onset gastric cancer often lacks specific symptoms, and patients with gastric cancer often present in advanced stages when seeking medical attention. The overall 5-year survival rate of patients with advanced gastric cancer is relatively low, and the overall prognosis is poor. The degree of tumor infiltration, lymph node metastasis and tumor size are the most critical factors affecting the postoperative survival rate of gastric cancer, but many related studies have shown that the prognoses of gastric cancer patients are also related to LVI and PNI [22,23]. Currently, relevant studies have shown that LVI is one of the independent factors affecting the prognoses of gastric cancer patients, while there is relatively little research on PNI [24]. PNI is one of the biological features of gastric cancer. Patients with PNI always have worse outcomes compared to patients without PNI [25]. Whether PNI is an independent risk factor affecting the prognoses of gastric cancer patients is still under debate. At the same time, the current criteria of PNI mainly rely on postoperative pathology, which is an invasive method with a time lag that has little significance in the selection of a treatment mode. In this research, we developed a radiomics model to preoperatively predict the PNI of gastric cancer, which has more guiding significance for the treatment mode and surgical approach. Therefore, this study attempted to use artificial intelligence methods to non-invasively predict the PNI of gastric cancer before surgery, and to determine its value in predicting the prognoses of gastric cancer patients.

In terms of the specific mechanism of the PNI of gastric cancer, Kai Yin et al. confirmed that the axon guidance molecule promoted gastric cancer cell navigation along peripheral neuritis [26]. Jia X et al. analyzed the expression of PNI gene signatures in gastric cancer via a meta-analysis of gene expression profiling, and found that genes regulating cell adhesion molecules were upregulated and associated with poor survival in gastric cancer [27]. Currently, there is relatively little research on the mechanism of the PNI of gastric cancer, and more basic experiments are still needed to elucidate this mechanism. At the same time, there are a few studies that have focused on the preoperative assessment of PNI in patients with gastric cancer via radiomics. Yardımcı analyzed texture features of gastric cancer based on multiple machine learning models. However, this study had a small sample size and did not analyze clinical data, and the models had poor efficacy [28]. In this study, we included a cohort of 204 patients and enough clinicopathological risk factors to obtain more convincing results.

In our study, we found that advanced T stage and N stage classifications were closely associated with the PNI of gastric cancer, which was consistent with the results of a previous study [29]. There were no other covariates associated with PNI, and studies with large sample sizes need to be further developed in the future. The AUC of the clinical model (including the T stage and N stage) reached 0.820 in the training set and 0.768 in the testing set, indicating that the above clinical factors provided a high level of accuracy when predicting PNI. Adding radiomics features to the clinical model could help improve performance, as the combined model had AUC values of 0.851 and 0.842 in the training set and testing set, respectively. The combined model was also validated using an external validation dataset, achieving an AUC value of 0.813. The relatively inferior performance of the combined model on the validation set compared to the training set and testing set may be attributed to the smaller sample size and the variable image quality across datasets. At the same time, our study also found that gastric cancer patients with PNI had poorer prognoses than those without PNI, which was in line with a previous study [30].

In recent years, the integration of radiomics and machine learning has shown great potential to advance our understanding of various gastric cancer conditions, especially PNI. In our research, we focused on exploring the mechanisms of PNI using radiomics and machine learning methods. By analyzing various shape and texture features in radiomics images, we could analyze the PNI of tumors from both macro and micro perspectives. Among the radiomics features we included, there were three wavelet features. These wavelet features can provide information about the internal vascular structure, tissue density and microscopic environments of tumors, and can thereby help us better understand the mechanisms of PNI and improve the model's predictive ability [31]. By synthesizing these multi-dimensional data, we can obtain more comprehensive and accurate tumor features and establish more reliable models to predict the PNI of gastric cancer. Furthermore, we plan to prospectively validate the developed model's effectiveness and evaluate its potential impact on clinical decision making and personalized treatment strategies.

Our study also has many limitations. Firstly, this was a single-center study with relatively small sample sizes; a multicenter study with large samples needs to be conducted. Secondly, our study did not investigate the specific mechanism of PNI through basic experiments. Thirdly, the manual delineation of ROIs presented a certain degree of heterogeneity, and a deep learning model without manual delineation needs to be further developed.

5. Conclusions

Gastric cancer patients with PNI have relatively poor prognoses; radiomics–clinicopathological models can effectively predict the PNI of gastric cancer, which could help clinicians in diagnosis and decision making.

Author Contributions: H.J. and R.L. wrote the manuscript. R.L. and Y.L. (Yawei Liu) analyzed clinical data. T.Z. reviewed and edited the manuscript. Y.L. (Yuan Li) contributed to the study design. J.Z. was responsible for submission of the manuscript. All authors have read and agreed to the published version of the manuscript.

Funding: This research was supported by the National Natural Science Foundation of China; (Grant No. 81874058).

Institutional Review Board Statement: Approval was received from the ethics committee of the Second Affiliated Hospital of Nanjing Medical University (No. 2023-KY-162-01). The approval date was 2 August 2023. The patients' or their family members' informed consent was acquired. All private patient information was removed.

Informed Consent Statement: Approval was received from the ethics committee. The patients' or their family members' informed consent was acquired.

Data Availability Statement: The data supporting this study are available from the corresponding author upon reasonable request.

Acknowledgments: The authors thank all their colleagues who contributed to this research.

Conflicts of Interest: The authors declare no potential conflicts of interest.

References

1. Ning, F.L.; Zhang, N.N.; Zhao, Z.M.; Du, W.Y.; Zeng, Y.J.; Abe, M.; Pei, J.P.; Zhang, C.D. Global, Regional, and National Burdens with Temporal Trends of Early-, Intermediate-, and Later-Onset Gastric Cancer from 1990 to 2019 and Predictions up to 2035. *Cancers* **2022**, *14*, 5417. [CrossRef] [PubMed]
2. Van Cutsem, E.; Sagaert, X.; Topal, B.; Haustermans, K.; Prenen, H. Gastric cancer. *Lancet* **2016**, *388*, 2654–2664. [CrossRef] [PubMed]
3. Riihimäki, M.; Hemminki, A.; Sundquist, K.; Sundquist, J.; Hemminki, K. Metastatic spread in patients with gastric cancer. *Oncotarget* **2016**, *7*, 52307–52316. [CrossRef]
4. Yu, X.; Hu, F.; Li, C.; Yao, Q.; Zhang, H.; Xue, Y. Clinicopathologic characteristics and prognosis of proximal and distal gastric cancer. *OncoTargets Ther.* **2018**, *11*, 1037–1044. [CrossRef]
5. Liu, L.; Wang, Z.W.; Ji, J.; Zhang, J.N.; Yan, M.; Zhang, J.; Liu, B.Y.; Zhu, Z.G.; Yu, Y.Y. A cohort study and meta-analysis between histopathological classification and prognosis of gastric carcinoma. *Anti-Cancer Agents Med. Chem.* **2013**, *13*, 227–234. [CrossRef]
6. Wang, W.; Yang, Q.; Huang, Q.; Zhang, H.; Zhang, Z.; Gao, J.; Ren, W.; Hu, Y.; Lin, Y.; Dang, Y.; et al. The rs2839698 Single Nucleotide Polymorphism of lncRNA H19 is Associated with Post-Operative Prognosis in T3 Gastric Adenocarcinoma. *Clin. Lab.* **2018**, *64*, 105–112. [CrossRef] [PubMed]
7. Yang, K.; Dan, Y.Q.; Choi, Y.Y.; Zhou, Z.G.; Hyung, W.J.; Hu, J.K.; Noh, S.H. The Impact of Nerve Involvement on the Prognosis of Gastric Cancer Patients with Curative Gastrectomy: An International Multicenter Analysis. *Dis. Markers* **2021**, *2021*, 8870562. [CrossRef]
8. Liu, Q.; Ma, Z.; Cao, Q.; Zhao, H.; Guo, Y.; Liu, T.; Li, J. Perineural invasion-associated biomarkers for tumor development. *Biomed. Pharmacother.* **2022**, *155*, 113691. [CrossRef]
9. Liebig, C.; Ayala, G.; Wilks, J.A.; Berger, D.H.; Albo, D. Perineural invasion in cancer: A review of the literature. *Cancer* **2009**, *115*, 3379–3391. [CrossRef]
10. Zhao, B.; Lv, W.; Mei, D.; Luo, R.; Bao, S.; Huang, B.; Lin, J. Perineural invasion as a predictive factor for survival outcome in gastric cancer patients: A systematic review and meta-analysis. *J. Clin. Pathol.* **2020**, *73*, 544–551. [CrossRef]
11. Li, C.; Wang, M.; Cheng, X.; Jiang, Y.; Xiao, H. Nerve invasion as an independent predictor of poor prognosis in gastric cancer after curative resection. *Medicine* **2022**, *101*, e30084. [CrossRef]
12. Mayerhoefer, M.E.; Materka, A.; Langs, G.; Häggström, I.; Szczypiński, P.; Gibbs, P.; Cook, G. Introduction to Radiomics. *J. Nucl. Med. Off. Publ. Soc. Nucl. Med.* **2020**, *61*, 488–495. [CrossRef] [PubMed]
13. Du, G.; Zeng, Y.; Chen, D.; Zhan, W.; Zhan, Y. Application of radiomics in precision prediction of diagnosis and treatment of gastric cancer. *Jpn. J. Radiol.* **2023**, *41*, 245–257. [CrossRef]
14. Jiang, Y.; Chen, C.; Xie, J.; Wang, W.; Zha, X.; Lv, W.; Chen, H.; Hu, Y.; Li, T.; Yu, J.; et al. Radiomics signature of computed tomography imaging for prediction of survival and chemotherapeutic benefits in gastric cancer. *EBioMedicine* **2018**, *36*, 171–182. [CrossRef] [PubMed]
15. Liu, S.; He, J.; Liu, S.; Ji, C.; Guan, W.; Chen, L.; Guan, Y.; Yang, X.; Zhou, Z. Radiomics analysis using contrast-enhanced CT for preoperative prediction of occult peritoneal metastasis in advanced gastric cancer. *Eur. Radiol.* **2020**, *30*, 239–246. [CrossRef]
16. Zheng, H.; Zheng, Q.; Jiang, M.; Han, C.; Yi, J.; Ai, Y.; Xie, C.; Jin, X. Contrast-enhanced CT based radiomics in the preoperative prediction of perineural invasion for patients with gastric cancer. *Eur. J. Radiol.* **2022**, *154*, 110393. [CrossRef] [PubMed]
17. Zwanenburg, A.; Vallières, M.; Abdalah, M.A.; Aerts, H.; Andrearczyk, V.; Apte, A.; Ashrafinia, S.; Bakas, S.; Beukinga, R.J.; Boellaard, R.; et al. The Image Biomarker Standardization Initiative: Standardized Quantitative Radiomics for High-Throughput Image-based Phenotyping. *Radiology* **2020**, *295*, 328–338. [CrossRef]
18. Liu, S.; Liu, S.; Ji, C.; Zheng, H.; Pan, X.; Zhang, Y.; Guan, W.; Chen, L.; Guan, Y.; Li, W.; et al. Application of CT texture analysis in predicting histopathological characteristics of gastric cancers. *Eur. Radiol.* **2017**, *27*, 4951–4959. [CrossRef]
19. Yushkevich, P.A.; Piven, J.; Hazlett, H.C.; Smith, R.G.; Ho, S.; Gee, J.C.; Gerig, G. User-guided 3D active contour segmentation of anatomical structures: Significantly improved efficiency and reliability. *Neuroimage* **2006**, *31*, 1116–1128. [CrossRef]
20. van Griethuysen, J.J.M.; Fedorov, A.; Parmar, C.; Hosny, A.; Aucoin, N.; Narayan, V.; Beets-Tan, R.G.H.; Fillion-Robin, J.C.; Pieper, S.; Aerts, H. Computational Radiomics System to Decode the Radiographic Phenotype. *Cancer Res.* **2017**, *77*, e104–e107. [CrossRef]
21. Weir, J.P. Quantifying test-retest reliability using the intraclass correlation coefficient and the SEM. *J. Strength Cond. Res.* **2005**, *19*, 231–240. [CrossRef] [PubMed]
22. Orman, S.; Cayci, H.M. Gastric cancer: Factors affecting survival. *Acta Chir. Belg.* **2019**, *119*, 24–30. [CrossRef] [PubMed]
23. Zhang, F.; Chen, H.; Luo, D.; Xiong, Z.; Li, X.; Yin, S.; Jin, L.; Chen, S.; Peng, J.; Lian, L. Lymphovascular or perineural invasion is associated with lymph node metastasis and survival outcomes in patients with gastric cancer. *Cancer Med.* **2023**, *12*, 9401–9408. [CrossRef] [PubMed]

24. Chen, Y.F.; Wang, S.Y.; Le, P.H.; Chen, T.H.; Kuo, C.J.; Lin, C.J.; Chou, W.C.; Yeh, T.S.; Hsu, J.T. Prognostic Significance of Perineural Invasion in Patients with Stage II/III Gastric Cancer Undergoing Radical Surgery. *J. Pers. Med.* **2022**, *12*, 962. [CrossRef] [PubMed]
25. Uzun, O.; Gülmez, S.; Senger, A.S.; Ömeroğlu, S.; Polat, E.; Duman, M. Prognostic effect of perineural invasion in successive years in patients with locally advanced gastric cancer. *Indian J. Pathol. Microbiol.* **2021**, *64*, 479–483. [CrossRef] [PubMed]
26. Yin, K.; Wang, L.; Xia, Y.; Dang, S.; Zhang, X.; He, Z.; Xu, J.; Shang, M.; Xu, Z. Netrin-1 promotes cell neural invasion in gastric cancer via its receptor neogenin. *J. Cancer* **2019**, *10*, 3197–3207. [CrossRef]
27. Jia, X.; Lu, M.; Rui, C.; Xiao, Y. Consensus-Expressed CXCL8 and MMP9 Identified by Meta-Analyzed Perineural Invasion Gene Signature in Gastric Cancer Microarray Data. *Front. Genet.* **2019**, *10*, 851. [CrossRef]
28. Yardimci, A.H.; Kocak, B.; Bektas, C.T.; Sel, I.; Yarikkaya, E.; Dursun, N.; Bektas, H.; Afsarriz, C.U.; Gursu, R.U.; Kilickesmez, O. Tubular gastric adenocarcinoma: Machine learning-based CT texture analysis for predicting lymphovascular and perineural invasion. *Diagn. Interv. Radiol.* **2020**, *26*, 515–522. [CrossRef]
29. Liu, J.; Huang, X.; Chen, S.; Wu, G.; Xie, W.; Franco, J.P.C.; Zhang, C.; Huang, L.; Tian, C.; Tang, W. Nomogram based on clinical characteristics for preoperative prediction of perineural invasion in gastric cancer. *J. Int. Med. Res.* **2020**, *48*, 300060519895131. [CrossRef]
30. Deng, J.; You, Q.; Gao, Y.; Yu, Q.; Zhao, P.; Zheng, Y.; Fang, W.; Xu, N.; Teng, L. Prognostic value of perineural invasion in gastric cancer: A systematic review and meta-analysis. *PLoS ONE* **2014**, *9*, e88907. [CrossRef]
31. Demircioğlu, A. The effect of preprocessing filters on predictive performance in radiomics. *Eur. Radiol. Exp.* **2022**, *6*, 40. [CrossRef] [PubMed]

MDPI AG
Grosspeteranlage 5
4052 Basel
Switzerland
Tel.: +41 61 683 77 34
www.mdpi.com

Cancers Editorial Office
E-mail: cancers@mdpi.com
www.mdpi.com/journal/cancers

www.ingramcontent.com/pod-product-compliance
Lightning Source LLC
Chambersburg PA
CBHW061609210926
51850CB00009B/2265